Functionalized Polymers

Synthesis, Characterization and Applications

Edited by

Narendra Pal Singh Chauhan

Department of Chemistry
Bhupal Nobles' University, Udaipur
Rajasthan, India

CRC Press is an imprint of the
Taylor & Francis Group, an **informa** business
A SCIENCE PUBLISHERS BOOK

Cover credit: Image reproduced by kind courtesy of Prof. Sharon Gerecht, Johns Hopkins University.

CRC Press
Taylor & Francis Group
6000 Broken Sound Parkway NW, Suite 300
Boca Raton, FL 33487-2742

Version Date: 20201130

International Standard Book Number-13: 978-0-367-42061-1 (Hardback)

Library of Congress Cataloging-in-Publication Data

Names: Chauhan, Narendra Pal Singh, editor.
Title: Functionalized polymers : synthesis, characterization and applications / edited by Narendra Pal Singh Chauhan, Department of Chemistry, Bhupal Nobles University, Udaipur Rajasthan, India.
Description: Boca Raton : CRC Press, [2021] | Includes bibliographical references and index.
Identifiers: LCCN 2020043459 | ISBN 9780367420611 (hardcover)
Subjects: LCSH: Polymers. | Polymerization.
Classification: LCC QD381 .F865 2021 | DDC 547/.7--dc23
LC record available at https://lccn.loc.gov/2020043459

Visit the Taylor & Francis Web site at
http://www.taylorandfrancis.com

and the CRC Press Web site at
http://www.routledge.com

Preface

The field of functional polymer research comprises the synthesis of functional polymer materials and the study and tuning of their properties. On the one hand, polymers with specific functions are exciting in their underlying principles of physics and chemistry, which result in outstanding tunability of the resulting properties. Functional polymers are and will continue to be in demand as they expand the property and range of application of their non-functional counterparts.

On the other hand, functional polymers govern our lives in a variety of applications, ranging from surface-specific coatings to polymer foams that are indispensable in modern construction and building.

There are very few books in the market today that deal with practical polymers. The existing books focus primarily on functional polymers for versatile applications. In recent years, however, significant progress has been made in the synthesis of functional polymers, and a number of national and international academic conferences were held to address these developments.

This book consists of fourteen chapters. The first provides an introduction as well as a view of future prospects. The subsequent chapters explain conjugated polymers, amphiphilic hyperbranched polymers, biodegradable polymers, pseudo-proteins, functionalized cellulose, 3D-printed polymers, poly(vinylcarbazole), functional plastics and rubbers, polyurethane, biopolymeric sensors, stimuli-responsive polymers and polysiloxane and polycarbosilanes.

I sincerely hope that this book will be of interest to polymer scientists and to all those who deal with the interdisciplinary branches of related subjects. I believe this book offers a balanced, informative and innovative perspective that is applicable to academics and industries.

I wish to thank all of the authors for their contributions and their sincere efforts, without which the book would not be in its present form. Suggestions from critics are always welcome. I am extremely thankful to Prof. Sharon Gerecht, Johns Hopkins University for her valuable suggestions and also for her kind consent of cover image.

Dr. Narendra Pal Singh Chauhan

Contents

CHAPTER

1

Introduction and Future Prospects

Narendra Pal Singh Chauhan
Department of Chemistry, Faculty of Science, Bhupal Nobles' University, Udaipur 313001, Rajasthan, India

1. Introduction

Polymers with atomic groups with higher polarity or reactivity than traditional hydrocarbon chains are functional polymers (Chauhan 2019a). Such materials display improved properties in contrast to their non-functional counterparts in terms of stronger contact, separation or reactivity. Living anion polymerization is a versatile and popular method to change well-designed polymers with an in-chain or chain-end structure group or classes (Vana and Yagci 2013). Such functional groups may be controlled by reversible ionic activity, chain extension, branching or cross-linking reactions. In some oligomeric or polymeric chains, these functional macromolecular materials can often be combined with reactive copolymers (Chauhan et al. 2015). Polymerization or copolymerization of other monomers of sufficient functional polymerization groups is also feasible.

The processing of bulk polymers results in chemical heterogeneity, resulting in various advantages, such as improved reactivity, phase isolation, and improved compatibility or association. The ability to construct self-assemblies and supramolecular structures of functionalized polymers with other architectures is another advantage (Chauhan 2019a). The creation or dissociation of self-assembled materials can lead to "smart" materials in response to chemical or physical stimuli. Many of the functional polymers are used in single linear backbones, including block, grafted, chain-end and in-chain polymer (Vana and Yagci 2013). For technical applications such as optics, electronics or catalysis, functional polymers are essential. These materials are often widely used for the synthesis of solid-phase oligonucleotides and membranes (e.g., column chromatography). In rapidly evolving medical environments, specially formulated functional polymer products are essential, including suture protection issues, dental fillings,

Email: narendrapalsingh14@gmail.com

wound dressings, bone cement and hollow fiber dialysis (Chauhan 2019a). Hydrogels and inhibitors that are the basis of ophthalmic surgery are usually among these products.

Polymer functionalization is defined by a functional polymer synthesis process consisting of three main categories: direct-functional polymer synthesis (in situ), post-functionalization (post-polymer modification), and functional group transformation.

2. Direct polymerization

2.1. End-functionalized polymerization

End-functional polymers are most often anionically generated using the post-polymerization reactions of anionic living polymers with the appropriate electrophiles to create the desired (possibly masked) functionality (Konkolewicz et al. 2009). End-functional polymers have been produced with carboxylic acid, hydroxyl, amine, thiol, aldehyde, acetyl, ring opening of epoxy group and vinyl moieties having functionalized pendant groups (Coessens et al. 2001).

Many variables, including chain-end structure, solvent, temperature, concentration, stoichiometry, and the order and nature of the addition of reagents (such as polar additives), have a dramatic effect on yield and functionality. Side reactions can be minimized by optimizing the termination procedure. Many of the functional groups can be utilized in the termination reactions without protection, as the reactions with living chain polymers are much quicker with haloalkyl moieties than in functional groups.

2.2. In-chain functionalized polymerization

Since most functional groups are deleterious for living anionic polymerization, the use of protective groups in the production of polymeric materials has become increasingly important. Monomers can provide protected functionality that can be translated into reactive or associative pendant groups. The anionic preparation of poly(methacrylic acid) is an important example (Donini et al. 2002). The carboxylic acid groups in the backbone are usually masked as tert-butyl esters and carboxylic acid is formed during polymerization hydrolysis, a substance that cannot be produced by direct anionic polymerization of the corresponding monomer (acrylic acid) (Morton and Fetters 1975).

Similarly, a broad variety of designed polymers and copolymers can be synthesized by sulfonation of homopolymers or styrene copolymers (Yang and Mays 2002). A different route is used by a different protection system. 1-Alkoxyethyl methacrylate is anionically polymerized to produce a well-being substance and polyacids are formed by acetal group hydrolysis (Zhang and Ruckenstein 1998). In particular, protective groups are needed for functional groups with active hydrogen-containing organic compounds such as alcohol, and primary and secondary amines.

Dimethyl-tert-butylsilyl will also protect alcohols (ROHs), with primary and secondary amines (RNH_2 and R_2NH) typically protected by trimethylsilyl groups. Reactive polymer sites such as halogen may be used as sites for attaching preformed side chains or sites for initiation. Various functionalized monomers based on methacrylates, styrene and conjugated dienes with different pendant groups have been made of smart functionalized polymers for various applications from electronics to biomedical applications.

3. Post-polymerization

Sometimes it is difficult to polymerize directly, since certain functional groups may interfere with polymerization; it is preferable to polymerize by post-polymerization or post-modification (Theato and Klok 2013), a technique that has been extensively used for many years (Gauthier et al. 2009). Thiol-ene click reaction, sulphonylation, chloromethylation, carboxylation, formylation, esterification, alkylation, acylation or amine functionality may be incorporated into the polymer to achieve the desired functionalized polymer.

4. Transformation of functional group

Functional group transformation can be used if functionalization of polymer is difficult using direct and post-polymerization technique. In this method, the synthesis of the undesired functional polymer is performed by direct or post-functionalization and subsequent modification to the desired functional polymer using a functional group transformation such as oxidation or reduction.

5. Future prospects

The development of functional polymers has remained an area of considerable interest and importance in electronics, industrial applications and rapidly expanding biomedical subfields such as tissue engineering, drug delivery, and gene delivery (Fig. 1).

Conducting polymers (CPs) such as polyacetylene, polyaniline, polypyrrole and polythiophene provide semiconductive molecular architectures and interesting sensing and biomedical properties (Chauhan 2019b). They have brought about significant progress in the field of sensing and biomedical applications.

Unfortunately, the most important requirements for sensors, such as analytical selectivity and detection of a particular analysis in a complex environment, are not easily achieved by pristine CPs. Such shortcomings in pristine CPs and processability necessitate the creation of functional CPs by means of intelligent structural changes to pristine CPs or the modification of CPs by incorporating functionalized moieties. The features

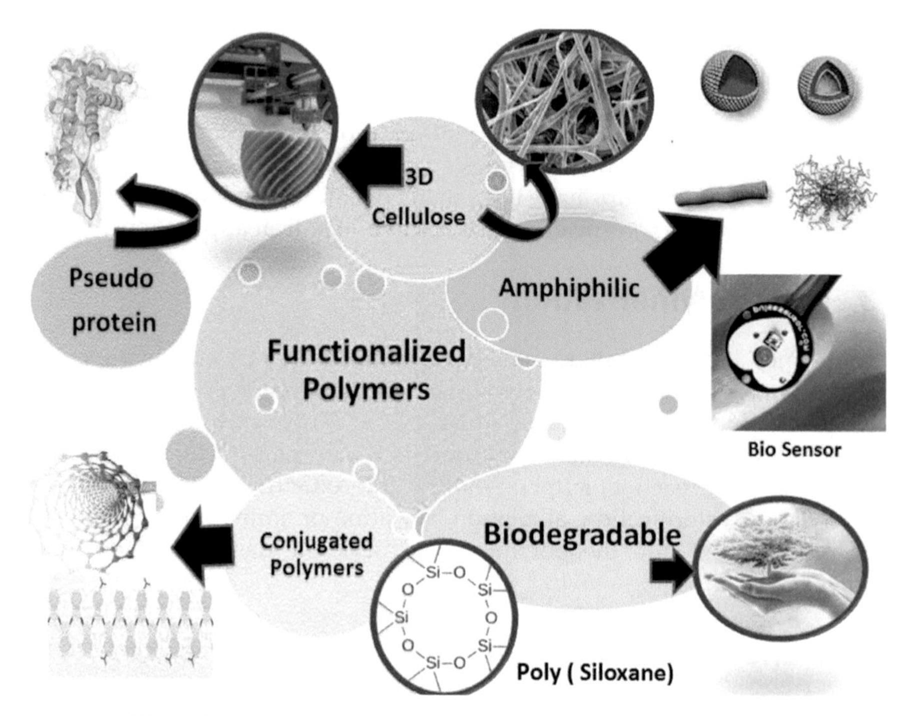

Fig. 1. Some important applications of functionalized polymers.

of functionalized conductive polymers (FCPs) are important for the design of molecular imprinted polymers and provide a variety of perspectives for sensor applications. FCPs, biofunctionalized FCPs, nanostructured FCPs and multi-components are used in sensing products (Chauhan 2019a, Chauhan and Mozafari 2019). Organic electro-active materials can be used for the production of simple alcohols, e.g., ethanol and 1-butanol, because these solvents come from biomass and have lower health risks, and can easily be extended to additional applications in organic electronics and bioelectronics.

Modification of functional groups added to the conjugated backbone polymer changes structural properties substantially. A highly promising development is the oxidation of electron-thioalkyl substitutions for electron donation sulfoxides or sulfones; however, this reaction has not been investigated in modification of the conjugated polymers in organic electronics. Questions on critical selectivity and response time remain to be answered. The use of dimethyldioxirane to oxidize thioalkyl compounds bound to conjugate polymers could be an important alternative to critical selectivity.

Organic donor-acceptor hybrid polymers are widely developed and used in electronic devices (Chauhan et al. 2016). Polymers based on diketopyrrolopyrrole (DPP) have gained the most popularity because of their high efficiency. Extension of conjugation may result in a variety of optical, electrochemical and system performance polymers (Chauhan and

Chundawat 2019). Polymers containing such high-performance electron-deficient pigments are expected to be potentially higher in electronic device applications than DPP-based ones.

Multimetallic gold-plated nanoalloys decorated with conjugated polymer nanofibers may exhibit superior catalytic activity for the production of solar hydrogen under visible light irradiation. This can provide new opportunities for photocatalysts using polymer-based hybrids to generate efficient solar hydrogen.

Photocrosslinked biodegradable synthetic polymer networks are highly interesting materials for a wide range of applications such as drug delivery, cell encapsulation, and fabrics engineering.

Differing from macromer precursor molecules in their composition, chemical, functional degree and molecular weight, they result in networks with a broad variety of physical and mechanical properties, crosslinking densities, degradation characteristics and thus possible applications.

Electro-active biomaterials are intrinsically conductive polymers based on a new generation of "smart" biomaterials. Poly(3,4-ethylenedioxythiophene), polypyrrole and polyaniline are well-conducting polymers with excellent electrical and optical characteristics that are key candidates for possible biomedical applications. In addition, biodegradability of biomaterials is very useful and attractive.

In this sense, biodegradable polyester-based polymers such as poly(D, L lactic acid), polycaprolactone and poly(glycolic acid) tend to be promising for good biocompatibility and thus become more sensitive as sustainable alternatives to medical applications. The use of polymers as biomaterials is a key issue for poor molecular interactions with cell biocompatibility. The key approaches to enhancing their own biocompatibility are whether monomers with different molecules are doped with unique counter ions (bio-doping) or chemically modified.

New methods have been introduced to solve the problem of conductive polymer insomnia, primarily by (1) the synthesis of modified electromones associated with block copolymers by the combination of degraded ester and (2) polyesters used in the second stage of copolymerization with conductive monomers.

Biological molecular engineering is a key issue in the production of highly functional polymers. Computing of physical (non-covalent) interactions with path-dependent, hierarchical covalent "press" to produce peptide-based synthetic hybrid polymers could be optimized to ensure maximum stability, size and spatial display control of chemical properties. The relationship of proteins to polymers changes the properties of each unit and paves the way for a number of new applications. The behavior of protein attachment in recent years has been very cautious and mainly affected by surface characteristics. There are three major routes for protein polymer conjugate synthesis and the forces interacting between protein molecules and polymer particles include hydrophobic contact, electrostatic bonding, van der Waals and hydrogen bonding.

A variety of chemical strategies have been developed to bind certain synthetic polymers to proteins. Surface changes play a key role in the production of activated materials and are achieved by physical and chemical processes. The functional groups on the protein surfaces, which indicate different potential reactions, are derived from the selected amino acid residues.

Protein attachment to polymer chains is based on ten natural amino acids found in the field of ligation chemistry and modern non-natural compounds (Tribet 1998). The synergistic combination of properties offers several advantages for bio-applications (Moghaddam et al. 2018). Related combinations of biomedical protein solutions such as bovine serum albumin or immunoglobulin G are very important biomedical strategies. The development of highly functional structures is an exciting new limit for living organisms. In particular, protein-based engineered living materials (ELMs) benefit from basic molecular biology to regulate their properties.

Caf1 (chromatin assembly factor-1) is an important ELM protein that, because of its unusual combination of properties, is exported by bacterial cells as a large, light-weight, non-covalent polymer, resistant to thermal and chemical degradation (Shibahara and Stillman 1999). As a result of complex protein structures and diverse interactions with polymers, soya protein based biomaterial is one of the promising candidates used as a functional modifier for polymers.

Specific protein-polymer interactions may have different property-modification effects on poly(vinylidenefluoride) (PVDF) (Martins et al. 2014). For instance, the powerful interaction between PVDF and the protein molecule that is denatured without disulfide bonds significantly improves the dipole moment of PVDF in the interface and improves polarization on external electric fields.

Nanocellulose refers to nano-scale rigid cellulose materials that are non-toxic, biodegradable and biocompatible without harmful health and environmental effects (Kargarzadeh et al. 2018). Because of their low thermal expansion coefficient, high aspect ratio, higher tensile strength, and excellent mechanical and optical properties, they have many applications in thermo-reversible and robust hydrogels, coating additives, food packaging, flexible displays and biopharmaceuticals. Cellulose nanocrystals are also very important because of their higher aspect ratios, low cost, durability and ability to modify the surface.

Three-dimensional extrusion printing is widely explored for the manufacture of patient-specific scaffolds made from biodegradable polyesters such as poly(lactic acid) (PLA) (Chauhan 2019c). PLA scaffolds are not bioactive for promoting bone regeneration given the required mechanical support. This research provides a technique to amend the surface to enhance the bioactivity of 3D-printed scaffold bone tissue regeneration.

Biodegradable polymeric scaffolds are widely used to promote early regeneration and can be used to induce the desired cell behavior using various chemical groups or bioactive signals. Nevertheless, these scaffolds are often modified after construction, resulting in undesirable changes in the biochemical composition of the native tissue and homogenously dispersed chemistries. To address these challenges, the deposition of preferred chemistries can be monitored by using 3D printing with peptide-polymer conjugates.

Multiphoton 3D lithography is becoming a method of choice in various fields. Its true 3D structuring capabilities beyond diffraction can be exploited to create diverse functionality structures. Such artifacts may also be made from specific materials enabling increased output.

Reversible addition-fragmentation chain transfer (RAFT) polymerization is an important method for synthesizing macromolecules with regulated topologies and various chemical functions (Perrier et al. 2005). However, slow polymerization rates of traditional systems hindered the application of RAFT polymerization to additives. A rapid visible (green) light-induced RAFT polymerization process could be generated and optimized for an open-air 3D printer. The reaction elements should be non-toxic, metal-free and eco-friendly, ideal for biomaterial processing.

The RAFT agent in photosensitive resin was used to monitor the mechanical characteristics of 3D-printed materials after printing. This RAFT-mediated 3D printing method can provide a variety of new, responsive materials.

The theory is based on supramolecular hydrogen-polymers that form nanomicellar clusters. The printability of the supramolecular polymer network is based on reversible thermal and shear-induced decoupling, which generates stable structures after printing, as monitored by x-ray scattering and melt rheology.

Hydrophobic functional polyurethane resins by Diels-Alder reaction combining N-alkyl maleimides with linear furan-modified polyurethane can be used to functionalize polyurethane by post-polymerization. The reaction of furan-containing diol, polyethylene glycol (PEG) and various diisocyanates provides access to furan-bearing polyurethanes. Maleimide and furan function undergo a Diels-Alder reaction that allows the hydrophobic side chains to be covalently bound to the polyurethane backbone (Pandey et al. 2019). Kraft lignin-powered isocyanate is commonly used as a reactive macromonomer for the preparation of polyurethane foam prepolymers to absorb bio macromolecules into a high-value polymer material. Polyurethane resin of three different carbon nanotube (CNT) forms, non-functioning CNT, carboxylic-functioning CNT and amine-functional CNT were studied (Koerner et al. 2005). The results showed that the CNT worked with the amine (NH_2) group, creating a good polyurethane resin interaction, forming a level and uniform film.

Polymers that can respond to external stimuli, especially as controlled drug release vehicles, are of great interest in medicine. The broad variety of possibilities for biomedical use of these polymers, from drug delivery mechanisms and cell adhesion mediators to enzyme activity regulation mechanisms and gene expression, are of special concern.

Temperature-responsive polymers are smart polymers that exhibit phase separation and cloud points in response to change in temperature of solution (Okano 1993). The main aim of this study is to track the thermo-responsiveness of dual light- and temperature-responsive polymers using light illumination that is useful for the development of new dual or multi-responsive polymers with potential applications in drug and gene delivery systems. Thermo-responsive and pH-responsive properties have been shown by bio-based polymers such as gelatine and chitosan and offer the ability to function differently in a wide range of temperature and pH.

Silicone products are widely used in everyday contexts and in military applications (Muzafarov 2010). With advances in science and technology and growing consumer demands, the need for high-precision structural silicone materials has increased. A breakthrough in synthetic methods is therefore the most important factor in this field.

References

Chauhan, N., N.S. Hosmane and M. Mozafari. 2019. Boron-based polymers: opportunities and challenges. *Materials Today Chemistry* 14: 100184.

Chauhan, N.P.S., M. Mozafari, R. Ameta, P.B. Punjabi and S.C. Ameta. 2015. Spectral and thermal characterization of halogen-bonded novel crystalline oligo (p-bromoacetophenone formaldehyde). *The Journal of Physical Chemistry B* 119(7): 3223-3230.

Chauhan, N.P.S., M. Mozafari, N.S. Chundawat, K. Meghwal, R. Ameta and S.C. Ameta. 2016. High-performance supercapacitors based on polyaniline–graphene nanocomposites: Some approaches, challenges and opportunities. *Journal of Industrial and Engineering Chemistry* 36: 13-29.

Chauhan, N.P.S. 2019a. Biocidal Polymers, Walter de Gruyter GmbH & Co KG.

Chauhan, N.P.S. 2019b. Functionalized polyaniline and composites. *Fundamentals and Emerging Applications of Polyaniline*, 177-201. Elsevier.

Chauhan, N.P.S. and N.S. Chundawat. 2019. Inorganic and Organometallic Polymers, Walter de Gruyter GmbH & Co KG.

Chauhan, N.P.S., K. Meghwal, D. Singh and M. Mozafari. 2019. Heterotelechelic multiblock polymers using click chemistry. *Advanced Functional Polymers for Biomedical Applications*, 129-142. Elsevier.

Chauhan, N.P.S. and M. Mozafari. 2019. Synthetic route of PANI (II): Enzymatic method. *Fundamentals and Emerging Applications of Polyaniline*, 43-65. Elsevier.

Coessens, V., T. Pintauer and K. Matyjaszewsk. 2001. Functional polymers by atom transfer radical polymerization. *Progress in Polymer Science* 26(3): 337-377.

Donini, C., D.N. Robinson, P. Colombo, F. Giordano and N.A. Peppas. 2002. Preparation of poly (methacrylic acid-g-poly (ethylene glycol)) nanospheres from methacrylic monomers for pharmaceutical applications. *International Journal of Pharmaceutics* 245(1-2): 83-91.

Gauthier, M.A., M.I. Gibson and H.A. Klok. 2009. Synthesis of functional polymers by post-polymerization modification. *Angewandte Chemie International Edition* 48(1): 48-58.

Kargarzadeh, H., M. Mariano, D. Gopakumar, I. Ahmad, S. Thomas, A. Dufresne, J. Huang and N. Lin. 2018. Advances in cellulose nanomaterials. *Cellulose* 25(4): 2151-2189.

Koerner, H., W. Liua, M. Alexander, P. Mirau, H. Dowty and R.A. Vaia. 2005. Deformation–morphology correlations in electrically conductive carbon nanotube—thermoplastic polyurethane nanocomposites. *Polymer* 46(12): 4405-4420.

Konkolewicz, D., A. Gray-Weale and S. Perrier. 2009. Hyperbranched polymers by thiol– yne chemistry: From small molecules to functional polymers. *Journal of the American Chemical Society* 131(50): 18075-18077.

Martins, P., A.C. Lopes and S. Lanceros-Mendez. 2014. Electroactive phases of poly (vinylidene fluoride): Determination, processing and applications. *Progress in Polymer Science* 39(4): 683-706.

Moghaddam, M.M., M. Eftekhary, S. Erfanimanesh, A. Hashemi, V.F. Omrani, B. Farhadihosseinabadi, Z. Lasjerdi, M. Mossahebi-Mohammadi, N.P.S. Chauhan, A.M. Seifalian and M. Gholipourmalekabadi. 2018. Comparison of the antibacterial effects of a short cationic peptide and 1% silver bioactive glass against extensively drug-resistant bacteria, Pseudomonas aeruginosa and Acinetobacter baumannii, isolated from burn patients. *Amino Acids* 50(11): 1617-1628.

Morton, M. and L.J. Fetters. 1975. Anionic polymerization of vinyl monomers. *Rubber Chemistry and Technology* 48(3): 359-409.

Mozafari, M. and N.P.S. Chauhan. 2019a. Advanced Functional Polymers for Biomedical Applications. Elsevier.

Mozafari, M. and N.P.S. Chauhan. 2019b. Fundamentals and Emerging Applications of Polyaniline. Elsevier.

Muzafarov, A.M. 2010. Silicon Polymers. Springer Science & Business Media.

Okano, T. 1993. Molecular design of temperature-responsive polymers as intelligent materials. *Responsive Gels: Volume Transitions II* 179-197. Springer.

Pandey, A., N.S. Chundawat and N.P.S. Chauhan. 2019. Maleimide and acrylate based functionalized polymers. *Advanced Functional Polymers for Biomedical Applications* 167-189. Elsevier.

Perrier, S., P. Takolpuckdee and C.A. Mars. 2005. Reversible addition–fragmentation chain transfer polymerization: End group modification for functionalized polymers and chain transfer agent recovery. *Macromolecules* 38(6): 2033-2036.

Shibahara, K.-i. and B. Stillman. 1999. Replication-dependent marking of DNA by PCNA facilitates CAF-1-coupled inheritance of chromatin. *Cell* 96(4): 575-585.

Theato, P. and H.-A. Klok. 2013. Functional Polymers by Post-polymerization Modification: Concepts, Guidelines and Applications. John Wiley & Sons.

Tribet, C. 1998. Hydrophobically driven attachments of synthetic polymers onto surfaces of biological interest: Lipid bilayers and globular proteins. *Biochimie* 80(5-6): 461-473.

Vana, P. and Y. Yagci. 2013. Fundamentals of Controlled/Living Radical Polymerization. Royal Society of Chemistry.

Yang, J.C. and J.W. Mays. 2002. Synthesis and characterization of neutral/ionic block copolymers of various architectures. *Macromolecules* 35(9): 3433-3438.

Zhang, H. and E. Ruckenstein. 1998. Graft, block–graft and star-shaped copolymers by an in situ coupling reaction. *Macromolecules* 31(15): 4753-4759.

CHAPTER

2

Conjugated Polymers

Ebad Asadi[1] and Noushin Ezzati[2]*

[1] Department of Chemistry, Amirkabir University of Technology, Tehran, Iran

[2] Young Researchers and Elites Club, Science and Research Branch, Islamic Azad University, Tehran, Iran

1. Introduction

Conjugated polymers have been the subject of many studies (Hu et al. 2013). Conjugated polymers have a structure similar to that of metals and semiconductors (Kausar 2016). This class of materials consists of amazing electronic and photonic functional materials originating from reversible redox/de-doping mechanism and of unusual conjugated and π-conjugated length (Srinivasan et al. 2019). In the structure of conjugated polymers, besides the backbone, each atom has a π bond, which is much weaker than σ bonds and holds the atoms in a polymer chain together (Kausar 2016). Because of the charge mobility along their p-electron backbones, conjugated polymers show unusual electronic properties such as electrical conductivity, low-energy optical transitions, good mechanical properties, high affinity, suitable flexibility and high surface area (Naseri et al. 2018). Investigations in the field of conjugated polymers have generated outstanding scientific concepts and new aspects in advanced technologies in many areas (Kumar et al. 2015). The change in redox behavior that alters optical and electrical properties is adopted to develop conjugated polymer membranes for use in actuators, sensors and energy storage devices (Nautiyal et al. 2019). Additionally, the availability of various preparation methods makes it possible to design and tune conjugated polymers for new technologies (Pavase et al. 2018). Owing to these promising advantages, conjugated polymers have a wide range of applications. These polymers are used as supercapacitors (Afzal et al. 2017, Bouldin et al. 2010), lithium ion batteries (Luo et al. 2019), transistors (bin Mohd Yusoff and Shuib 2011), and sensors (Lai et al. 2016). Furthermore, conjugated polymers are widely used in biomimetic devices such as artificial

*Corresponding author: s416em@gmail.com

muscles, smart windows and mirrors, polymeric batteries, smart membranes, smart drug delivery systems, nervous interfaces, sensors and biological sensors (Otero et al. 2012). This chapter presents a brief introduction of conjugated polymers and their application in advanced materials. The main preparation methods are also briefly discussed.

2. Conjugated polymer classification

Various types of conducting polymers have been produced and investigated but according to ease of preparation, cost, applications and properties, some conducting polymers such as polypyrrole, polyaniline and polythiophene have become more widespread (Pavase et al. 2018).

2.1. Polyacetylene

Polyacetylene was first synthesized by Shirakawa in the 1970s (Awuzie 2017). It is one of the most widely known conjugated polymers with a linear polymer structure and the simplest member of the conjugated polymer family. The two hydrogen atoms in a basic unit of polyacetylene can be replaced by functional groups in order to prepare subtitled polyacetylene (Zhang et al. 2019, Stowell et al. 1989).

The origin of helicity in conjugated polymers is the aggregated form but substituted polyacetylenes have demonstrated the ability to twist the main chain in solutions. This ability refers to the steric repulsion among pendant groups (Nomura et al. 2002). Both *cis* and *trans* polyacetylene can be prepared in the form of flexible, shiny, semi-conducting films (Stowell et al. 1989). The structure of polyacetylene is depicted in Fig. 1.

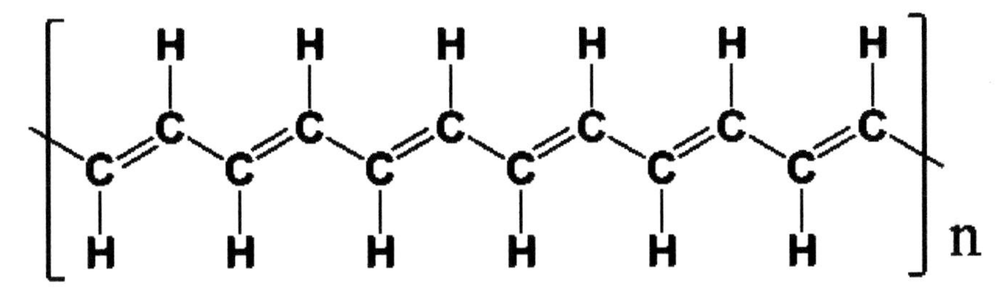

Fig. 1. Polyacetylene structure.

2.2. Polyaniline

Polyaniline (PANi) as the most common conducting polymer exists in various forms with different chemical, electrical and physical properties in addition to desirable environmental stability in the vicinity of oxygen and water (Kausar 2016). Owing to its promising tunable conductivity through doping and protonation mechanisms, PANi has a large number of potential applications compared to other conducting polymers (Le et al. 2013). The three main forms of PANi include fully oxidized state (pernigraniline), fully reduced (leucoemeraldine), and conducting emeraldine base, which is half oxidized (Naseri et al. 2018). The polymerization includes the oxidation of aniline monomer to form oligomers as depicted in Fig. 2. During the procedure, the dimers are first formed and then oxidized immediately to react with other aniline monomers via an electrophilic aromatic substitution and further oxidation followed by deprotonation to form trimmers occurs (Kumar et al. 2015). Polyaniline is prepared on the basis of aniline monomer via chemical, electrochemical, emulsion and other methods. The synthesis procedure seems to be simple, but the mechanism is complex (Nezakati et al. 2018, Kryszewski 1991). The preparation pathways give specific characterization and uses to a polymer. For example, chemical method is useful for mass production. The electrochemical method is used to prepare a fine film with high purity and polyaniline with desired optical and electrochemical sensing properties can be obtained via enzymatic method (Bouldin et al. 2010).

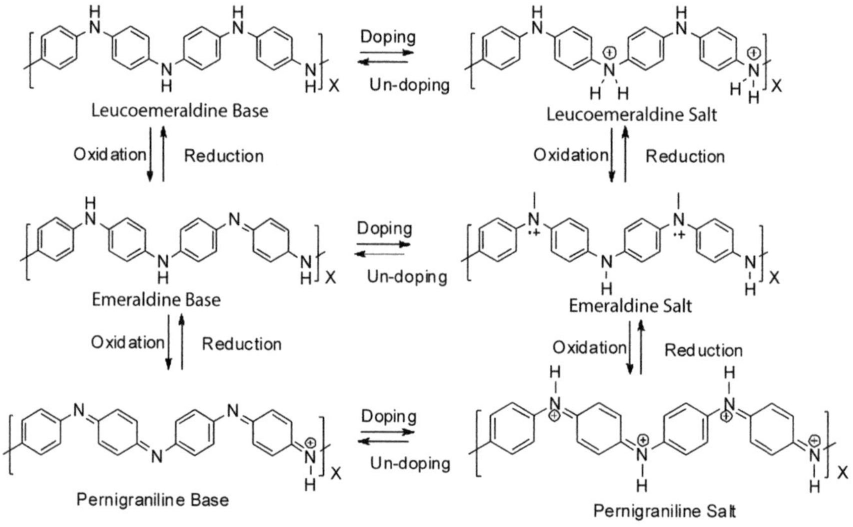

Fig. 2. Polyaniline structure.

2.3. Polypyrrole

Polypyrrole is obtained through polymerization of pyrrole as monomer (Awuzie 2017). This polymer is a heteroaromatic conducting polymer that was discovered in 1968 (Al-Mashat et al. 2012). Because of its suitable capacitance, facile synthesis, environmental stability, high conductivity, good chemical/ thermal stability, aqueous solubility, desirable biocompatibility, and low cost, it is considered a commercially viable polymer with a wide range of uses (Hosseini et al. 2015, Afzal et al. 2017, Jain et al. 2017, Antony and Mohanan 2019, Apetrei et al. 2018, Bhadra and Sarkar 2010). Owing to a rigid conjugated backbone structure, this polymer suffers from brittle, insoluble and infusible nature. To overcome these drawbacks many researches have proposed the addition of various fillers to the polymer matrix (Tarmizi et al. 2018).

Electrochemical polymerization of pyrrole is the most common method for preparation of polypyrrole but, depending on the target application, both chemical and electrochemical pathways are used to obtain this polymer. In general, freestanding films with high conductivity and suitable mechanical characteristics are obtained by electrochemical route but large-scale polymers can be prepared through chemical method (Shen and Wan 1998).

Polypyrrole is prepared via one electron oxidation of pyrrole to a radical cation that couples with other radical cations to produce bi-pyrrole and then this process is repeated to form polypyrrole (Fig. 3) (Kumar et al. 2015). The electrical conductivity of polypyrrole is influenced by doping ratio, the dopant nature and preparation technique (Kausar 2016).

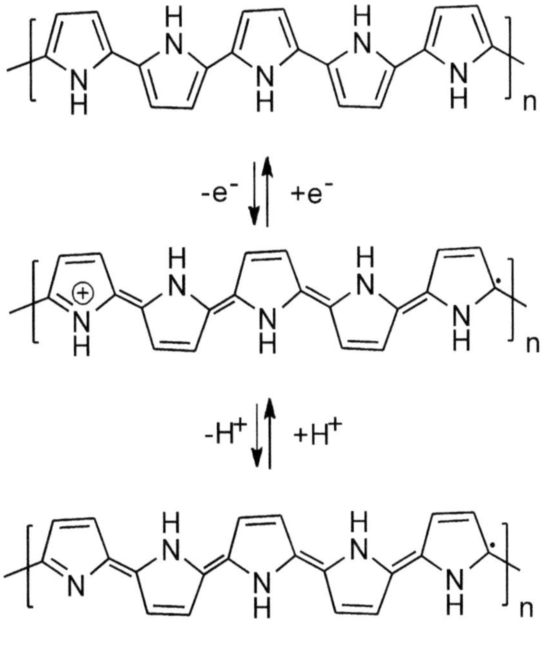

Fig. 3. Polypyrrole structure.

2.4. Polythiophene

Polythiophene was first developed by two researchers in 1980 (Awuzie 2017). This polymer belongs to the family of conjugated aromatic ring structure polymers (Fig. 4). Polythiophene derivatives such as poly (3-methylthiophene), poly(3-hexylthiophene) (P3HT) and poly (3,4-propylenedioxythiophene) are obtained easily through functionalization in 3 or 4 positions (Dutta and Rana 2019). In molecular scale, polythiophene demonstrates high charge density, suitable surface area, appropriate molecular structure, manageable electrochemical behavior and potential sites to interact with pollutant molecules. Also, the easy processing and scaling up, reasonable cost, and environmental stability of doped and un-doped form have given this polymer many uses (Jaymand et al. 2015, Husain et al. 2020). By electrochemical synthesis of polythiophene, highly conductive polythiophene with conductivity of 10-100 S/cm is obtained (Kausar 2016).

Depending on the preparation situation the charge on polythiophene surface can be regulated; for example, in pH below 8.5 it demonstrates positive potential, while negative zeta potential is observed in pH above 8.5 (Dutta and Rana 2019).

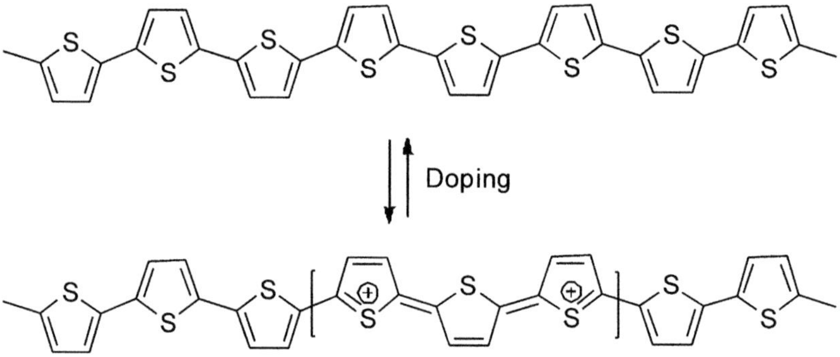

S S S S S S S S

Doping

S S S S S S S S

Fig. 4. Polythiophene structure.

2.5. Polyphenylene and polyparaphenylene

The first attempt to obtain a non-acetylene hydrocarbon polymer with doping process resulted in the development of polyparaphenylene in 1980 (Awuzie 2017). Polyparaphenylene possesses outstanding thermal and thermo-oxidation stability, good optical characteristics, and very wide conductive range. In view of the low solubility of this polymer in common organic solvents, its applications have been limited. However, this polymer can be used as light-emitting diodes, filler in rubber stock, and other products (Chen et al. 2006). The highly crystalline polyparaphenylene can be prepared through electrochemical route (Awuzie 2017).

3. Synthesis methods

3.1 Electro-polymerization

The electrochemical route consists of placing the counter and reference electrode into the solution comprising monomer and electrolyte as dopant. By applying voltage the polymer is produced onto the working electrode (Harun et al. 2007). Through electro-polymerization method by means of potential variation, the thickness and characterization of the film can be controlled. Among different types of electrochemical polymerization including potentiostatic, galvanostatic, sweeping, and pulse technique, the potentiodynamic route is preferable because a homogenous film forms and there is strong adherence of film on the electrode surface (Naseri et al. 2018). Electrochemical synthesis is the most common technique for preparation of some conjugated polymers such as polypyrrole. Additionally, it was proved that, though via electro-polymerization the nanostructure polymer is formed onto the electrode and the morphology of polymer is controllable, electro-polymerization is not a suitable method for mass production (Nezakati et al. 2018). Through electrochemical method there is no need for purification but the process could be reversible, so in some cases chemical synthesis is the better choice (Nezakati et al. 2018). The main limit of the electrochemical route is that the polymer is produced on the conducting substrate (Apetrei et al. 2018). Also, conducting polymers obtained by electro-polymerization are not processable or soluble (Stelmach et al. 2019). Another challenge of using this pathway is that via electrochemical synthesis the polymer can irreversibly oxidize and decompose on the electrode surface during the procedure. To solve this problem, the use of hydrazine or ammonium hydroxide has been proposed (Pavase et al. 2018). By electrochemical pathway, highly conductive films with desirable mechanical properties are obtained that are useful especially in electronic devices (Harun et al. 2007, Waltman and Bargon 1986).

3.2. Chemical method

Chemical polymerization is carried out through oxidation of monomer in the vicinity of dopant ions as oxidant or catalyst (Guimard et al. 2007). Linear conducting polymers can be obtained through chemical polymerization (Wang et al. 2002). The major advantage of chemical polymerization compared to electrochemical pathway is that it is cost efficient, whereas the amount of polymer obtained is obviously much greater than with the electrochemical method. But the inability of preparing thin films is one of the drawbacks defined for chemical polymerization (Awuzie 2017, Iqbal and Ahmad 2018).

3.3. Template polymerization

Template polymerization is mostly proposed for preparation of nanostructure conjugated polymers (Gangopadhyay and De 2000). Depending on the nature of monomers, chemistry of the template in addition to the type of adopted monomer and template interactions, various micro/nano morphologies are obtained (Sołoducho and Cabaj 2016). Also, the type of hard or soft template can influence the structure of the polymer (Jang 2006). Hard templates such as V_2O_5 or MnO_2 are desirable choices for fabrication of nano-size materials such as nanorods, nanotubes, and nanocapsules. In the soft template method, surfactants are used (Sołoducho and Cabaj 2016). Typically, microemulsion and reversed-microemulsion polymerization are considered as soft polymerization pathways. It has been proved that by changing the morphology of template, various kinds of nanostructures are obtained (Pan et al. 2010).

3.4. Vapor phase polymerization

Vapor phase polymerization method provides the possibility of simple preparation of conjugated polymers thin film in soluble and insoluble forms (Hanif et al. 2020). In this method, the polymerization is carried out at the liquid-vapor interface, which is prepared via casting of an oxidant solution on the suitable substrate. This substrate is exposed to monomer vapor. The following equation demonstrates the reaction that occurs during vapor phase polymerization: Fe^{3+} as the oxidant, X as the doping anion and M as the monomer (Brooke et al. 2017).

$$Fe^{3+}X^{-}_{3} + M \rightarrow Fe^{2+}X^{-}_{2} + M^{*} + X^{-}$$

The most common oxidants used in vapor phase polymerization are Fe(III) chloride and Fe $(Tosylate)_3$ and the usual dopants include 2,4,6 trimethylbenzene, paratoluensulfonate, tiron, polyvinyl alcohol, methanesulfonate, octanesulfonate, butanesulfonate and polystyrenesulfonate (Lawal and Wallace 2014).

The monomer concentration, temperature of substrate and humidity level are important factors for better conductivity, but the vital element in vapor phase polymerization is the humidity level (Hanif et al. 2020). The conjugated polymers prepared via vapor phase polymerization demonstrated applications in piezoelectric, supercapacitor, electronic devices, LEDs, biomedicine, artificial devices and other fields (Lawal and Wallace 2014, Hanif et al. 2020).

3.5. Photochemical method

Photochemical method is an important technique of preparation of conjugated polymers with advantages of rapid and simple processability at room temperature, suitable thermal stability, and environment-friendly process, and applications for coating insulate substrates. Additionally, with some changes in process such as excitation wavelength, desirable morphology can be obtained by this method. Two main approaches are commonly used for photochemical polymerization of known conducting polymers such as polyaniline, polythiophene and polypyrrole including direct photopolymerization through photoexcitation of the monomer and photopolymerization by using photocatalytic systems (Heydarnezhad et al. 2018, Kasisomayajula et al. 2010). This process is started via light emission of a photo-synthesizer, which leads to initiating species and continues through transfer of active sites to monomers. The photosynthesizer plays an important role in this process. This molecule transfers the absorbed energy to adjacent molecules in the photopolymerization process (Heydarnezhad et al. 2018, Sharif et al. 2016, Geetha et al. 2006).

3.6. Emulsion method

Emulsion polymerization is performed in a homogenous system (Yamak 2013). This outstanding method is used for preparation of conjugated polymers with desirable processability (Palaniappan and John 2008). A micelle-forming surfactant, a water-soluble initiator in a mixture, and a water-insoluble monomer are used (Kumar et al. 2015). Two types of emulsion polymerization are common: water in oil emulsion consisting of small drops of water in oil and oil in water emulsion consisting of small drops of oil in water. The procedure is conducted by addition of initiator, monomer and surfactant all together to the solvent to form micelles. Afterwards, the polymerization starts and well-dispersed colloids are obtained (Iqbal and Ahmad 2018, John et al. 2008).

3.7. Plasma polymerization

Plasma polymerization is considered a simple, inexpensive, solvent-free, room-temperature and oxidant-free method for fabrication of conjugated polymer ultra-thin films. This method offers the possibility of controlling thin films of thickness in nano-meter range onto the substrate (Tamirisa et al. 2004, Wang et al. 2002). During this pathway, polymerization occurs in a single step. The product obtained by this method possesses unique properties that cannot obtained with common wet polymerization techniques (Mathai et al. 2002). According to the type of substrate, various kinds of polymer are produced; for example, on a glass surface a continuous film is formed but on Pt substrate spongy spherical particles are obtained (Wang et al. 2002, Bhat and Joshi 1994). Polymers prepared through this method are applicable for electrochromic display, corrosion protection and sensors (Tamirisa et al. 2004, Stilwell and Park 1988).

3.8. Enzyme synthesis

Bio-catalytic and enzymatic synthesis pathways of polymer preparation have demonstrated environment-friendly characteristics. Oxidoreductases are the most common enzymes for preparing conducting polymers. These enzymes catalyse oxidation-reduction reactions.

The reaction starts with two-electron oxidation of enzyme and provides an oxidized intermediate using hydrogen peroxide. Afterwards, dimer is produced by oxidizing the monomer. Then the enzyme is reduced to another intermediate state that provides oxidation of another monomer. This process is repeated to form the polymer (Aizawa et al. 1990).

3.9. Solid state method

Solid state polymerization is an effective way to prepare conducting polymers. Low cost, simplicity and reduced pollution are some advantages of this pathway (Abdiryim et al. 2012). It has been proved that by polymerization of aniline in solid state, due to the restricted diffusion of reagents that results in less contact, the reaction will be slowed down. As a result, the nucleation and growth of polymer must be carefully controlled to obtain the target morphology (Konyushenko et al. 2008, Bhadra et al. 2009). In solid state method the reagents are brought into intimate contact using ball-milling or grinding process. Consequently, little or no solvent is used in the reaction process (Shao et al. 2012).

3.10. Sono-chemical route

The chemical effect of the sono-chemical method refers to acoustic cavitation including the formation, growth and implosive collapse of bubble in the liquid (Kumar et al. 2000). In sono-chemical method, because of the intense collapse effect, the mechanical and electrical properties of polymer improve considerably (Sivakumar and Gedanken 2005). The ultrasound radiation effects on the chemical reaction rate, destruction of the agglomerated sites, and improved dispersion reduce the size of the particles formed. The method has been demonstrated to result in long and thin polyaniline nanofibers (Abdelraheem et al. 2018).

4. Application

4.1. Gas sensing

The most important characteristic of conjugated polymers for sensor application is their conductivity. Moreover, the flexibility, low cost, room-temperature applicability and desirable mechanical properties make them good candidates for gas sensing devices. Compared with other common sensors, polymer-based sensors have the advantages of easy preparation pathway and the possibility of shape and morphology control to prepare films with micro/nanostructures (Sanjuán Cortázar et al. 2018, Rawal and Kaur 2013). Metal oxide particles, owing to their large and fast change in conductivity, are another candidate for gas sensing applications but they demonstrate some drawbacks, such as non-stable nature, high cost and morphological limit (Subbaiah et al. 2019).

The responses obtained from a sensor are presented in different ways such as changes in shape, solubility, or color, or modification of surface and fluorescence properties (Sanjuán Cortázar et al. 2018). Conjugated polymers can be doped and de-doped through redox reactions so their doping level is changed via transformation of electrons to or from analytes (Karmakar et al. 2017). Conjugated polymers are sensitive to a broad range of volatile spices and in the vicinity of volatile spices demonstrate changes in some parameters, of which the main one is conductivity (Stussi et al. 1996).

Among the known conjugated polymers (polyaniline, polythiophene and polypyrrole), polyaniline has attracted much attention for gas sensing applications and currently is applied in electronic nose systems. Polyaniline is considered a potential sensing material for detection of ammonia (Guo et al. 2019). Some disadvantages such as low mechanical strength and low reproducibility limit the use of polyaniline. The incorporation of some metal filler in polymer matrix has been proposed to overcome these drawbacks. The addition of metal fillers improves gas sensing and also the selectivity and stability of the sensor (Subbaiah et al. 2019, Ram et al. 2005b). Conducting polymer films or blends are used as air-borne volatiles such as alcohol, NH_3, NO_2, CO, and ether (Ram et al. 2005a). Polythiophene film has demonstrated the ability to detect ppb amount of hydralazine gases, and polypyrrole film doped with Cu and Pd presented CO gas detection ability (Ram et al. 2005b). Polypyrrole possesses suitable stability in aqueous media and air. To improve the gas sensing efficiency, the combination of polypyrrole with other inorganic materials is proposed. Also, different dopants can be used to change its properties and structure (Karmakar et al. 2017). Generally, by increment in surface to volume ratio, the gas sensing behavior of materials increases obviously. For example, by decreasing the particle size of polypyrrole (100-20 nm), 8-12% increment in sensitivity at 200 ppm ammonia is observed. Moreover, considering the further enhancement in surface to volume ratio, it has been suggested that 1D nanostructures such as nano-rod, nanowire, and nanotubes are more efficient morphologies than nano-spheres or nano-cubes in gas sensing applications (Rawal and Kaur 2013).

4.2. Biosensors

Today, biosensors are in demand in many fields of application such as biosecurity, medical science, agriculture, environment monitoring and food. They are categorized as immune biosensor, nuclear biosensor, enzymatic biosensor, receptor biosensor, and others. They are also classified as semiconductor, electrochemical, photoelectrical and piezoelectric biosensors. Biosensors are a series of analytical apparatus containing a connected detector and a biological sensing agent. In this kind of sensor the bio-element response is changed to a measurable signal such as voltage of current (Zheng et al. 2020). Among various types of biosensors, the enzymatic biosensors are the most common because of their high specificity and sensitivity (Lai et al. 2016). The most important challenge in electrochemical systems is transmitting electrical signals between signal transducers and biological recognition elements. The application of biosensors depends on the fast and simple electron transfer among analyte and electrode surface (Zamani et al. 2019, Jain et al. 2017). The easy doping de-doping ability, which obviously changes the electrical, optical and magnetic properties, together with high surface area and biocompatibility are some of the main characteristics of conjugated polymers in the field of biosensors (Lai et al. 2016, Antony and Mohanan 2019, Stelmach et al. 2019). Biosensors based on conjugated polymers have presented improvement in the flexibility and quantity of charge carrier, which modifies selectivity, conductivity and sensitivity (Zamani et al. 2019).

Polypyrrole and its composites with inorganic fillers have been studied as one of the most important candidates for designing bio-analytical sensor and biosensors such as immune sensors and DNA biosensors (Jain et al. 2017).

To improve the electrochemical performance and sensitivity of conducting polymers for biosensing applications the incorporation of carbon-based and metallic filler has been proposed (Zheng et al. 2019, Soni et al. 2019, Hansen et al. 2016). Other studies have proposed the conjugated polymer blends for better processability and improved mechanical strength of target sensor (Hansen et al. 2016). Along with increasing the available surface area the sensitivity of sensor for detection of analyst obviously improves (Mahmoudian et al. 2016). The film thickness is the most important factor in enhancing the biosensing characteristic, speed of response and sensitivity. Thinner films present a larger working area, which results in better interaction of target analyte. The presence of mediators can also improve the efficiency of biosensor. The most commonly used mediator is ferrocence and its derivatives (Hansen et al. 2016, Heller and Feldman 2008, Lawrence 2014).

4.3. Supercapacitor

Energy storage is presently one of the most essential requirements in urban life and industries. To fulfill this demand, various types of efficient energy storage such as fuel cells, batteries, and electrochemical supercapacitors have been fabricated (Li et al. 2019). Supercapacitors, which can be charged and discharged in just a few moments and have the advantages of high power density and long cycle life, have the greatest potential in energy storage devices (Meng et al. 2017, Bello et al. 2016). The performance of supercapacitors is dependent on cyclic voltammetry, galvanostatic charge/discharge, and electrochemical impedance spectroscopy. Also, the mechanism of energy storage in supercapacitors includes quick and reversible Faradaic redox reaction on the surface and bulk in the vicinity of electrode materials and electrolyte ions (Wu et al. 2019).

Conjugated polymers and transition metal oxide are the common materials used in supercapacitors (Wang et al. 2015). Pseudocapacitance ability provided by conducting polymers is attributed to doping, de-doping of the polymer backbone. This procedure results in intercalation and de-intercalation of electrolyte ions within the polymer electrode to maintain charge neutrality (Fong et al. 2017). The conducting polymers provide high specific capacitance in comparison with double-layered supercapacitors and also show faster kinetics than most inorganic material-based capacitors (Wang et al. 2016). The use of pure conducting polymer as electrodes does not provide desirable efficiency, and therefore binary and ternary conducting polymer composites with other materials such as metal oxide or carbon-based materials have been proposed (Meng et al. 2017, Wang et al. 2016).

It has been proved that conducting polymer nanostructures show higher efficiency as supercapacitors. Their advantages include higher charge/discharge rates caused by higher electrode/electrolyte contact area, and short path lengths for electronic transport and ion transport (Pan et al. 2010).

4.4. Corrosion protection

In 1985, MacDiarmid suggested the corrosion protection application of conducting polymers (Pehkonen and Yuan 2018). The correction mechanism via organic compounds depends on their barrier properties. Conducting polymers, as a new class of organic coatings, have offered outstanding properties in corrosion protection (Eslami et al. 2019). Polyaniline and polypyrrole are two common conductive polymers applied for corrosion protection of aluminum, steel and magnesium (Volpi et al. 2012, Ascencio et al. 2018, Iroh and Su 2000). The mechanism of protection is based on the electrochemical interaction of these polymers with target surface. The redox activity results in anion exchange with electrolyte to compensate for the positive charge on polymer branches. A conjugated polymer coat onto the metal surface could tolerate the reduction reaction and serve as oxidizer according to "passivation protection mechanism" (Eslami et al. 2019, Volpi et al. 2012).

The studies have proved that in NaCl solution the non-conducting form causes the best protection, while in HCl solution the conducting form is responsible for the best protection (Tan and Blackwood 2003). The corrosion protection mechanism of polyaniline on metal is based on the formation of passive oxide layer of polyaniline on the target surface through cyclic reduction of emeraldine salt form of polyaniline to leucoemeraldine form with subsequent reoxidation of the latter species by oxygen (Tavandashti et al. 2016, Baldissera and Ferreira 2012).

It has been proved that metal oxide nanoparticles and carbon-based nanomaterials incorporated in conjugated polymer matrix influence the morphology and electrochemical properties of polymer and enhance corrosion protection ability (El Jaouhari et al. 2019, Breslin et al. 2005). For example, polyaniline/TiO_2 particles demonstrated obvious corrosion protection in aggressive situation compared with polyaniline. This efficiency was attributed to improvement of barrier characteristics, redox property of polyaniline, and formation of p-n junctions, which do not permit easy charge transport when the coating is scratched or damaged (Baldissera and Ferreira 2012).

Use of conducting polymers as corrosion protection has some drawbacks. For example, continuous coating of conducting polymers is not possible because they may cause corrosive attack in the vicinity of large defects. Consequently, they may be more useful as additives in composite coating for corrosion protection (Bai et al. 2015).

4.5. Electromagnetic shielding

With the extensive use of wireless communication and network, the demand for electromagnetic shielding materials has been high. The first candidate for electromagnetic shielding was metal sheets, but metals have the disadvantages of corrosion, non-flexibility, high price and heavy weight (Gahlout and Choudhary 2019, Yan et al. 2017). Conducting polymers found many applications in various fields as soon as they were discovered. Conducting polymers as synthetic metals have been considered as alternatives for metals in many fields including electromagnetic shielding (Gahlout and Choudhary 2019, Hosseini et al. 2014, 2015).

Many studies have concentrated on microwave-absorbing materials. It has been found that addition of some conductive or magnetic filler through magnetic chain could improve shielding performance (Jafarian et al. 2019). A three-phase composite of conducting polymers, magnetic particles and graphene is considered a suitable choice for dielectric loss (Majdzadeh-Ardakani and Holl 2017). The heterogeneous interface in organic-inorganic composite improves the microwave properties due to dielectric loss as a consequence of multi-interfacial polarization. Additionally, the interface causes reflection and microwave scattering (Jiang et al. 2016). Conducting polymers such as polyaniline and polypyrrole do not reflect the electromagnetic waves and with absorption mechanism diminish these waves (Farrokhi et al. 2016).

4.6. Electrochromic devices

Electrochromism is defined as the ability to change optical characteristics within the whole electromagnetic spectrum under an applied voltage. In other words, these materials are used to change from opaque to transmissive at certain regions of electromagnetic spectrum (Harun et al. 2007). In electrochromic devices, because the system needs to switch from a fully bleached state to a fully colored state, two redox couples with two conditions are mandatory, i.e., colorless bleached state and the complementary coloration.

The conduction procedure in conjugated polymers involves formation of polaron and bipolaron. The extensive delocalization of π electrons has made conducting polymers a promising candidate for electrochromic devices (Zhan et al. 2017, Vaia et al. 1993, Fukushima et al. 1988).

4.7. Organic light-emitting diodes (LEDs)

The development of organic light-emitting diodes has attracted a great deal of interest (Jang et al. 2008). The first investigation of conjugated polymer application as light-emitting diodes dates from 1990 (Alonso et al. 2009). Polyaniline, polypyrrole and poly(3,4-ethylenedioxythiophene) have been studied as transparent anodic material for use in OLEDs (Jang et al. 2008, Mohsennia et al. 2015).

Polyaniline as the best-known conjugated polymer demonstrates barrier properties against oxygen and can act as planarizing layer to prevent electrical shorts and extend device lifetime. This characteristic improves the efficiency of a device by enhancing the hole injection layer. In other words, it acts as a buffer layer that inhibits the diffusion of oxygen from the anode into the device (Mohsennia et al. 2015). Additionally, polyaniline has shown increased transparency and conductivity, which lead to lower voltage operation (Bejbouj et al. 2010, Huh et al. 2007). The greatest challenge posed by use of conjugated polymers such as polyaniline in LED devices is their poor solubility. Functionalized polyaniline has been proposed with improved processability in organic solvents (Mohsennia et al. 2015).

4.8. Drug delivery

Following recent progress in conducting polymer applications, they have been introduced as potential materials for drug delivery. The performance of conducting polymers in drug delivery has been proven through *in vitro* and *in vivo* analyses (Boehler et al. 2019). The drug loading mechanism refers to the electrostatic interactions between the drug and the polymer. During preparation of conjugated polymer films, in order to counterbalance the positive charge, the dopant anions are incorporated into the polymer associated with the oxidative polymerization. In general, when a drug is incorporated directly as a dopant into the conducting polymer, loading is limited to 1 drug molecule per 3-5 monomer units (Otero et al. 2012). Variation in charge, volume, molecular permeability and hydrophobic/hydrophilic balance changes the drug release efficiency (Uppalapati et al. 2016).

Most conjugated polymers are not biodegradable and this is one of the main concerns with using conducting polymers as drug carrier. However, it has been demonstrated that biodegradability can be achieved via surface modification (Svirskis et al. 2010). Polymer properties including surface energy (hydrophobicity), surface roughness, conductivity, mechanical actuation and dopant retention influence the biocompatibility. Generally, polypyrrole is known as the most biocompatible conjugated polymer (Svirskis et al. 2010). Polyaniline, owing to some advantages, has been a candidate in drug delivery applications. The possibility of transformation from insulator to high conductor in various pH environments, easy processing and suitable thermal and mechanical stability have made polyaniline useful in drug delivery systems (Fan et al. 2008). Additionally, poly(3,4 ethylenedioxythiophene) (PEDOT) has demonstrated promising potential in drug delivery (Boehler et al. 2019).

In the case of drug delivery, the electrochemical method is preferable because it allows control of the quantity and final electrical properties of the polymer (Uppalapati et al. 2016). Also, preparation of the conducting polymers in micro/nanostructures provides them greater loading capacity and improved efficiency (Uppalapati et al. 2016, Svirskis et al. 2010).

5. Nanotechnology on conjugated polymers

The characteristics of bulk polymers change at nano-scale due to quantum confinement effect, obviously (Rawal and Kaur 2013). Various methods are adapted to prepared nanostructure conducting polymers such as interfacial technique, template-assisted, well-controlled solution synthesis, and seeding polymerization (Liu et al. 2010). Also, the fabrication strategies for conjugated polymer nanomaterials are categorized into hard template synthesis, soft template synthesis, and template-free technique (Sołoducho and Cabaj 2016, Pan et al. 2010). In the template-free approach, special polymerization situations are used to manage the initial nucleation reactions, which results in final microstructure of the obtained polymer. The hard template method offers convenience, simplicity and controllability in preparation. The structural properties of target polymer can be controlled by hard template structure and other features. The challenge is the removal of the template from the final polymer. Template polymerization can be applied in chemical and electrochemical polymerization (Uppalapati et al. 2016). Surfactants are thermodynamically stable materials adopted as soft template molecules for preparation of nanostructure conjugated polymers. Among surfactants used as soft templates, reverse micro-emulsion is a common route to prepare nanomaterials. The surface micelles, liquid crystalline phase and soap bubbles are other new soft templates (Israelachvili and Adams 1976). The polymerization medium and condition directly influence the morphology of polymer particles. For example, colloidal polyaniline particles are fabricated in the presence of hydroxypropylcellulose, while under similar conditions, when the polymerization was done in ice, sponge-like macroporous paticles were observed (Gangopadhyay and De 2000).

Electrospinning method has been used in many studies to prepare electro-conductive polymers. Ultra-thin film and nano/microfibers are produced using this method (Li et al. 2014). Electrospinning, bubble electrospinning, centrifugal spinning, melt-blowing, and bio-component spinning are common methods to develop nanofibers (Yanılmaz and Sarac 2014). In electrospinning process, external electrical force is applied for polymerization (Yanılmaz and Sarac 2014). This technique is a viable method for preparing 3D ultrafine nanofibers (Razak et al. 2015).

6. Conclusion and outlook

The tremendous array of applications of conjugated polymers in various fields was discovered gradually. Conjugated polymers demonstrated characteristics of plastics and semiconductor materials. Various methods are proposed to prepare conjugated polymers. It has been proved that preparation method and condition directly influence the final characteristics of the polymer. The type of materials used and the mode of fabrication can be regulated according to the target applications. Also, owing to carbon-based structure of these materials, many approaches are taken to control the organic synthesis process. Conjugated polymers also have inherent defects that limit their applications. To overcome these drawbacks the incorporation of some fillers such as metal particles or carbon-based materials is proposed. As a result, many conjugated polymer composites with outstanding properties have been prepared and investigated. Nanotechnology is a widespread approach in many scientific fields. In the synthesis of conjugated polymers the micro/nanostructure of materials is very important, whereas the advanced applications of these materials directly depend on their morphology, shape and size. Many studies have discovered important factors in the field of conducting polymers and nanomaterials, but many challenges remain to be met. With increased focus on the synthesis mechanisms, it appears possible that greater control will be achieved over the features of conducting polymers.

References

Abdelraheem, A., A.H. El-Shazly and M.F. Elkady. 2018. Characterization of atypical polyaniline nano-structures prepared via advanced techniques. *Alexandria Engineering Journal* 57(4): 3291-3297.

Abdiryim, Tursun, Aminam Ubul, Ruxangul Jamal and Adalet Rahman. 2012. Solid-state synthesis of polyaniline/single-walled carbon nanotubes: A comparative study with polyaniline/multi-walled carbon nanotubes. *Materials* 5(7): 1219-1231.

Afzal, Adeel, Faraj A. Abuilaiwi, Amir Habib, Muhammad Awais, Samaila B. Waje and Muataz A. Atieh. 2017. Polypyrrole/carbon nanotube supercapacitors: Technological advances and challenges. *Journal of Power Sources* 352: 174-186.

Aizawa, Masuo, Lili Wang, Hiroaki Shinohara and Yoshihito Ikariyama. 1990. Enzymatic synthesis of polyaniline film using a copper-containing oxidoreductase: Bilirubin oxidase. *Journal of Biotechnology* 14(3-4): 301-309.

Al-Mashat, Laith, Catherine Debiemme-Chouvy, Stephan Borensztajn and Wojtek Wlodarski. 2012. Electropolymerized polypyrrole nanowires for hydrogen gas sensing. *The Journal of Physical Chemistry C* 116 (24): 13388-13394.

Alonso, J.L., J.C. Ferrer, M.A. Cotarelo, F. Montilla and S. Fernández de Ávila. 2009. Influence of the thickness of electrochemically deposited polyaniline used as hole transporting layer on the behaviour of polymer light-emitting diodes. *Thin Solid Films* 517(8): 2729-2735.

Antony, Navya and P.V. Mohanan. 2019. Template synthesized polypyrroles as a carrier for diastase alpha amylase immobilization. *Biocatalysis and Agricultural Biotechnology* 19: 101164.

Apetrei, Roxana-Mihaela, Geta Carac, Gabriela Bahrim, Almira Ramanaviciene and Arunas Ramanavicius. 2018. Modification of Aspergillus niger by conducting polymer, Polypyrrole, and the evaluation of electrochemical properties of modified cells. *Bioelectrochemistry* 121: 46-55.

Ascencio, M., M. Pekguleryuz and S. Omanovic. 2018. Corrosion behaviour of polypyrrole-coated WE43 Mg alloy in a modified simulated body fluid solution. *Corrosion Science* 133: 261-275.

Awuzie, C.I. 2017. Conducting polymers. *Materials Today: Proceedings* 4(4): 5721-5726.

Bai, Xiaoxia, The Hai Tran, Demei Yu, Ashokanand Vimalanandan, Xiujie Hu and Michael Rohwerder. 2015. Novel conducting polymer based composite coatings for corrosion protection of zinc. *Corrosion Science* 95: 110-116.

Baldissera, A.F. and Carlos A. Ferreira. 2012. Coatings based on electronic conducting polymers for corrosion protection of metals. *Progress in Organic Coatings* 75(3): 241-247.

Bejbouj, Habiba, Laurence Vignau, Jean Louis Miane, Thomas Olinga, Guillaume Wantz, Azzedine Mouhsen, El Moustafa Oualim and Mouhammed Harmouchi 2010. Influence of the nature of polyaniline-based hole-injecting layer on polymer light emitting diode performances. *Materials Science and Engineering: B* 166(3): 185-189.

Bello, A., F. Barzegar, M.J. Madito, D.Y. Momodu, A.A. Khaleed, T.M. Masikhwa, J.K. Dangbegnon and N. Manyala. 2016. Stability studies of polypyrole-derived carbon based symmetric supercapacitor via potentiostatic floating test. *Electrochimica Acta* 213: 107-114.

Bhadra, J. and D. Sarkar. 2010. Polypyrrole nanocomposite made by polypyrrole dispersion in poly (vinyl alcohol) matrix. *Indian Journal of Physics* 84(10): 1321-1325.

Bhadra, Sambhu, Nam Hoon Kim, Kyong Yop Rhee and Joong Hee Lee. 2009. Preparation of nanosize polyaniline by solid-state polymerization and determination of crystal structure. *Polymer International* 58(10): 1173-1180.

Bhat, N.V. and N.V. Joshi. 1994. Structure and properties of plasma-polymerized thin films of polyaniline. *Plasma Chemistry and Plasma Processing* 14(2): 151-161.

bin Mohd Yusoff, Abd Rashid and Saiful Anuar Shuib. 2011. Metal-base transistor based on simple polyaniline electropolymerization. *Electrochimica Acta* 58: 417-421.

Boehler, Christian, Felix Oberueber and Maria Asplund. 2019. Tuning drug delivery from conducting polymer films for accurately controlled release of charged molecules. *Journal of Controlled Release* 304: 173-180.

Bouldin, Ryan, Akshay Kokil, Sethumadhavan Ravichandran, Subhalakshmi Nagarajan, Jayant Kumar, Lynne A. Samuelson, Ferdinando F. Bruno and Ramaswamy Nagarajan. 2010. Enzymatic synthesis of electrically conducting polymers. Chapter 23, pp. 315-341. *In*: Green Polymer Chemistry: Biocatalysis and Biomaterials. Volume 1043. ACS Publications. DOI: 10.1021/bk-2010-1043.ch023

Breslin, Carmel B., Anna M. Fenelon and Kenneth G. Conroy. 2005. Surface engineering: Corrosion protection using conducting polymers. *Materials & Design* 26(3): 233-237.

Brooke, Robert, Philip Cottis, Pejman Talemi, Manrico Fabretto, Peter Murphy and Drew Evans. 2017. Recent advances in the synthesis of conducting polymers from the vapour phase. *Progress in Materials Science* 86: 127-146.

Chen, Hongxiang, Hongying Sun, Maosheng Zheng, Qingming Jia and Wanhai Dang. 2006. Characterization and thermo-stability of alkylated polyparaphenylene. *Polymer Bulletin* 56(2-3): 221-227.

Dutta, Kingshuk and Dipak Rana. 2019. Polythiophenes: An emerging class of promising water purifying materials. *European Polymer Journal* 116: 370-385.

El Jaouhari, A., A. Chennah, S. Ben Jaddi, H. Ait Ahsaine, Z. Anfar, Y. Tahiri Alaoui, Y. Naciri, A. Benlhachemi and M. Bazzaoui. 2019. Electrosynthesis of zinc phosphate-polypyrrole coatings for improved corrosion resistance of steel. *Surfaces and Interfaces* 15: 224-231.

Eslami, Maryam, Flavio Deflorian and Caterina Zanella. 2019. Electrochemical performance of polypyrrole coatings electrodeposited on rheocast aluminum-silicon components. *Progress in Organic Coatings* 137: 105307.

Fan, Qiuxi, Kamalesh K. Sirkar and Bozena Michniak. 2008. Iontophoretic transdermal drug delivery system using a conducting polymeric membrane. *Journal of Membrane Science* 321(2): 240-249.

Farrokhi, Hadi, Omid Khani, Firouzeh Nemati and Mohammad Jazirehpour. 2016. Synthesis and investigation of microwave characteristics of polypyrrole nanostructures prepared via self-reactive flower-like MnO_2 template. *Synthetic Metals* 215: 142-149.

Fong, Kara D., Tiesheng Wang and Stoyan K. Smoukov. 2017. Multidimensional performance optimization of conducting polymer-based supercapacitor electrodes. *Sustainable Energy & Fuels* 1(9): 1857-1874.

Fukushima, YI., A. Okada, M. Kawasumi, T. Kurauchi and O. Kamigaito. 1988. Swelling behaviour of montmorillonite by poly-6-amide. *Clay Minerals* 23(1): 27-34.

Gahlout, Pragati and Veena Choudhary. 2019. Microwave shielding behaviour of polypyrrole impregnated fabrics. *Composites Part B: Engineering* 175: 107093.

Gangopadhyay, Rupali and Amitabha De. 2000. Conducting polymer nanocomposites: A brief overview. *Chemistry of Materials* 12(3): 608-622.

Geetha, S., Chepuri R.K. Rao, M. Vijayan and D.C. Trivedi. 2006. Biosensing and drug delivery by polypyrrole. *Analytica Chimica Acta* 568(1-2): 119-125.

Guimard, Nathalie K., Natalia Gomez and Christine E. Schmidt. 2007. Conducting polymers in biomedical engineering. *Progress in Polymer Science* 32(8-9): 876-921.

Guo, Zhi, Ningbo Liao, Miao Zhang and Aixin Feng. 2019. Enhanced gas sensing performance of polyaniline incorporated with graphene: A first-principles study. *Physics Letters A* 383(23): 2751-2754.

Hanif, Zahid, Daeyong Shin, Dongwhi Choi and Sung Jea Park. 2020. Development of a vapor phase polymerization method using a wet-on-wet process to coat polypyrrole on never-dried nanocellulose crystals for fabrication of compression strain sensor. *Chemical Engineering Journal* 381: 122700.

Hansen, Betina, Marcele A. Hocevar and Carlos A. Ferreira. 2016. A facile and simple polyaniline-poly (ethylene oxide) based glucose biosensor. *Synthetic Metals* 222: 224-231.

Harun, Mohd Hamzah, Elias Saion, Anuar Kassim, Noorhana Yahya and Ekramul Mahmud. 2007. Conjugated conducting polymers: A brief overview. *UCSI Academic Journal: Journal for the Advancement of Science & Arts* 2: 63-68.

Heller, Adam and Ben Feldman. 2008. Electrochemical glucose sensors and their applications in diabetes management. *Chemical Reviews* 108(7): 2482-2505.

Heydarnezhad, Hamid Reza, Behzad Pourabbas and Masoud Tayefi. 2018. Conducting Electroactive Polymers via Photopolymerization: A Review on Synthesis and Applications. *Polymer-Plastics Technology and Engineering* 57(11): 1093-1109.

Hosseini, Seyed Hossein, M. Askari and S. Noushin Ezzati. 2014. X-ray attenuating nanocomposite based on polyaniline using Pb nanoparticles. *Synthetic Metals* 196: 68-75.

Hosseini, Seyed Hossein, S. Noushin Ezzati and M. Askari. 2015. Synthesis, characterization and X-ray shielding properties of polypyrrole/lead nanocomposites. *Polymers for Advanced Technologies* 26(6): 561-568.

Hu, Peng, Lei Han and Shaojun Dong. 2013. A facile one-pot method to synthesize a polypyrrole/hemin nanocomposite and its application in biosensor, dye removal, and photothermal therapy. *ACS Applied Materials & Interfaces* 6(1): 500-506.

Huh, Dal Ho, Miyoung Chae, Woo Jin Bae, Won Ho Jo and Tae-Woo Lee. 2007. A soluble self-doped conducting polyaniline graft copolymer as a hole injection layer in polymer light-emitting diodes. *Polymer* 48(25): 7236-7240.

Husain, Ahmad, Sharique Ahmad and Faiz Mohammad. 2020. Synthesis, characterisation and ethanol sensing application of polythiophene/graphene nanocomposite. *Materials Chemistry and Physics* 239: 122324.

Iqbal, Sajid and Sharif Ahmad. 2018. Recent development in hybrid conducting polymers: Synthesis, applications and future prospects. *Journal of Industrial and Engineering Chemistry* 60: 53-84.

Iroh, Jude O. and Wencheng Su. 2000. Corrosion performance of polypyrrole coating applied to low carbon steel by an electrochemical process. *Electrochimica Acta* 46(1): 15-24.

Israelachvili, J.N. and G.E. Adams. 1976. Theory of self-assembly of hydrocarbon amphiphiles into micelles and bilayers. *J. Chem. Soc., Faraday Trans.* 2 72: 1525-1568. DOI: https://doi.org/10.1039/F29767201525

Jafarian, Mojtaba, Seyyed Salman Seyyed Afghahi, Yomen Atassi and Mohsen Salehi. 2019. Insights on the design of a novel multicomponent microwave absorber based on $SrFe_{10}A_{12}O_{19}$ and $Ni_{0.5}Zn_{0.5}Fe_2O_4$/MWCNTs/polypyrrole. *Journal of Magnetism and Magnetic Materials* 471: 30-38.

Jain, Rajeev, Nimisha Jadon and Archana Pawaiya. 2017. Polypyrrole based next generation electrochemical sensors and biosensors: A review. *TrAC Trends in Analytical Chemistry* 97: 363-373.

Jang, Jyongsik. 2006. Conducting polymer nanomaterials and their applications. pp. 189-260. *In*: Emissive Materials Nanomaterials. Volume 199. Springer. DOI: https://doi.org/10.1007/12_075

Jang, Jyongsik, Jungseok Ha and Kyungho Kim. 2008. Organic light-emitting diode with polyaniline-poly (styrene sulfonate) as a hole injection layer. *Thin Solid Films* 516(10): 3152-3156.

Jaymand, Mehdi, Maryam Hatamzadeh and Yadollah Omidi. 2015. Modification of polythiophene by the incorporation of processable polymeric chains: Recent progress in synthesis and applications. *Progress in Polymer Science* 47: 26-69.

Jiang, Linwen, Zhenhua Wang, Dianyu Geng, Yu Wang, Jing An, Jun He, Da Li, Wei Liu and Zhidong Zhang. 2016. Carbon-encapsulated Fe nanoparticles embedded in organic polypyrrole polymer as a high performance microwave absorber. *The Journal of Physical Chemistry C* 120(49): 28320-28329.

John, Amalraj, Srinivasan Palaniappan, David Djurado and Adam Pron. 2008. One-step preparation of solution processable conducting polyaniline by inverted emulsion polymerization using didecyl ester of 4-sulfophthalic acid as multifunctional dopant. *Journal of Polymer Science Part A: Polymer Chemistry* 46(3): 1051-1057.

Karmakar, N., R. Fernandes, Shilpa Jain, U.V. Patil, Navinchandra G. Shimpi, N.V. Bhat and D.C. Kothari. 2017. Room temperature NO_2 gas sensing properties of p-toluenesulfonic acid doped silver-polypyrrole nanocomposite. *Sensors and Actuators B: Chemical* 242: 118-126.

Kasisomayajula, Subramanyam V., Xiaoning Qi, Chris Vetter, Kenneth Croes, Drew Pavlacky and Victoria J. Gelling. 2010. A structural and morphological comparative study between chemically synthesized and photopolymerized poly (pyrrole). *Journal of Coatings Technology and Research* 7(2): 145-158.

Kausar, Ayesha. 2016. Review on structure, properties and appliance of essential conjugated polymers. *American Journal of Polymer Science & Engineering* 4(1): 91-102.

Konyushenko, Elena N., Jaroslav Stejskal, Miroslava Trchova, Natalia V. Blinova and Petr Holler. 2008. Polymerization of aniline in ice. *Synthetic Metals* 158(21-24): 927-933.

Kryszewski, M. 1991. Heterogeneous conducting polymeric systems: Dispersions, blends, crystalline conducting networks—An introductory presentation. *Synthetic Metals* 45(3): 289-296.

Kumar, R. Vijaya, Y. Mastai and A. Gedanken. 2000. Sonochemical synthesis and characterization of nanocrystalline paramelaconite in polyaniline matrix. *Chemistry of Materials* 12(12): 3892-3895.

Kumar, Ravindra, Satyendra Singh and B.C. Yadav. 2015. Conducting polymers: Synthesis, properties and applications. *International Advanced Research Journal in Science, Engineering and Technology* 2(11): 110-124.

Lai, Jiahui, Yingchun Yi, Ping Zhu, Jing Shen, Kesen Wu, Lili Zhang and Jian Liu. 2016. Polyaniline-based glucose biosensor: A review. *Journal of Electroanalytical Chemistry* 782: 138-153.

Lawal, Abdulazeez T. and Gordon G. Wallace. 2014. Vapour phase polymerisation of conducting and non-conducting polymers: A review. *Talanta* 119: 133-143.

Lawrence, Jeffrey. 2014. "I Read even the Scraps of Paper I Find on the Street": A Thesis on the Contemporary Literatures of the Americas. *American Literary History* 26(3): 536-558.

Le, Trong Huyen, Ngoc Thang Trinh, Le Huy Nguyen, Hai Binh Nguyen, Van Anh Nguyen and Tuan Dung Nguyen. 2013. Electrosynthesis of polyaniline–mutilwalled carbon nanotube nanocomposite films in the presence of sodium dodecyl sulfate for glucose biosensing. *Advances in Natural Sciences: Nanoscience and Nanotechnology* 4(2): 025014.

Li, Jak, Jinli Qiao and Keryn Lian. 2019. Hydroxide ion conducting polymer electrolytes and their applications in solid supercapacitors: A review. *Energy Storage Materials* 24: 6-21. January 2020. DOI: https://doi.org/10.1016/j.ensm.2019.08.012

Li, Xiao-Qiang, Wan-Wan Liu, Shui-Ping Liu, Meng-Juan Li, Yong-Gui Li and Ming-Qiao Ge. 2014. In situ polymerization of aniline in electrospun microfibers. *Chinese Chemical Letters* 25(1): 83-86.

Liu, Zhen, Xinyu Zhang, Selcuk Poyraz, Sumedh P. Surwade and Sanjeev K. Manohar. 2010. Oxidative template for conducting polymer nanoclips. *Journal of the American Chemical Society* 132(38): 13158-13159.

Luo, Yani, Ruisong Guo, Tingting Li, Fuyun Li, Zhichao Liu, Mei Zheng, Baoyu Wang, Zhiwei Yang, Honglin Luo and Yizao Wan. 2019. Application of Polyaniline for Li-Ion Batteries, Lithium–Sulfur Batteries, and Supercapacitors. *ChemSusChem* 12(8): 1591-1611.

Mahmoudian, Mohammad Reza, Wan Jeffrey Basirun and Yatimah Binti Alias. 2016. Sensitive dopamine biosensor based on polypyrrole-coated palladium silver nanospherical composites. *Industrial & Engineering Chemistry Research* 55(25): 6943-6951.

Majdzadeh-Ardakani, Kazem and Mark M. Banaszak Holl. 2017. Nanostructured materials for microwave receptors. *Progress in Materials Science* 87: 221-245.

Mathai, C. Joseph, S. Saravanan, M.R. Anantharaman, S. Venkitachalam and S. Jayalekshmi. 2002. Characterization of low dielectric constant polyaniline thin film synthesized by ac plasma polymerization technique. *Journal of Physics D: Applied Physics* 35(3): 240.

Meng, Qiufeng, Kefeng Cai, Yuanxun Chen and Lidong Chen. 2017. Research progress on conducting polymer based supercapacitor electrode materials. *Nano Energy* 36: 268-285.

Mohsennia, M., M. Massah Bidgoli, F. Akbari Boroumand and A. Mohsen Nia. 2015. Electrically conductive polyaniline as hole-injection layer for MEH-PPV: BT based polymer light emitting diodes. *Materials Science and Engineering: B* 197: 25-30.

Naseri, Maryam, Lida Fotouhi and Ali Ehsani. 2018. Recent progress in the development of conducting polymer-based nanocomposites for electrochemical biosensors applications: A mini-review. *The Chemical Record* 18(6): 599-618.

Nautiyal, Amit, Jonathan E. Cook and Xinyu Zhang. 2019. Tunable electrochemical performance of polyaniline coating via facile ion exchanges. *Progress in Organic Coatings* 136: 105309.

Nezakati, Toktam, Amelia Seifalian, Aaron Tan and Alexander M. Seifalian. 2018. Conductive polymers: Opportunities and challenges in biomedical applications. *Chemical Reviews* 118(14): 6766-6843.
Nomura, Ryoji, Hideo Nakako and Toshio Masuda. 2002. Design and synthesis of semiflexible substituted polyacetylenes with helical conformation. *Journal of Molecular Catalysis A: Chemical* 190(1-2): 197-205.
Otero, T.F., J.G. Martinez and J. Arias-Pardilla. 2012. Biomimetic electrochemistry from conducting polymers. A review: Artificial muscles, smart membranes, smart drug delivery and computer/neuron interfaces. *Electrochimica Acta* 84: 112-128.
Palaniappan, Srinivasan and Amalraj John. 2008. Polyaniline materials by emulsion polymerization pathway. *Progress in Polymer Science* 33(7): 732-758.
Pan, Lijia, Hao Qiu, Chunmeng Dou, Yun Li, Lin Pu, Jianbin Xu and Yi Shi. 2010. Conducting polymer nanostructures: Template synthesis and applications in energy storage. *International Journal of Molecular Sciences* 11(7): 2636-2657.
Pavase, Tushar Ramesh, Hong Lin, Qurat-ul-ain Shaikh, Sameer Hussain, Zhenxing Li, Ishfaq Ahmed, Liangtao Lv, Lirui Sun, Syed Babar Hussain Shah and Muhammad Talib Kalhoro. 2018. Recent advances of conjugated polymer (CP) nanocomposite-based chemical sensors and their applications in food spoilage detection: A comprehensive review. *Sensors and Actuators B: Chemical* 273: 1113-1138.
Pehkonen, Simo Olavi and Shaojun Yuan. 2018. Conducting polymer coatings as effective barrier to corrosion. pp. 23-61. *In*: Interface Science and Technology. Volume 23. Elsevier. DOI: https://doi.org/10.1016/B978-0-12-813584-6.00003-X
Ram, Manoj K., Ozlem Yavuz and Matt Aldissi. 2005a. NO_2 gas sensing based on ordered ultrathin films of conducting polymer and its nanocomposite. *Synthetic Metals* 151(1): 77-84.
Ram, Manoj Kumar, Özlem Yavuz, Vitawat Lahsangah and Matt Aldissi. 2005b. CO gas sensing from ultrathin nano-composite conducting polymer film. *Sensors and Actuators B: Chemical* 106(2): 750-757.
Rawal, Ishpal and Amarjeet Kaur. 2013. Synthesis of mesoporous polypyrrole nanowires/nanoparticles for ammonia gas sensing application. *Sensors and Actuators A: Physical* 203: 92-102.
Sanjuán Cortázar, Ana María, José A. Reglero Ruiz, Félix Clemente García García and José Miguel García Pérez. 2018. Recent developments in sensing devices based on polymeric systems. *Reactive and Functional Polymers* 133: 103-125.
Shao, Weiwei, Ruxangul Jamal, Feng Xu, Aminam Ubul and Tursun Abdiryim. 2012. The effect of a small amount of water on the structure and electrochemical properties of solid-state synthesized polyaniline. *Materials* 5(10): 1811-1825.
Sharif, Mehdi, Behzad Pourabas and Ali Fazli. 2016. Photo-reduction of graphene oxide during photo-polymerization of graphene oxide/epoxy-novolac nanocomposite coatings. *Journal of Photopolymer Science and Technology* 29(5): 769-773.
Shen, Youqing and Meixiang Wan. 1998. Soluble conducting polypyrrole doped with DBSA–CSA mixed acid. *Journal of Applied Polymer Science* 68(8): 1277-1284.
Sivakumar, Manickam and Aharon Gedanken. 2005. A sonochemical method for the synthesis of polyaniline and Au–polyaniline composites using H_2O_2 for enhancing rate and yield. *Synthetic Metals* 148(3): 301-306.
Sołoducho, Jadwiga and Joanna Cabaj. 2016. Conducting polymers in sensor design. *InTech: Rijeka, Croatia*: 27-48.
Soni, Amrita, Chandra Mouli Pandey, Manoj Kumar Pandey and Gajjala Sumana. 2019. Highly efficient polyaniline-MoS_2 hybrid nanostructures based biosensor for cancer biomarker detection. *Analytica Chimica Acta* 1055: 26-35.

Srinivasan, Sumithra Y., Kavita R. Gajbhiye, Kishore M. Paknikar and Virendra Gajbhiye. 2019. Conjugated polymer nanoparticles as a promising tool for anticancer therapeutics. pp. 257-280. *In*: Polymeric Nanoparticles as a Promising Tool for Anti-cancer Therapeutics, Academic Press. Elsevier. DOI: https://doi.org/10.1016/B978-0-12-816963-6.00012-1

Stelmach, Emilia, Ewa Jaworska, Vijay D. Bhatt, Markus Becherer, Paolo Lugli, Agata Michalska and Krzysztof Maksymiuk. 2019. Electrolyte gated transistors modified by polypyrrole nanoparticles. *Electrochimica Acta* 309: 65-73.

Stilwell, David E. and Su-Moon Park. 1988. Electrochemistry of conductive polymers II. Electrochemical studies on growth properties of polyaniline. *Journal of the Electrochemical Society* 135(9): 2254-2262.

Stowell, Jonathan A., Allan J. Amass, Martin S. Beevers and Trevor R. Farren. 1989. Synthesis of block and graft copolymers containing polyacetylene segments. *Polymer* 30(2): 195-201.

Stussi, Elisa, Sara Cella, Giorgio Serra and Giusto Stoppato Venier. 1996. Fabrication of conducting polymer patterns for gas sensing by a dry technique. *Materials Science and Engineering C* 4(1): 27-33.

Subbaiah, K.M., K.S. Nithin, B.M. Jagajeevan Raj and Sachhidananda Shivanna. 2019. Highly flexible and conducting polymer nanocomposite films for selective gas sensing applications. *Materials Today: Proceedings* 9: 515-521.

Svirskis, Darren, Jadranka Travas-Sejdic, Anthony Rodgers and Sanjay Garg. 2010. Electrochemically controlled drug delivery based on intrinsically conducting polymers. *Journal of Controlled Release* 146(1): 6-15.

Tamirisa, Prabhakar A., Knona C. Liddell, Patrick D. Pedrow and Mohamed A. Osman. 2004. Pulsed-plasma-polymerized aniline thin films. *Journal of Applied Polymer Science* 93(3): 1317-1325.

Tan, C.K. and D.J. Blackwood. 2003. Corrosion protection by multilayered conducting polymer coatings. *Corrosion Science* 45(3): 545-557.

Tarmizi, Emma Ziezie Mohd, Hussein Baqiah, Zainal Abidin Talib and Halimah Mohamed Kamari. 2018. Preparation and physical properties of polypyrrole/zeolite composites. *Results in Physics* 11: 793-800.

Tavandashti, N. Pirhady, M. Ghorbani, A. Shojaei, J.M.C. Mol, H. Terryn, K. Baert and Y. Gonzalez-Garcia. 2016. Inhibitor-loaded conducting polymer capsules for active corrosion protection of coating defects. *Corrosion Science* 112: 138-149.

Uppalapati, Dedeepya, Ben J. Boyd, Sanjay Garg, Jadranka Travas-Sejdic and Darren Svirskis. 2016. Conducting polymers with defined micro- or nanostructures for drug delivery. *Biomaterials* 111: 149-162.

Vaia, Richard A., Hope Ishii and Emmanuel P. Giannelis. 1993. Synthesis and properties of two-dimensional nanostructures by direct intercalation of polymer melts in layered silicates. *Chemistry of Materials* 5(12): 1694-1696.

Volpi, E., M. Trueba and S.P. Trasatti. 2012. Electrochemical investigation of conformational rearrangements of polypyrrole deposited on Al alloys. *Progress in Organic Coatings* 74(2): 376-384.

Wahab, Izzati Fatimah, Fatirah Fadil, Farah Nuruljannah Dahli, Ahmad Zahran Md Khudzari and Hassan Adeli. 2015. A review of electrospun conductive polyaniline based nanofiber composites and blends: Processing features, applications, and future directions. *Advances in Materials Science and Engineering* DOI: https://doi.org/10.1155/2015/356286

Waltman, R.J. and J. Bargon. 1986. Electrically conducting polymers: A review of the electropolymerization reaction, of the effects of chemical structure on polymer film properties and of applications towards technology. *Canadian Journal of Chemistry* 64(1): 76-95.

Wang, Huanhuan, Jianyi Lin and Ze Xiang Shen. 2016. Polyaniline (PANi) based electrode materials for energy storage and conversion. *Journal of Science: Advanced Materials and Devices* 1(3): 225-255.

Wang, Jinggong, K.G. Neoh, Luping Zhao and E.T. Kang. 2002. Plasma polymerization of aniline on different surface functionalized substrates. *Journal of Colloid and Interface Science* 251(1): 214-224.

Wang, Yan, Jin Guo, Tingfeng Wang, Junfeng Shao, Dong Wang and Ying-Wei Yang. 2015. Mesoporous transition metal oxides for supercapacitors. *Nanomaterials* 5(4): 1667-1689.

Wu, Xinming, Bin Huang, Qiguan Wang and Yan Wang. 2019. Thermally chargeable supercapacitor using a conjugated conducting polymer: Insight into the mechanism of charge-discharge cycle. *Chemical Engineering Journal* 373: 493-500.

Yamak, Hale Berber. 2013. Emulsion polymerization: Effects of polymerization variables on the properties of vinyl acetate based emulsion polymers. *In*: Polymer Science, IntechOpen.

Yan, Lili, Xixi Wang, Shichao Zhao, Yunqin Li, Zhe Gao, Bin Zhang, Maosheng Cao and Yong Qin. 2017. Highly efficient microwave absorption of magnetic nanospindle–conductive polymer hybrids by molecular layer deposition. *ACS Applied Materials & Interfaces* 9(12): 11116-11125.

Yanılmaz, Meltem and A. Sezai Sarac. 2014. A review: Effect of conductive polymers on the conductivities of electrospun mats. *Textile Research Journal* 84(12): 1325-1342.

Zamani, Faezeh Ghorbani, Hichem Moulahoum, Metin Ak, Dilek Odaci Demirkol and Suna Timur. 2019. Current trends in the development of conducting polymers-based biosensors. *TrAC Trends in Analytical Chemistry* 118: 264-276.

Zhan, Chuanxing, Guoqiang Yu, Yang Lu, Luyan Wang, Evan Wujcik and Suying Wei. 2017. Conductive polymer nanocomposites: A critical review of modern advanced devices. *Journal of Materials Chemistry C* 5(7): 1569-1585.

Zhang, Chunhong, Lijia Liu and Yoshio Okamoto. 2019. Enantioseparation using helical polyacetylene derivatives. *TrAC Trends in Analytical Chemistry* 115762.

Zheng, Hao, Maokun Liu, Zupeng Yan and Jianfang Chen. 2020. Highly selective and stable glucose biosensor based on incorporation of platinum nanoparticles into polyaniline-montmorillonite hybrid composites. *Microchemical Journal* 152: 104266.

Zheng, Hao, Zupeng Yan, Minghui Wang, Jianfang Chen and Xinzheng Zhang. 2019. Biosensor based on polyaniline-polyacrylonitrile-graphene hybrid assemblies for the determination of phenolic compounds in water samples. *Journal of Hazardous Materials* 378. doi: https://doi.org/10.1016/j.jhazmat.2019.05.107

CHAPTER

3

Amphiphilic Hyperbranched Polymers

Srijoni Sengupta[1], Prashant Gupta[2], Priyanka Sengupta[3] and Ayan Dey[3*]

[1] Department of Polymer Science and Technology, University of Calcutta, 92, A.P.C. Road, Kolkata - 700009

[2] Department of Plastic and Polymer Engineering, Maharashtra Institute of Technology, Beed Bypass Road, Aurangabad - 431010

[3] Indian Institute of Packaging, Plot E-2, MIDC Area, Andheri East, Mumbai - 400093

1. Introduction

In the last few years, many research articles have been reported on the synthesis of dendrimers and hyperbranched polymers (HBPs), as size and structure are important aspects of a polymer. Though dendrimers are symmetrically branched and mono-disperse macromolecules, large-scale commercial production was hindered by their stringent synthesis procedure (Caminade et al. 2015). Thus, many scientists aimed to synthesize an alternative polymer with properties comparable to dendrimers that can be synthesized in one step. This led to the development of randomly branched, unsymmetrically structured HBPs. The properties of HBPs which make them superior to their linear analog are high solubility, high degree of functionality, lower viscosity (which leads to better rate of processing), high thermal stability, and high segment density (Kim and Webster 1998). Degree of branching (DB) is an important aspect for characterizing these classes of polymers. It ranges from DB = 1 (for dendrimers) to DB < 1 (for HBPs). The wide variation of architecture of these special classes of polymers makes them ideal for application in diverse fields. The alterable end group functionality of HBPs has led to a varying range of property modifications. Because of high solubility and low viscosity, HBPs find wide applications in blend components, additives, adhesives, delivery devices, sensing, catalysts and coating components. Hence, an understanding of the structure and the methods of synthesis of these polymers are of primary interest to many researchers.

*Corresponding author: deyayanchem@yahoo.co.in

Flory designed a theoretical approach to synthesize branched and hyperbranched polymers by polycondensation technique (Flory 1941). In 1995, Fréchet and co-workers studied living cationic and living free radical techniques and introduced the term "self-condensing vinyl polymerization" or SCVP (Frechet et al. 1995). The techniques were further modified by Matyjaszewski and co-workers who proposed the concept of atom transfer radical polymerization or ATRP. Specialized monomers and catalysts were required to make the process a sophisticated one (Tsarevsky et al. 2009). But with the introduction of the Strathclyde method, a simpler approach for generating HBPs from vinyl monomers was put into practice (Chisholm et al. 2009).

Amphiphilic HBPs can be synthesized by polycondensation, SCVP, nitrogen-mediated polymerization (NMP), living-radical polymerization such as ATRP and reversible addition-fragmentation chain-transfer (RAFT), ring-opening polymerization, free radical polymerization using multifunctional vinyl monomers, and chain transfer agent (known as Strathclyde method).

Amphiphilic polymers are macromolecules that can be assembled into a nanostructured entity in aqueous environment. The hydrophilic and hydrophobic parts control the self-assembling ability of the polymer. The generation of the assembly is generally stimuli responsive, i.e., it gets initiated under the influence of pH, temperature, and other variables; thus, the supramolecular structure formed by the self-assembling nature of these amphiphilic polymers acts as a nano-reactor (Barriau et al. 2005). The core-shell structure thus generated exhibits excellent encapsulation efficiency.

In this chapter we present the various synthesis procedures for the generation of amphiphilic HBPs, their structure-property relationship, and their various applications.

2. Synthesis and properties of amphiphilic hyperbranched polymers

As the demand for architectural polymers increases, simultaneously there is a rise in the requirement for developing different synthesis procedures. Three-dimensional globular structured dendrimers and hyperbranched polymers are among the major classes of architectural polymers. With exclusive properties, this class of polymers has gained much attention from researchers and scientists. On a commercial scale, HBPs are more acceptable because of their easy "one-pot" synthesis technique, unlike the stringent procedure for dendrimers (Gao and Yan 2004). Hyperbranched polymers are endowed with some unique properties: they are irregular highly branched structures with a large number of terminal functional groups, and they have excellent solubility, low viscosity, and a non-entangled compact structure. Here, we focus mainly on amphiphilic HBPs whose self-assembling nature has paved a new path in the field of polymers. With excellent stability, unique mechanical properties and a highly stimuli-responsive nature, self-assembling amphiphilic HBPs find a wide range of applications, especially in biomedicine. Different synthesis procedures are employed to obtain different grades of amphiphilic HBPs with varying hydrophobic and hydrophilic parts. As the constituents are varied, their self-assembling nature also varies, along with their stimuli-responsive nature.

Here we discuss condensation polymerization, free radical polymerization (i.e., Strathclyde method), and self-condensing vinyl polymerization. Structure-property relationship of these polymers synthesized from different procedures is subsequently elaborated.

2.1. ATRP

Atom transfer radical polymerization or ATRP has gained popularity in synthesis of multicyclic polymers with hyperbranched structure. Liu and co-workers reported a synthesis of hyperbranched polystyrene using this process (Liu et al. 2018). They reported that in living radical techniques all polymer chains are (ideally) initiated simultaneously and grow at an equal rate to reach a mean degree of polymerization. This is governed by the initiator-to-monomer ratio. Use of (2,2,6,6-tetramethylpiperidin-1-yl) oxyl (TEMPO) as co-catalyst for the polymerization of styrene has opened a new route to control radical polymerization procedures. ATRP is effectively used to prepare block copolymers. In the first step of ATRP, reactivation of the alkyl halide adduct with the prepared alkene (monomer) occurs, and subsequently there is a reaction of the intermittently generated radical with additional monomer units (propagation). This is an example of living radical polymerization. Most successful ATRP processes have been developed using copper as catalyst (Matyjaszewski 2018). Bannister and co-workers reported a three-stage technique to prepare amphiphilic hyperbranched copolymer using tertiary–bromoalkane type molecules as an ATRP initiator for the polymerization of mono and divinyl monomers. Cu(I)Cl is used as an activator and the reaction is carried out using methanol at 20°C.

Matyjaszewski (2018) classified ATRP into nine categories:

(i) *Traditional ("normal") ATRP*
Mechanistically, ATRP is defined by Matyjaszewski as "an inner sphere electron transfer process, which involves a reversible (pseudo) halogen transfer from a dormant species (P_n-X) to a transition metal complex (M_t^m/L_n)". Here, M_t^m denotes the transition metal in the oxidation state m complexed with an appropriate ligand L_n. Propagating radicals (P_n^*) and the metal complex of higher oxidation state (compared to its original oxidation state), i.e., X-M_t^{m+1}, are obtained as a product of the reaction. This behaves as an intermittent radical that allows the formation of macromolecular chains. The radical reacts reversibly with the metal complex with higher oxidation state that is deactivated to form a dormant species and transition metal complex with lower oxidation state. The transition metal complex obtained is used again to reform the activator.

(ii) Reverse ATRP
The "Reverse ATRP" mechanism is activated by the addition of transition metal compound of comparatively more stable higher oxidation state, which is then converted to the activator by reacting the compound with standard thermal initiators such as AIBN (azobisisobutyronitrile) or BPO (benzoyl peroxide). Here, the metal ion used, mostly, should have the ability to undergo comproportionation reaction.

(iii) *Simultaneous reverse and normal ATRP and AGET*
Simultaneous reverse and normal ATRP, is a combination of the two aforementioned procedures that makes it possible to use more active, readily oxidized catalyst complexes. In this procedure, alkyl halides act as an initiator from which majority of polymer chains grow. Here, a transition metal complex with higher oxidation state is used at sub-stoichiometric proportion. A thermal initiator is used to reduce it. In the subsequent reaction, the transition metal halide is incorporated to convert the molecule by electron transfer (AGET) to activator, which helps in polymerization process activation.

(iv) *ARGET ATRP*
Matyjaszewski mentioned activator regenerated by electron transfer (ARGET) ATRP as another way to initiate ATRP, but it can be considered as a new way to carry out a reversible deactivation radical polymerization (RDRP). It is a recently developed green procedure that requires the catalyst in ppm level with an appropriate FDA-approved reducing agent such as tin(II) 2-ethylhexanoate ($Sn(EH)_2$), glucose, or ascorbic acid along with excess comparatively inexpensive ligands such as amines or nitrogen-containing monomers.

(v) *ICAR ATRP*
Initiators for continuous activator regeneration (ICAR) is also considered as a reverse ARGET ATRP, where organic free radicals are used to continuously produce very low concentration of Cu(I) activator (5–50 ppm).

(vi) SARA ATRP
Cu can also react with alkyl halide and act as supplemental activator. Here, the majority of alkyl halides are activated by Cu(I) formed by comproportionation or deactivation with Cu(II). Hence, it is named "supplemental activator and reducing agent (SARA) ATRP".

(vii) eATRP
Electrochemical process is used to offer better control over multiple parameters such as applied current, potential, and total charge passed in order to manipulate polymerization rates effectively and easily. As an example, $CuBr_2/Me_6TREN$ catalyst complexes can be used, which are electrochemically reduced to $CuBr/Me_6TREN$ activators to trigger a controlled polymerization process. Sacrificial aluminium anode and simple glassy carbon, stainless steel, or gold-based cathode are used for these processes.

(viii) Photo-ATRP

Photo-induced ATRP or photo-ATRP reaction is used to provide excellent control over tuning the process to synthesize HBPs. The reaction is initiated through irradiation with light beam of wavelength 450 nm. Both organic solvents and aqueous medium are reported to be used as solvent for performing photo-ATRP. This process is again classified into two categories based on the type of initiator: (a) transition metal ions/metal halides initiated photo-ATRP and (b) metal-free photo-ATRP. Photo-ATRP can be initiated with transition metals such as Fe, Co, Ni, or Cu. Ferric bromide itself on irradiation with UV light converts methacrylates to alkyl halides, the complex of which *in situ* activates the ATRP mechanism. A metal-free ATRP is reported to be carried out with phenazines, phenoxazines and phenothiazines, which are found to activate the ATRP mechanism for methacrylate monomers. X. Zhang and co-workers also developed HBPs using photo-ATRP (Zhang et al. 2020).

(ix) *Mechano ATRP*

In mechano ATRP, ultrasound bath is used to carry out the ATRP reaction with a low concentration (in ppm) of copper catalyst along with barium titanate or zinc oxide as piezoelectric catalyst.

2.2. RAFT

Brookhart et al. and Guan et al. reported the "formation of hyperbranched polyethylene at low pressure by a 'chain walking' process using Pd-α-diimine catalyst" (Johnson et al. 1995, Guan et al. 1999). Alfurhood and his team developed a complex structured HBP by copolymerizing HPMA (2-hydroxypropyl methylacrylamide) and a chain transfer monomer (CTM) via RAFT-based SCVP (Alfurhood et al. 2016).

Initially the CTM was reported to be prepared by reacting the chain transfer agent, butyl 2-cyanopropan-2-yl-carbonotrithioate, with HPMA. Then the synthesized CTM was copolymerized with HPMA to generate a highly branched structure.

Synthesis of pH-responsive hyperbranched poly(bis(N,N-propylacrylamide) (HPNPAM) was carried out by Zhou and his team. They opted for RAFT polymerization to synthesize a highly branched structure (Zhou et al. 2016). The reaction scheme is shown in Fig. 1.

The control of molecular architecture and the monomer conversion were directly affected by the reaction conditions. It was observed that, with increasing temperature, the molecular weight of the polymer increases along with the degree of branching. Again, with addition of more chain transfer agent the monomer conversion rate slows down, making the molecular weight lower, and the distribution becomes quite narrow. HPNPAM exhibits both temperature sensitivity and pH-responsive nature, along with lower cytotoxicity, making it a potential candidate for biomedical applications.

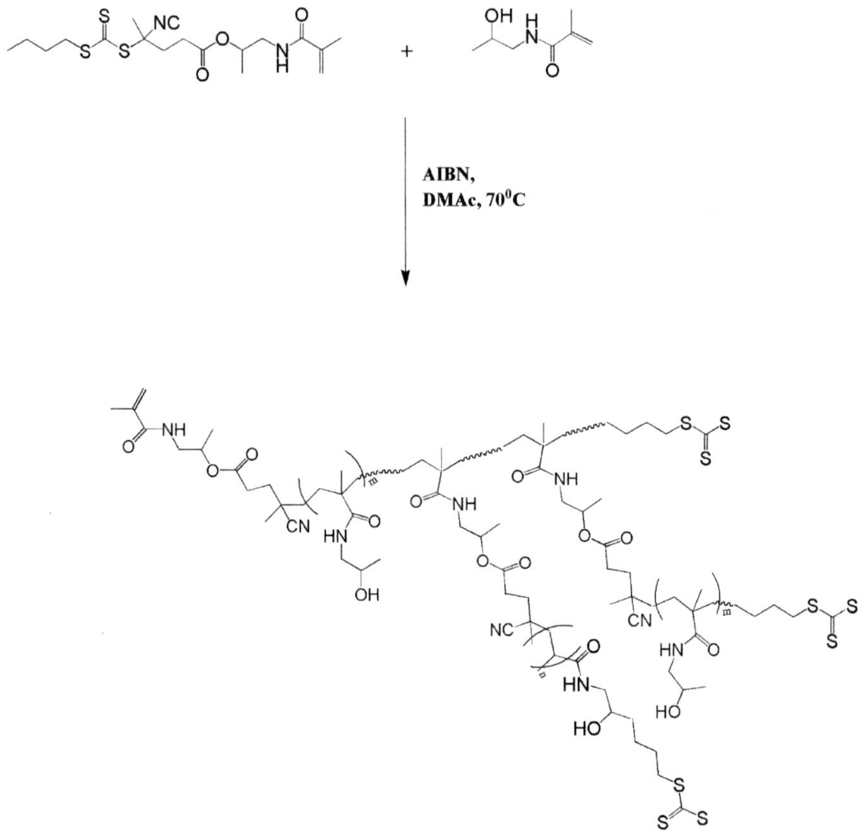

Fig. 1. Synthesis of hyperbranched PHPMA.

2.3. NMP

Hawker reported the preparation of star and hyperbranched polymers by self-condensing radical polymerization. The monomer (Hawker 1995) consists of styrene group and a polymerizable chemical moiety with a nitroxide linkage. Thermal oxidation is possible for polymerization by self-condensing radical route due to low dissociation energy of C-O bond with the nitroxide group (Kazmaier et al. 1995). The thermolysis of the monomer results in a benzylic radical that has the potential to add to the vinyl group of another molecule to facilitate propagation. A dimer thereby formed by thermal recombination (reversible) with TEMPO is a monomer with an unsaturation that is polymerizable and has two sites on which initiation/propagation can be carried out. This eventually leads to formation of oligomers and the hyperbranched polymer (3) having a vinyl group as shown in Fig. 2. The polymerization rate is dependent on the concentration of vinyl group. The product obtained by the polymerization as shown in Fig. 2 is living because of the retention of reactivity of terminal benzylic nitroxide

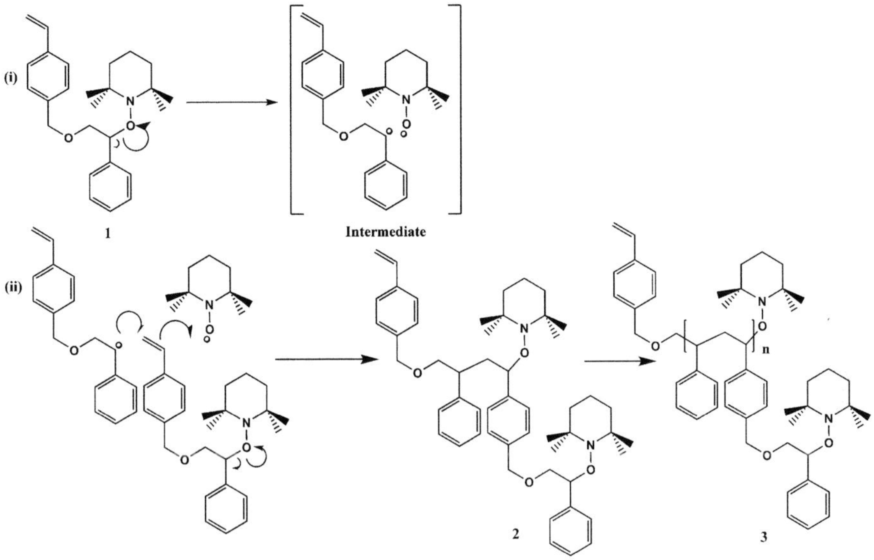

Fig. 2. Synthesis of hyperbranched polymer. Reprinted (adapted) with permission from Hawker (1995).

chemical moieties. The above polymerization was reported to be carried out by bulk technique at 130°C under inert argon atmosphere for a period of 72 h. Size exclusion chromatography of the reaction mixture at different times of reaction provided the evidence of the gradual conversion of monomers to dimer, trimer, tetramer and so on to polymer by exhibiting increase in the polymer concentration and the molecular weight. The polymer is obtained as a light yellow solid as a precipitate from the mixture after completion of 72 h. The yield of the reaction was 51% with Tg of 45°C, polystyrene equivalent molecular weight and polydispersity index (PDI) as shown in Table 1. The absence of termination reactions in NMP was supported by the absence of insoluble or crosslinked material. The presence of nitroxide functionality provides an opportunity for developing alternative architectures.

Multi-arm star polymers can be formed by the above synthesized polymers since they can function as initiating cores. The procedure has been demonstrated with the help of the same conditions of polymerization with much higher purification yield (88%). There is a substantial increase in the molecular weight and PDI. Furthermore, as the monomer has a benzyl ether group that is cleavable, the hyperbranched star-shaped molecule is dissociated by a reaction with trimethylsilyl iodide at its branch points to understand the presence and nature of branching.

Free radical living polymerization can also be employed to make polymers with controlled branch density and length by copolymerizing of vinyl monomers with specific monomers such as the starting material used by Hawker and co-workers, as they serve as both initiation and branch point in product macromolecule. The polymerization of the starting monomer with

25, 50 and 100 equivalents of styrene was performed, and the GPC results were analyzed for the polymerized macromolecule. The increase in styrene equivalent resulted in higher molecular weight with increasing repeating units within branch points. The cleavage results of copolymer formed are also tabulated in Table 1 (Hawker et al. 1995).

Table 1. GPC results of molecular weight (Mw) and PDI for various polymers formed by free radical living polymerization mediated by nitroxide (Hawker et al. 1995)

Sr. No.	*Sample*	*Molecular weight (Mw)*	*Polydispersity index (PDI)*
1	Hyperbranched polymer	6,000	1.4
2	Star-shaped hyperbranched (hyperstar) polymer	300,000	4.35
3	Cleavage of hyperstar at branch point	65,000	1.57
4	Copolymer formed by use of 100 equivalents of styrene	50,000	2.45
5	Cleavage of copolymer at branch point	11,000	1.17

Peleshanko and co-workers reported the synthesis of poly(ethylene oxide)-polystyrene (PEO-PS)n copolymers. The complex synthesis process as given in Fig. 3 deals with a primary alcohol (1-benzyloxymethyl triethylene glycol) and its selective functionalization by employing 4-dimethyl aminopyridine as a catalyst (Chaudhary and Hernandez 1979). Thereafter, 2-bromopropionyl bromide was used for the modification of secondary alcohol group to obtain polymer in the form of a clear, colorless liquid. NMP for this scenario can be performed in solution or bulk without the use of any other chemicals, whereas ATRP is to be done in solution with use of the catalyst. ATRP of styrene is carried out with the polymer obtained earlier, in the presence of copper (20%) as a catalyst to obtain polymer with a molecular weight below 3000 and PDI less than 1.1 as bulk NMP yielded higher molecular weight and highly polydisperse polymers. The obtained polymer was precipitated in ethanol solution. Subsequently, as reported by Matyjaszewski and Bon, the bromine group was substituted in the polymer by the use of 2,2,6,6-tetramethylpiperidin-1-yl)oxyl. (TEMPO) (Matyjaszewski et al. 1998, Bon et al. 2000). The modification of hydroxyl group in polymer obtained from previous reaction sequence was carried out with 4-vinylbenzoic acid after removing the protecting group, t-butyldimethylsilane. The polymer obtained was found to possess average molecular weight (Mn) of 2175 with a PDI of 1.15. It is noteworthy that RAFT polymerized macroinimer had low Mn of 1150 with PDI of 1.05. However, RAFT polymerized macroinimer resulted in HBP with higher molecular weight containing 3 ethylene oxide and 5 styrene units in comparison with 3 and 12 units respectively for NMP polymerized macroinimer. The synthesis of HBPs via macromonomers

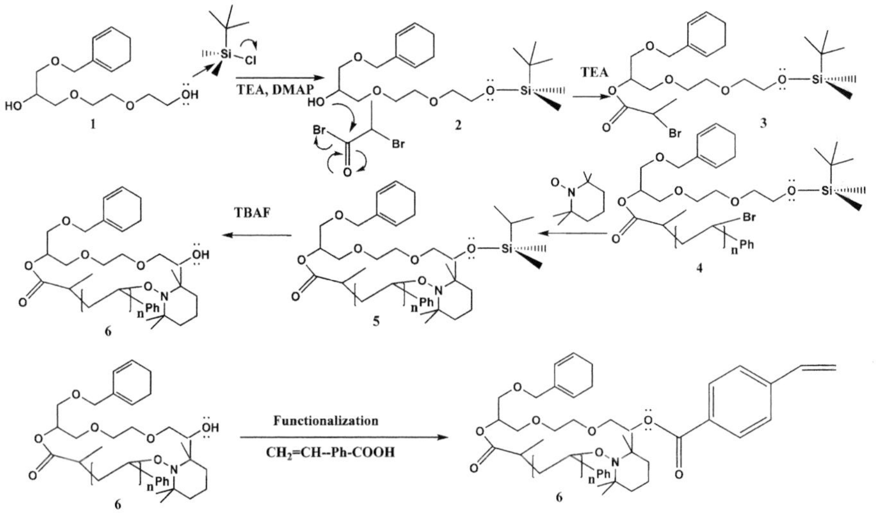

Fig. 3. Synthesis of macroinimer by NMP. Reprinted (adapted) with permission from Peleshanko et al. (2006).

was performed at 125°C by bulk/solution (with 60% DMF) techniques. The polymer obtained by bulk NMP technique possessed PDI in excess of 11 with wide molecular weight distribution. A combination of RAFT with NMP leads to formation of PEO-PS hyperbranched amphiphilic copolymer with mixed hydrophilic and hydrophobic components with strong amphiphilic behavior (Peleshanko et al. 2006).

Bian and co-workers reported a direct route of synthesis of poly hydroxyl ethyl acrylate (pHEA) using N-tert-butyl-N-(1-diethyl phosphono-2,2-dimethylpropyl) nitroxide (SG1, an Atofina product). It has proved more effective for controlled radical polymerization of acrylates (Benoit et al. 1998, Grimaldi et al. 2001) and acrylic acid (Couturier et al. 2003) than TEMPO as reported in other work (Peleshanko et al. 2006). The polymerization was initiated by MONAMS, which is an SG1-based alkoxyamine of methyl acrylate at 100-120°C in a Schlenk tube by bulk technique and solution technique in the presence of N,N-dimethylformamide (DMF) and water. The percentage conversion along with Mn and PDI are reported in Table. 2. The polymerization was controlled by addition of 6-12 mol% nitroxide with respect to initiator to achieve Mn of 90,000 and PDI around 1.3. Also, solution polymerization of pHEA in 50 wt% N,N-dimethylformamide (DMF) and water exhibited kinetics comparable to that of bulk technique at similar temperatures. The presence of DMF showed slight increase in the rate of acceleration. Similar results were also observed when water was used for bulk polymerization at 100°C. Macroinitiator poly(n-butyl acrylate) was used to synthesize amphiphilic copolymers of poly(BA-b-HEA) with varying molecular weight and composition of blocks. In bulk and aqueous media polymerizations ~1 mol% branches were found to form because of chain transfer to polymer, whereas 1.5-1.7 mol% branches were evident in DMF, which were either remaining the same or decreasing because of increase in long chain branching due to higher extent of conversion (Bian and Cunningham 2005).

Table 2. ^1H NMR and SEC results for bulk polymerized pHEA carried out with HEA (9.64 g), MONAMS (183 mg), and nitroxide SG1 (17 mg) (Bian and Cunningham 2005)

Time of sample (hours)	*Molecular weight (Mn)*	*Polydispersity index (PDI)*	*% Conversion*
0.4	3,600	1.18	7.9
1	3,800	1.25	13.9
1.75	5,600	1.20	21.1
2.75	7,000	1.19	31.4
3.75	9,200	1.15	41.0
4.75	11,500	1.17	48.7
6.25	14,200	1.13	59.2
7.8	17,800	1.17	69.1
9.3	23,200	1.26	78.4

2.4. Strathclyde method

Strathclyde method was introduced by D.C. Sherrington and his group to synthesize HBPs from vinyl monomers via free radical mechanism. The method is named for Strathclyde University, UK, where the experimental work was executed (Baudry and Sherrington 2006). In this method, vinyl polymerization reaction is carried out in the presence of a multivinyl monomer and a chain transfer agent where the multivinyl monomer acts as the brancher. Though multivinyl monomer acts as a crosslinking agent, the simultaneous addition of thiol-based chain transfer agents restricts the gelation leading into highly branched structure. Restricting the complete conversion of the monomers also prevents gelation. Being quite easy and cost effective, the Strathclyde method has been widely accepted. But a major shortcoming is that, owing to the presence of toxic thiol-based chain transfer agents along with vinyl monomers, and to the uncontrolled nature of free radical polymerization, HBPs synthesized via Strathclyde method cannot be used for biomedical purposes. The reaction scheme is shown in Fig. 4.

J. Weaver and his team developed an amphiphilic HBP via the easy and single-step Strathclyde method. They reported a reaction between a pH-responsive monomer ((2-diethylamino)ethyl methacrylate) and the hydrophilic polyethylene glycol methacrylate to generate a core-shell architecture in the resulting HBP (Weaver et al. 2008). The reaction scheme is explained in Fig. 5. The synthesized HBPs had a narrow and symmetrical molecular weight distribution.

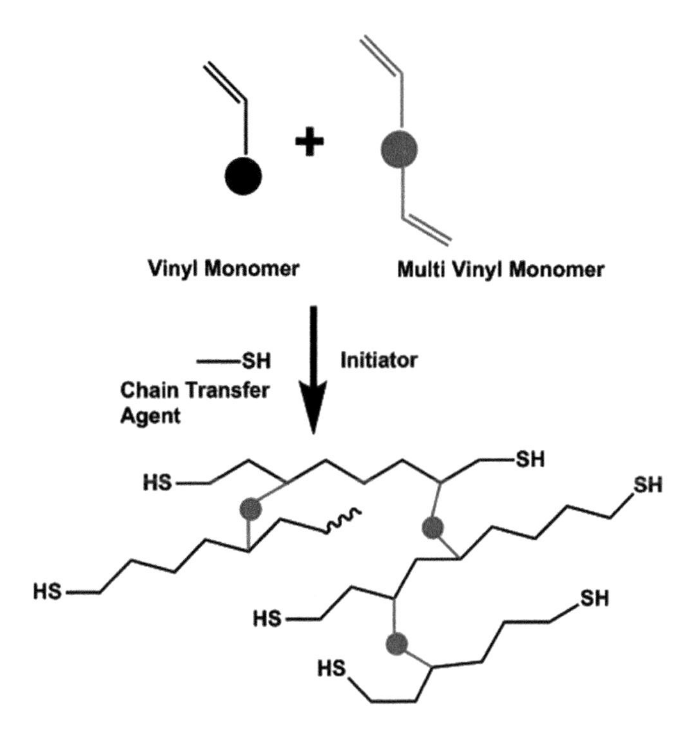

Fig. 4. Mechanism for the synthesis of hyperbranched polymers via Strathclyde method.

These polymers form a stable core-shell structured micelle starting from a fully hydrated form with increasing pH and it was also further observed that the amphiphilic characteristics depend on the extent of branching introduced by the brancher.

The above experiment was slightly modified by P. Chambon and his team to develop a temperature-responsive amphiphilic HBP. They synthesized a copolymer with N-isopropyl acrylamide and polyethylene glycol methacrylate, along with brancher ethylene glycol dimethacrylate and dodecanethiol as chain transfer agent (Chambon et al. 2011). The reaction scheme is shown in Fig. 6.

The size of the HBPs was in the nanometric range. As the distribution of individual NIPAAm was considerably less, there was a deviation from the conventional LCST behavior. However, opting for controlled synthesis technique can provide more control over the thermoresponsive behavior.

In another study, T. Das and her group synthesized an amphiphilic HBP using Strathclyde method and a post-polymerization modification was done by click reaction (Das et al. 2015). A comonomer was synthesized based on propargyl acrylate and acrylic acid in which divinyl benzene was used as the brancher and dodecanethiol as the chain transfer agent. Further, the alkyne ends of propargyl acrylate were modified by introduction of azide group via click reaction. The ratio of brancher and chain transfer agent was varied to

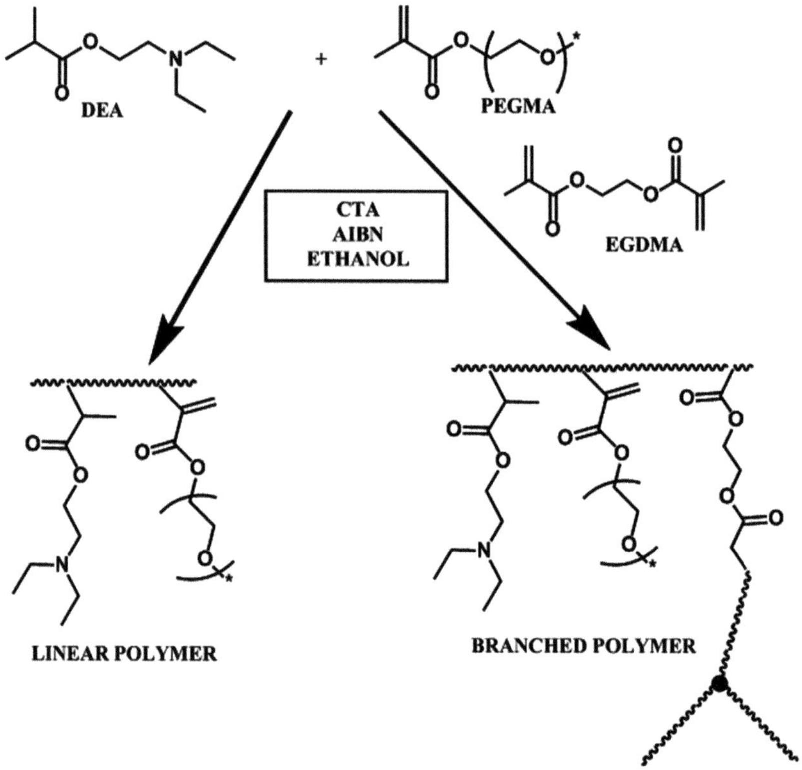

Fig. 5. Mechanism for the synthesis of the DEA-PEGMA-based linear and hyperbranched polymer.

generate HBPs with varying degree of branching and their properties were compared with their linear analogs. The synthesized HBP had an amphiphilic nature with a core-shell structure that is stimulated by pH. The amphiphilic HBPs were capable of encapsulating both hydrophobic and hydrophilic guest molecules.

Less control over the reaction and the end products restricts the acceptance of the Strathclyde method. More sophisticated techniques such as ATRP, RAFT, and NMP are more practicable for more advanced end applications that yield more symmetrical products than the simple free radical polymerization, but they have a high cost of execution. Low-cost applications require an easy process such as the Strathclyde method to make the products commercially viable. Though few studies on amphiphilic HBPs via Strathclyde method have been reported, this method still needs to be explored further for developing new set of vinyl-based HBPs.

2.5. Condensation polymerization

Synthesis of HBPs by step-growth polymerization generally results in uncontrolled molecular weight with broad molecular weight distribution,

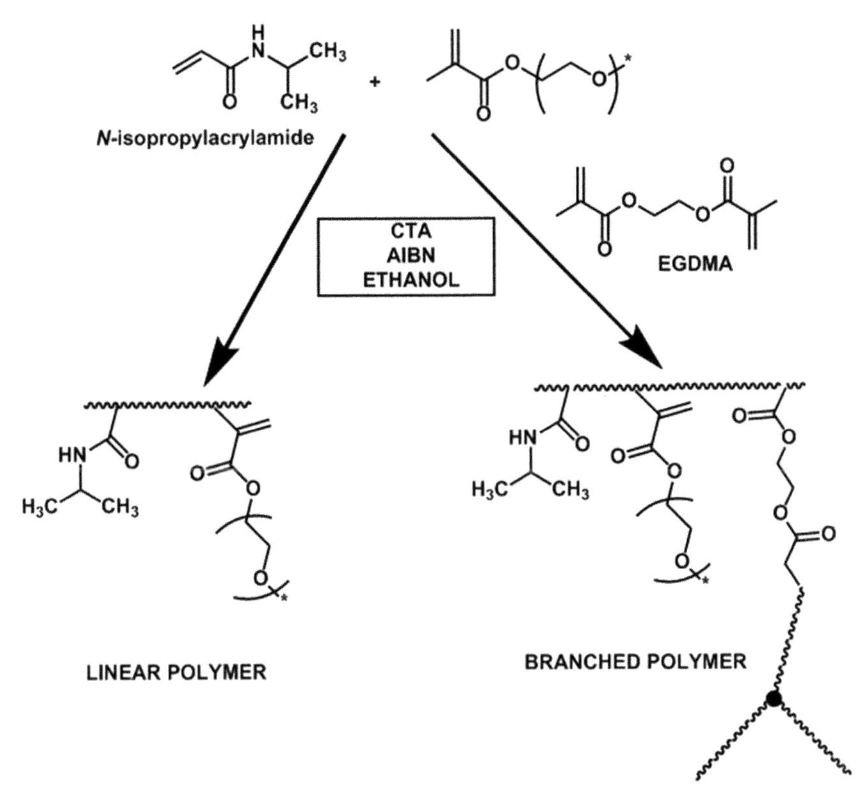

Fig. 6. Mechanism for the synthesis of the NIPAAm-PEGMA-based hyperbranched and linear polymer.

but due to the presence of numerous terminal functional groups, HBPs via condensation are easily tunable according to the need of the product. Though Flory designed the theoretical concept to develop HBPs a long time ago, it took a long time for it to be practically executed. The first commercial HBP prepared by Berzilius via condensation method was polyester-Boltorn. Generally, condensation needs AB_x or A_2+B_3 type monomers to yield highly branched three-dimensional structure.

For instance, Ohta and his group reported synthesis of amphiphilic copolymer, which consists of hydrophilic poly(ethylene glycol) (PEG) and hydrophobic hyperbranched polyamide (HBPA) (Ohta et al. 2013). The reaction scheme is presented in Fig. 7. In their reported process, they first performed chain-growth condensation polymerization with AB_2 monomer using PEG macroinitiator, which resulted in broad polydispersity; therefore, in the second step, condensation reaction was carried out with carboxylated PEG (at one end) and HBPA having hydroxyl groups at core.

Recently, Ohta and his group developed a janus amphiphilic diblock hyperbranched copolyamide by reacting hydrophobic carboxylic group of hyperbranched polyamide 1 with the hydrophilic hydroxyl group of hyperbranched polyamide 2 (Ohta et al. 2019).

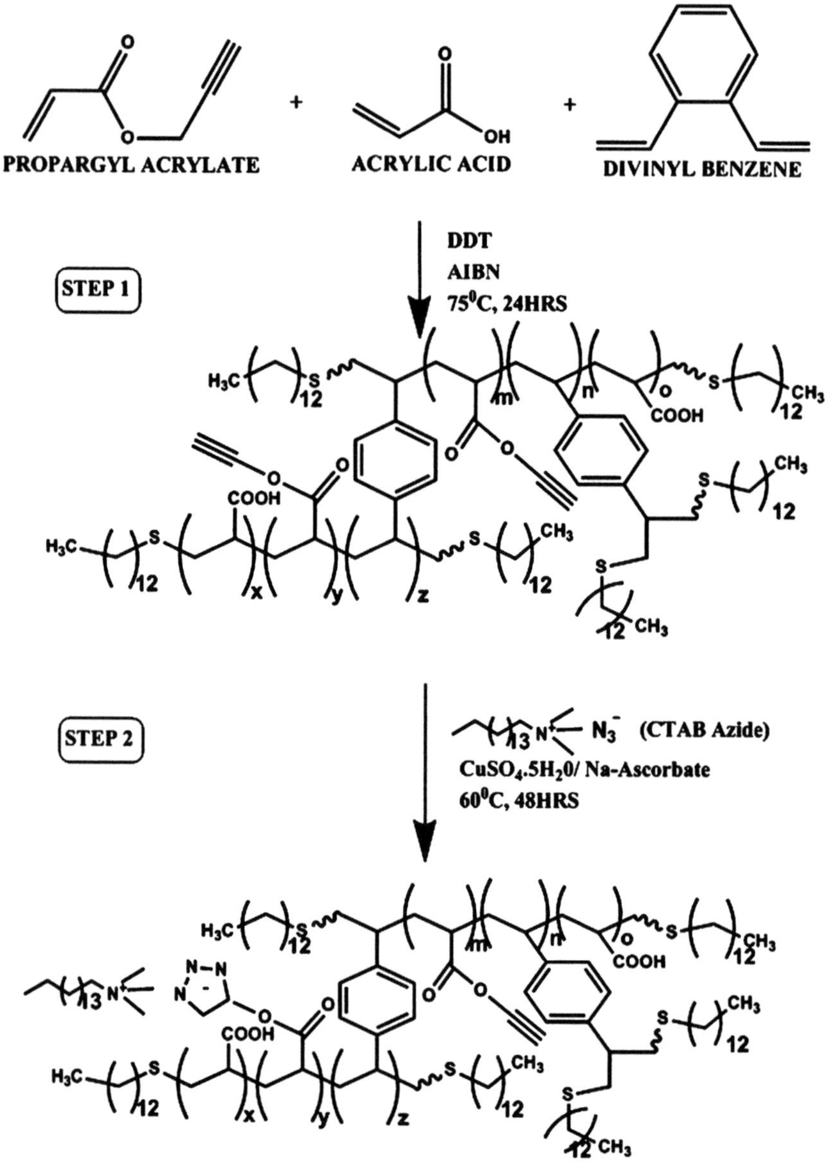

PROPARGYL ACRYLATE + ACRYLIC ACID + DIVINYL BENZENE

STEP 1

DDT
AIBN
75^0C, 24HRS

STEP 2

(CTAB Azide)
$CuSO_4.5H_2O$/ Na-Ascorbate
60^0C, 48HRS

Fig. 7. Mechanism for the synthesis of hyperbranched polymers from propargyl acrylate.

Amphiphilic HBPs can be applied as an excellent delivery vehicle, as demonstrated by Chen and his group. Initially, they used Boltorn H40 (with 64 terminal OH) and caprolactone to get H40-PCL. Then, H40-PCL was reacted to succinic anhydride in the presence of tin (II) chloride to give H40-PCL-COOH. H40-PCL-PEG was then obtained by reacting H40-PCL-COOH with PEG in the presence of DCC and DMAP in DMSO solvent. Finally, they incorporated folic acid as a targeting ligand to the H40-PCL-PEG to get H40-PCL-PEG-FA (Chen et al. 2008). The amphiphilic HBPs were nanometric in dimension. The reaction scheme is presented in Fig. 8.

H40-PCL forms the hydrophobic part, while PEG is the hydrophilic part used to form the amphiphilic nature (Fig. 9). Further attachment of folic acid improved the tumor-targeting properties and the drug release was controlled, as evident from the results. Paclitaxel, a highly hydrophobic drug

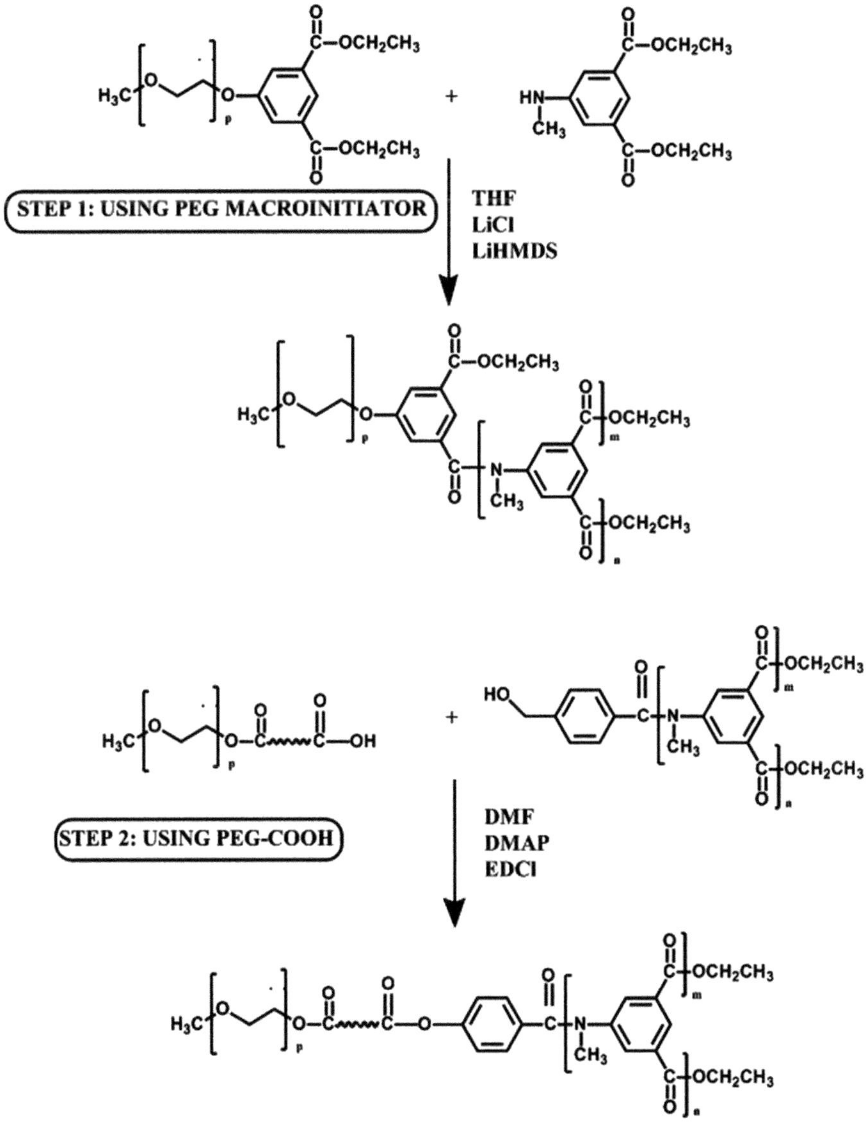

Fig. 8. Mechanism for the synthesis of hyperbranched polymers using PEG macroinitiator.

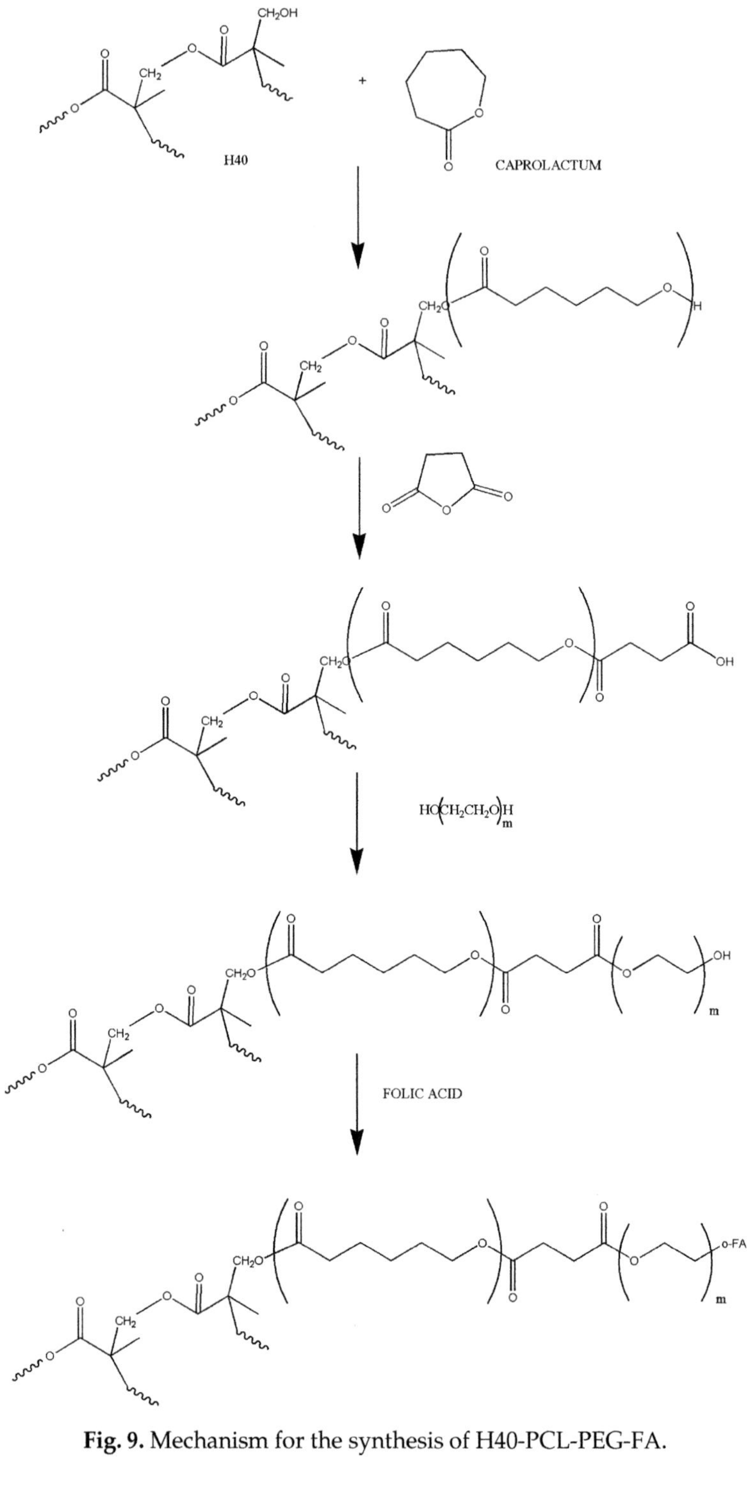

Fig. 9. Mechanism for the synthesis of H40-PCL-PEG-FA.

used to treat lung and ovarian cancer, was easily encapsulated within the hydrophobic core and sustained release profile was observed through the polymeric vehicle.

In another experimental work, Chen and his team developed an amphiphilic copolymer based on hydrophilic PEG and hydrophobic β-amino ester. First, in the presence of 4-(amino methyl) piperidine (AMP) and 1,6-hexanediol diacrylate (HDDA) β-amino ester was reacted by Michael Addition reaction at 60°C to generate a hyperbranched structure. In the second step, the residual acrylate of the hyperbranched β-amino ester groups reacted with PEG-NH_2 to obtain the amphiphilic structure (Chen et al. 2013). At pH ~7 the hyperbranched copolymer is able to self-assemble to form micelle with PEG at outward shell and hyperbranched amino ester at the core. Under UV irradiation, the micelles were crosslinked. The micelles were pH sensitive, highly sensitive, and proved to be highly efficient as an anti-cancerous drug delivery vehicle.

Amphiphilic HBPs synthesized via condensation polymerization varied widely in applications and properties. These types of polymers find extensive use in biomedicine.

2.6. Self-condensing vinyl polymerization

Self-condensing vinyl polymerization is preferred for generating highly branched structure with high number of functional groups. This method was first introduced by Frechet and his co-workers, who polymerized 3-((1-chloroethyl)ethylene)benzene with $SnCl_4$ as catalyst (Frechet et al. 1995). Vinyl monomers, which participate in SCVP, are called 'inimers' as they both have monomer unit and an initiator part that gets activated under the action of external stimuli.

Powell and his co-workers opted for atom transfer radical SCVP to develop amphiphilic fluoropolymer containing tri(ethylene glycol) (Powell et al. 2007). They used 4-(oxy(tri(ethyleneglycol))bromoisobutyryl)-2,3,5,6-tetrafluoro styrene as inimer for the synthesis of hyperbranched fluoro homopolymer, and in the next step they also developed a hyperbranched fluoro copolymer by reacting 2,3,4,5,6-pentafluorostyrene with the inimer. In all the cases, bipyridine along with CuCl/$CuCl_2$ was used as the catalyst. The reaction scheme is shown in Fig. 10.

The HBP was soluble in a wide range of organic solvents including tetrahydrofuran, acetone, chloroform, and toluene. Though incorporation of TEG did not improve the water solubility, the polymers were capable of forming complex micelles in aqueous solvent.

Another study was reported by Weng and his group, who combined RAFT with SCVP to develop a novel amphiphilic HBP. First, 2-(dimethyl amino)ethyl methacrylate (DMAEMA) and the chain-transfer monomer 2-((2-(((dodecylthio)carbonothioyl)thio)-2-methylpropanoyl)oxy) ethyl acrylate (ACDT) was reacted to form poly(2-dimethylamino) ethyl methacrylate). Then, in the presence of this polymer, styrene was

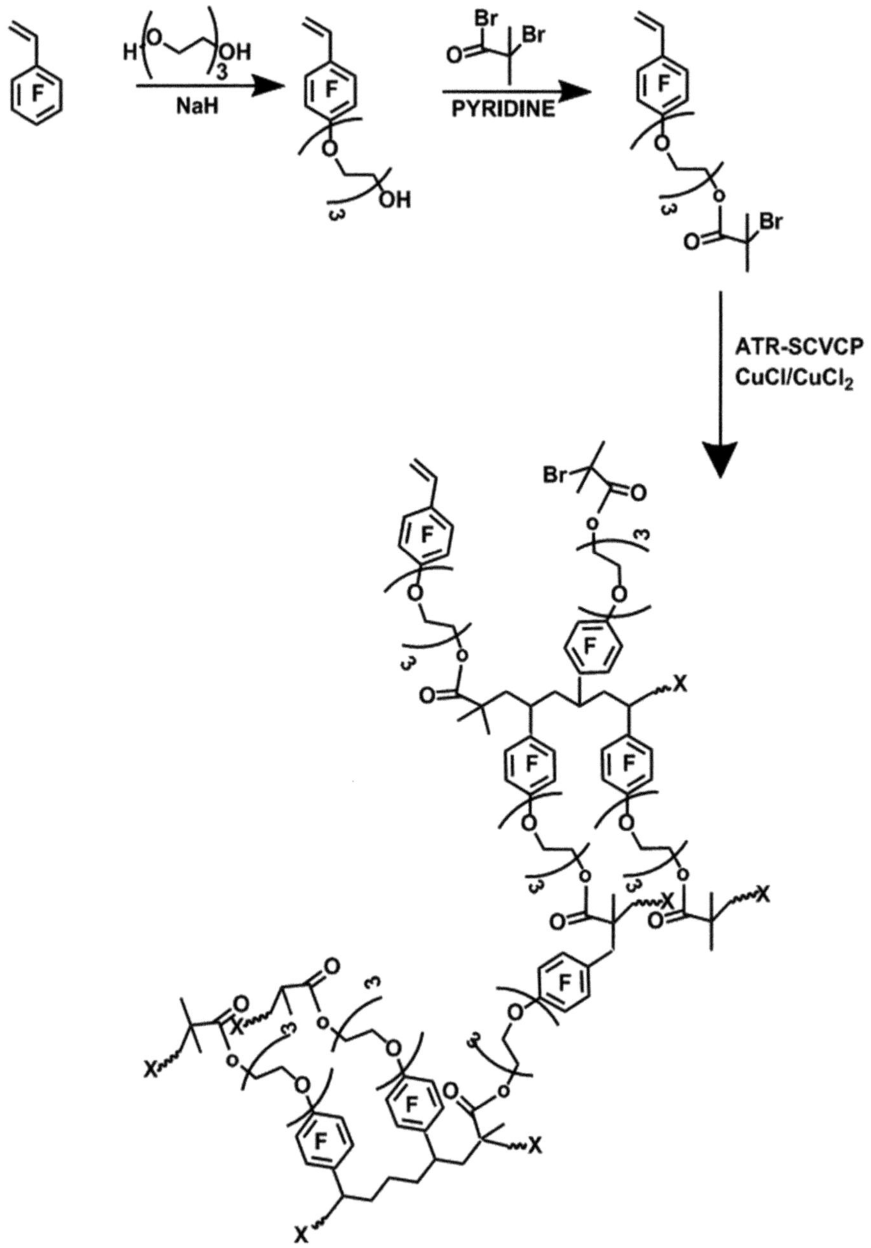

Fig. 10. Mechanism for the synthesis of amphiphilic hyperbranched fluoropolymer.

polymerized via RAFT polymerization to yield HPTAM-co-PS. It was further modified via Menshutkin click reaction, where the tertiary amine groups are functionalized by propargyl bromine (Fig. 11). The final product had a hydrophobic outer shell made of PS and a hydrophilic inner core made from

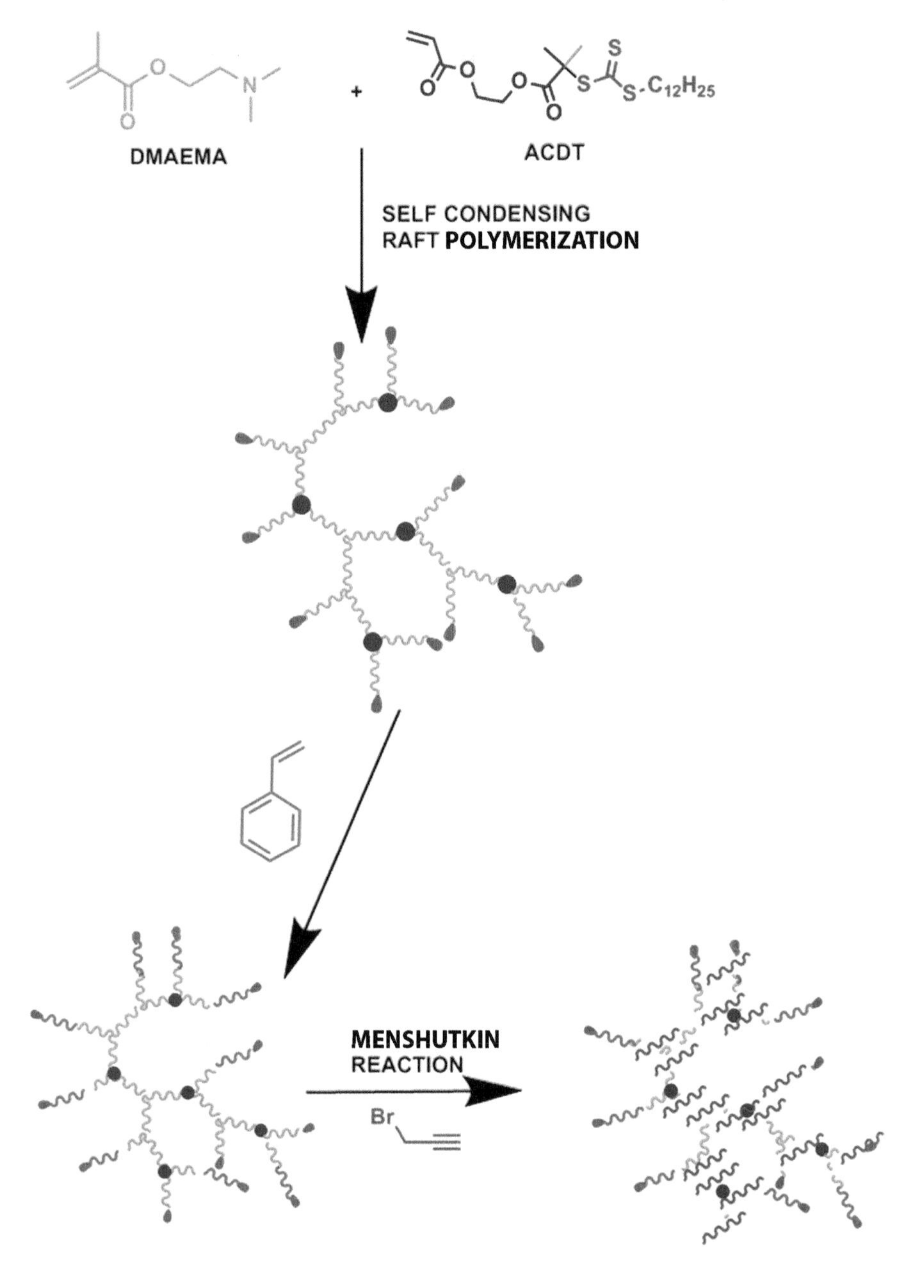

Fig. 11. Mechanism for the synthesis of amphiphilic hyperbranche polymer colorant.

quaternary ammonium salt. The resulting amphiphilic polymer exhibited an excellent dye-encapsulation property that was enhanced by the incorporation of the hydrophilic part. The polymers found application as coloring agents (Weng et al. 2014).

In a recent report, Aydogan and his group reported light-induced self-condensing vinyl copolymerization to develop amphiphilic HBPs. They polymerized methyl methacrylate (MMA) in the presence of 2-(2-bromoisobutryloxy)ethyl methacrylate (BIBEM) and poly(ethyl glycol) methyl ether methacrylate, where BIBEM was used as the inimer and the latter as the hydrophilic copolymer. $Mn_2(CO)_{10}$ was used as photo-initiator (Aydogan et al. 2018). The reaction scheme is shown in Fig. 12. There were excess unreacted bromine groups present at the terminal ends that could be further functionalized via click reaction. The process was simple, one step and rapid. Thus, this technique can be used further to develop more amphiphilic polymers with controlled hydrophilicity that can be applied in various fields.

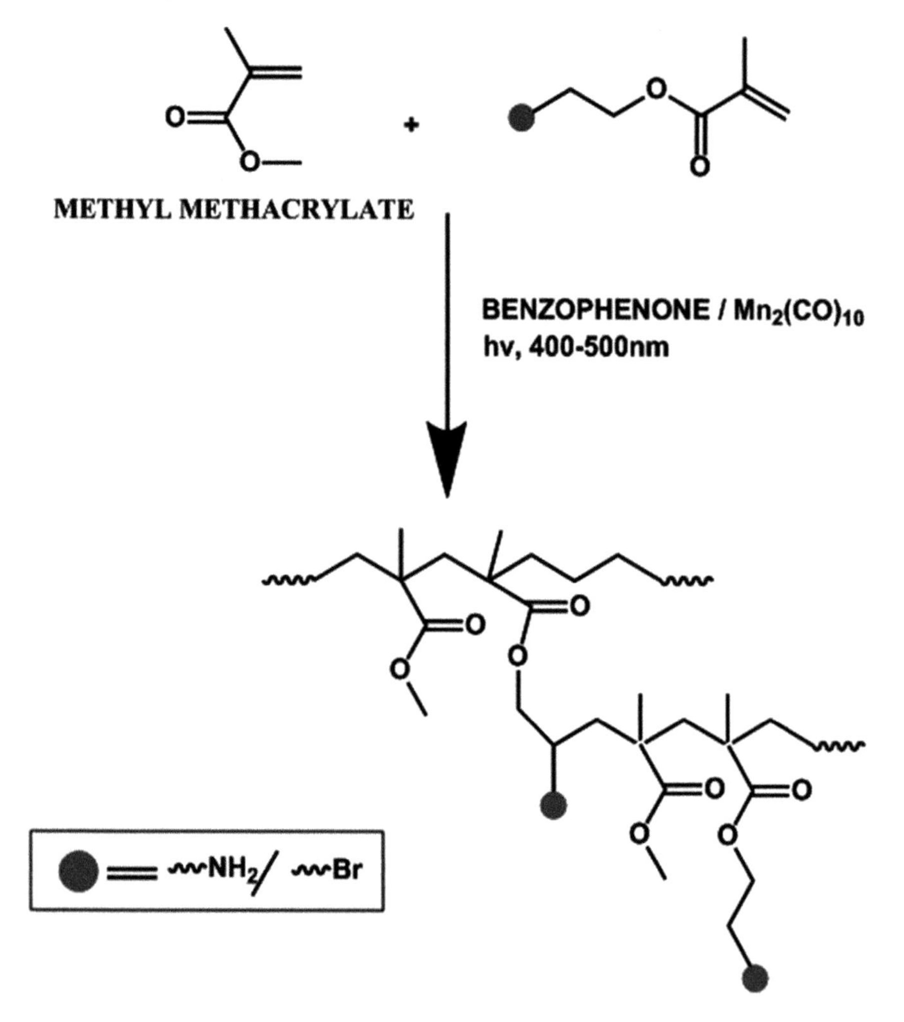

Fig. 12. Mechanism for the synthesis of light-induced amphiphilic hyperbranched polymer.

3. Structure-property relationship for amphiphilic hyperbranched polymers

In the Strathclyde method, the choice of chain transfer agent is an important factor as it finally determines the end property of the synthesized HBP. Even to restrict gelation in the random free radical mechanism, the chain transfer agent plays a major role (Weaver et al. 2008). With the introduction of chain transfer agent, the molar mass effectively decreases as the chain transfer agent controls the formation of number of primary polymer chain. The molar mass in HBPs synthesized via Strathclyde method is lower than in its linear counterpart. Again, molecular weight depends on the degree of branching in the HBPs (Isaure et al. 2006).

The compact globular structure of the HBPs was proved by P. Chambon and his team in their research work (Chambon et al. 2011). The absence of effective primary chain in the polymer even affects the thermoresponsive nature of poly(N-isopropyl acrylamide) (pNIPAAm).

Generally, branched polymers are synthesized by the Strathclyde method. They demonstrated broad and multimodal molecular weight distribution but due to the core-shell morphology within the amphiphilic HBPs, unimodal and narrow distribution is obtained (Weaver et al. 2008, Das et al. 2015). T. Das also showed that the intrinsic viscosities of the HBPs were lower than that of their linear analogs. From FESEM images (Fig. 13) it was seen that, with the introduction of branching within the structure, the

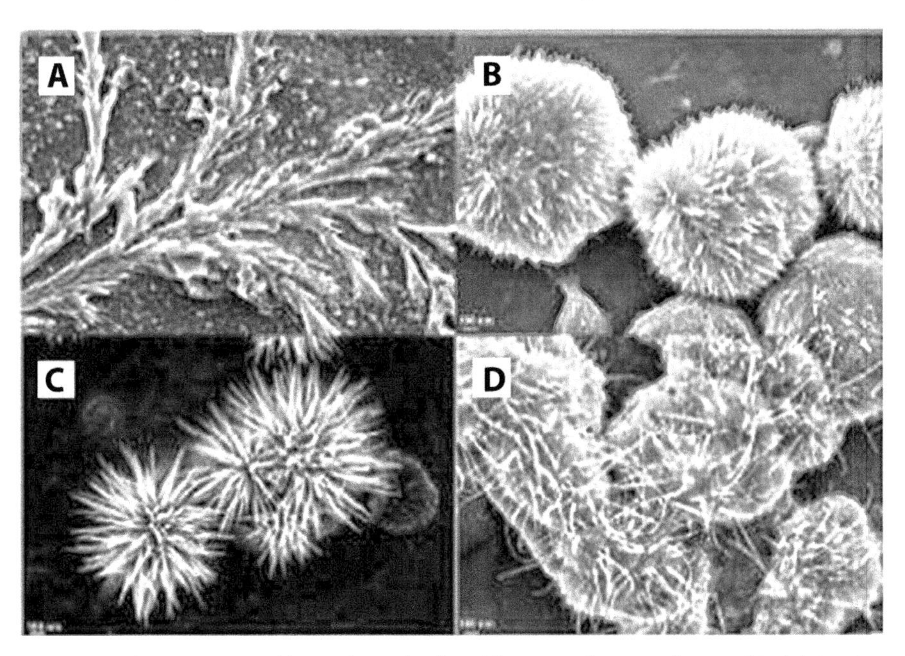

Fig. 13. FESEM images of hyperbranched and linear polymers above pK_a: (a) Linear; (b-d) HBP. Reprinted with permission from Das et al. (2015).

morphology of these amphiphilic copolymers also changed, leading to more stable aggregates with increasing branching (Das et al. 2015).

The amphiphilic HBPs when synthesized via condensation polymerization resulted in relatively narrow molecular weight distribution. In the work reported by Ohta and his group, the synthesized amphiphilic HBP, PEG-b-HBPA, demonstrated a narrow polydispersity (Ohta et al. 2013). When NMR characterization was carried out in D_2O, only the spectra of PEG were visible, which indicated that in water the polymer forms micelle with PEG in the outer shell and HBPA is present within the core. In a subsequent study, Ohta and co-workers developed janus diblock hyperbranched copolyamide, which shows unique self-assembling characteristics. With increasing temperature the shape of micelle changes from spherical at 25°C to dendritic aggregates at 70°C. This is due to the increasing hydrophobicity in the structure as well as the presence of triethylene glycol (Ohta et al. 2019).

Chen and his group developed a nanometric drug delivery vehicle based on amphiphilic HBPs. The sustained drug release profile exhibited by H40-PCL-PEG-FA is due to its unique architecture. The highly branched structure favors more attachment of targeting ligands, folic acid, which in turn enhances the specificity of the delivery vehicle. Even the hydrophobic core of the polymer helps the encapsulation of the highly hydrophobic drug, paclitaxel, and its proper administration within the specified site (Chen et al. 2008).

Self-condensing vinyl polymerization provides more controlled hyperbranched structures just by controlling the structures of the comonomers or inimers. Even HBPs by SCVP possess more stable –C-C- backbone (Powell et al. 2007).

Weng and his co-workers synthesized HPPrAM-co-PS; it showed excellent dye-encapsulating properties by forming stable polymer-dye complexes. Further, because of the presence of PS in the structure, the polymer gets compatible with other polymer matrices and thus acts as a staining agent. The number of cavities and binding sites present within the HBP regulates its encapsulation efficiency and also the polarity difference between the core part and the outer PS part. Hence, the amphiphilic polymers could possibly find application as coloring agents for plastics (Weng et al. 2014).

In the light-induced SCVP by Aydogan and his group, it was observed that as the inimer feed ratio is increased, the branching density in the amphiphilic copolymer also increases. Even the increment in irradiation time increases the molecular weight and the degree of branching in the resultant polymer.

4. Applications of amphiphilic hyperbranched polymers

Hyperbranched macromolecules have been studied since the 19th century, but the term "hyperbranched polymer" was coined by Kim and Webster

(1988). The first hyperbranched macromolecule (soluble polyphenylene) was prepared by Kim et al. in 1988 (Kim and Webster 1990, 1992). The presence of both hydrophobic and hydrophilic parts makes them versatile in terms of solubility, polar nature, crystallinity, solution viscosity, rigidity and thermal stability. They have reportedly been applied in areas such as nanotechnology to encapsulate guest molecules for aesthetic, therapeutic (drug and gene transfection), cosmetic, catalytic, and antimicrobial therapy. Furthermore, the applications as antifouling coatings and multifunctional crosslinkers have generated significant interest in the past few decades (Ho et al. 2012, Wan et al. 2009).

Their applications in encapsulation of guest molecules are reported above. Encapsulation occurs through the formation of a core-shell structure of the HBPs. The functional groups present in the core of amphiphilic HBP are critical for guest selection and guest release. The selective encapsulation is possible even in the case of host-guest system based on ionic interaction, since hydrogen bonding between the core molecule and the carrier can increase the strength of binding between them (Singh and Chauhan 2019).

The improvement in coloring of the polymers by using poly(propargyl quaternary ammonium methacrylate)-*co*-polymethyl methacrylate (HPPrAM-*co*-PMMA) as an amphiphilic HBP synthesized by SVCP-RAFT and Menschutkin reaction was synthesized which was found to exhibit high loading capacity for water soluble dyes. Here, the amphiphilic HBP behaves as a host molecule for anionic dye supermolecular encapsulation. A high loading capacity of methyl orange rose Bengal and fluorescein sodium (water-soluble dyes) is observed due to the powerful electrostatic interaction along with different polarity (large differences) between PPrAM and PMMA moieties. This amphiphilic HBP-dye complex is further employed as a colorant for some common polymers such as styrene butadiene styrene copolymer (SBS) and PMMA by solution casting technique, which has exhibited excellent dispersion characteristics of dye with uniformity, along with color stability. Transmission electron micrographs show that HPPrAM-*co*-PMMA exhibits little effect on the structure of SBS matrix (Figs. 14a and 14b), indicating a similar morphology. The compatibility between dye complex

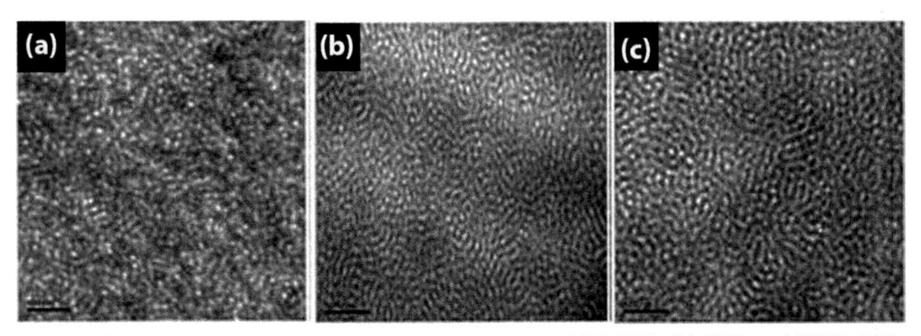

Fig. 14. Transmission electron microscopy images of (a) SBS, (b) SBS and HPPrAM-co-PMMA at 1% (w/w), and (c) SBS and HPPrAM-co-PMMA dye complex at 1% (w/w). Reprinted with permission from Zheng et al. (2016).

and SBS (Fig. 14c) was good, resulting in good dispersion characteristics in conjugation with high transparency of the colored sample analyzed by a spectrophotometer. A significant observation of the tests performed on the colored polymeric membranes that were immersed in deionized water for up to 40 d was the excellent stability of the membrane evidenced by little (~1%) escape of dye molecules (Zheng et al. 2016).

Also, hyperbranched PEI amidated with palmitic acid polymer was found to possess M_n of 10,000 and 2500 g/mol and was reported to transport hydrophilic anionic guest molecules such as eosin Y, methyl orange and fluorescein sodium. An interesting observation was noted that the dye-loading capacity to make an amphiphilic HBP-dye complex increased with an increase in quaternization of residual amines (amidation) (Kitajyo et al. 2007).

The use of nanocarriers in drug delivery (nanomedicine) for the controlled release of drug (guest) molecules was well studied in the last two decades. The encapsulation of water-soluble drug by hyperbranched D-glucancarbamation with *N*-carbonyl L-leucine in three different degrees of substitution of 46.0%, 68.7% and 93.7% was investigated and the investigation revealed that the degree of substitution governed the release of the entrapped drug molecules. The increase in hydrophobicity of the shell of amphiphilic HBP reduced the diffusion of water soluble molecules like RB (rose bengal) from the hydrophobic phase into the aqueous phase (Kitajyo et al. 2007). The effect of stereochemistry on transport behavior was studied, including functionalization of hyperbranched PG with PLA and PLLA (both differing only in stereochemistry) for making two nanocarriers with core-shell morphology. The hyperbranched PG-PLLA encapsulated drug molecule was six times as large as that of PG-PLA hyperbranched polymer (Adeli et al. 2007, Irfan and Seiler 2010).

The modification of dendritic polyglycerol to make it usable in biomedical applications such as scaffolds is shown by Calderón et al. The advantages of tunable functional end groups, bio-inertness for non-specified interactions and defined topological three-dimensional architecture in polyglycerol make it a suitable material for engineering to meet future demands (Calderón et al. 2010). The basic physical properties and drug delivery assisted by the polymer are dependent on the architecture of the polymer (Grayson and Godbey 2008). The synthesis of amphiphilic core-shell dendritic HBPs with targeting ligand as the folate group is reported. The core was made of Boltorn H40 (aliphatic polyester) scaffold. The hydrophobic poly(epsilon-caprolactone) chains were on the inner side and the hydrophilic PEG chains on the outer shell, which possesses folate moieties attached via terminal hydroxyl group in the amphiphilic polymer. These dendritic nanocarriers were loaded with two anticancer drugs, namely paclitaxel and 5-fluorouracil, and *in vitro* experiments were conducted to ascertain that the nanoparticles exhibited enhanced inhibition of cell growth. This was due to increase in cytotoxicity of loaded nanoparticles against tumor cells expressing reception to folates (Paleos et al. 2010). Drug polymer conjugates was evident to be the most effective way for drug release from a micelle system. It involves the formation of a conjugate with the amphiphilic polymer's hydrophobic section leading to micelle formation. The drug is retained in the micelle for a long period because of conjugation. The drug molecules can only be released through enzyme hydrolysis (or other means) of the polymer and diffusion of drug from the micelles. Some drugs that are delivered using micelles (normal, reverse or unimolecular) are presented in Table 3 (Trivedi and Kompella 2010).

Table 3. Micelle release of drugs (Forrest et al. 2008, Xiong et al. 2008, Yoo and Park 2001, Jeong and Park 2001, Gaber et al. 2006, Dong et al. 2008, Zeng et al. 2012, Chen et al. 2013)

Sr No	*Drug used*	*Drug application*	*Polymer applied*	*Duration of release (in vitro)*	*Size (nm)*
1	Paclitaxel	Anti-glioblastoma (tumor)	PEG–PCL	Up to 14 d	27-44
2	Doxorubicin	Breast cancer	PEG–PLGA	>15 d	61.4
3	Oligodinucleotide	Anti-inflammatory	ODN–PLGA	>50 d	65.2
4	Beclomethasone dipropionate	Lung disorders	PEO–DSPE	>6 d	22
5	Sulindac	Anti-inflammatory	PLA–PEO–PLA flower-like micelles	20 d	7-16
6	Tetracaine	Anesthesia	PLA–PEO–PLA flower-like micelles	10 d	7-16
7	Doxorubicin	Breast cancer	Boltorn-PEG	>2 d	110
8	5-Fu and paclitaxel	Antineoplastic (tumor)	Boltorn H40-PCL-PEG	>8 d	>100

The use of amphiphilic HBPs (outer shell of poly(ethylene glycol) (PEG) block and a core of poly(2-(4-vinylbenzyloxy)-N,N-diethyl nicotinamide)) in oxidation therapy of cancer (paclitaxel drug) was also reported. It was attempted by either inducting the reactive oxygen species generation (H_2O_2) in tumors by delivery or by inhibition of defense system (antioxidative enzyme such as catalase, SOD and heme oxygenase) of tumor cells (Boulikas and Tsogas 2008, Lee et al. 2007). The drug delivery for paclitaxel drug-oriented cancer treatment is reportedly done by amphiphilic copolymers of poly(ethyl ethylene phosphate) and poly(ε-caprolactone) (Forrest et al. 2008) and amphiphilic block co-polymer micelles of poly(ethylene glycol)-b-poly(ε-caprolactone) (PEG-b-PCL) (Liu et al. 2008).

Jensen and co-workers studied seven nanocarriers for gene transfection (LPS-activated macrophage cell line with siRNA) and demonstrated *in vitro* gene silencing and low toxicity of nanocarriers based on poly (amido amine) (PAMAM), PEG and PLGA. They reported that PEG and PLGA nanoparticles were less effective than their counterparts for carrying siRNA transfection. Furthermore, they studied five off-target genes to find only minor transcriptional changes, which indicated the effectiveness of PAMAM dendrimers for silencing of TNF-α gene expression (Jensen et al. 2012).

Ma et al. exhibited the synthesis of multifunctional fluorinated polymers with pyrrolidinone pendant groups and block copolymers with pyrrolidinone and alkenyl groups on different segments by RAFT. The formation of distinct fluorinated nanoparticles was observed upon micelle formation in amphiphilic block copolymers with alkenyl functionalized block copolymers inducing thiol-ene photolytic crosslinking. As a result, the formation of amphiphilic crosslinked networks having nanosurface features was obtained. The crosslinked networks formed were made up of robust materials with good mechanical strength, as upon exposure to seawater, it offered limited swelling due to the presence of shorter or equivalent length of hydrophilic block segments to that of hydrophobic ones (Ma et al. 2010). The presence of fluorine along with the well-defined, balanced amphiphilic characteristics and multiple functionalities aid in defining material morphology as well as topology. This further can help in exploring the application of this material in the marine and medical sectors as an antifouling and antimicrobial coating (Bartels et al. 2007, Dai et al. 2016). Bartels et al. further reported the hybridization of hydrophobic hyperbranched fluoropolymers (HBFP) with linear hydrophilic PEGs into complex networks of amphiphilic crosslinked copolymers with discrete nanostructures that demonstrated anti-biofouling abilities, good mechanical properties and atypical release behavior for guest molecules. They also reported that amphiphilic hyperbranched copolymer having HBFP and PEG-rich domains with nano and microscopic dimensions within the crosslinked networks can be employed as nanoscopic hosts for partitioning, packaging and releasing guest molecules with variations in volatility and hydrophilicity with the help of terpenegeraniol, thereby exhibiting superior host-guest behavior (Bartels et al. 2007).

Dai et al. reported the use of AHP based on PEG-PEI-PES (Boltorn) synthesized at room temperature with a mild and low-cost preparation route for stabilizing the gold catalyst nanoparticles (~10 nm). They reported research on the interlayer of core and shell of the nanoparticles in stabilization of catalyst nanoparticles by the absorption of PEI on Au ions and gold nanoparticles (AuNP) along with reduction of 4-nitrophenol (4-NP) to 4-aminophenol (4-AP). It can also be stored with a shelf life of one month. Furthermore, the most salient feature of the AHP was the increase in the catalytic activity of nanoparticles stabilized by AHP (Dai et al. 2016).

The future may hold exciting applications such as the use of AHP as a crosslinker to obtain amphiphilic co-networks that can serve as carriers for bio-catalysis in organic medium (Bruns et al. 2008). There may also be applications based on build-up of a membrane (Tobis et al. 2011) for separation of water with high salinity. The material development can also move towards biodegradable approach or biocompatibility for biomedical applications.

5. Conclusion and outlook

Amphiphilic HBPs have a vast range of applications in various fields. Different synthesis techniques can be selected for developing unique kinds of macromolecular architecture. On account of their special properties, these polymers are in high demand. Still, researchers are taking great effort to develop HBPs with more controlled structure and degree of branching. It is expected that these types of polymers with improved properties will be further explored in future and will be applied for various purposes.

Acknowledgement

The authors gratefully acknowledge the University of Calcutta (Kolkata), Indian Institute of Packaging (Mumbai) and Maharashtra Institute of Technology (Aurangabad) for providing support in writing this chapter.

References

Adeli M., R. Haag and Z. Zarnegar. 2007. Effect of the shell on the transport properties of poly (glycerol) and Poly (ethylene imine) nanoparticles. *Journal of Nanoparticle Research* 9(6): 1057-1065.

Alfurhood, J.A., H. Sun, P.R. Bachler and B.S. Sumerlin. 2016. Hyperbranched poly (N-(2-hydroxypropyl) methacrylamide) via RAFT self-condensing vinyl polymerization. *Polymer Chemistry* 7: 2099-2104.

Aydogan, C., M. Ciftci, A. Mohamed Asiri and Y. Yagci. 2018. Visible light induced one-pot synthesis of amphiphilichyperbranched copolymers. *Polymer* 158: 90-95.

Barriau, E., A.G. Marcos, H. Kautz and H. Frey. 2005. Linear-hyperbranched amphiphilic AB diblock copolymers based on polystyrene and hyperbranched polyglycerol. *Macromolecular Rapid Communications* 26(11): 862-867.

Bartels, J.W., C. Cheng, K.T. Powell, J. Xu and K.L. Wooley. 2007. Hyperbranched fluoropolymers and their hybridization into complex amphiphilic crosslinked copolymer networks. *Macromolecular Chemistry and Physics* 208: 1676-1687.

Baudry, R. and D.C. Sherrington. 2006. Synthesis of highly branched poly(methyl methacrylate)s using the "Strathclyde Methodology" in aqueous emulsion. *Macromolecules* 39: 5230-5237.

Benoit, D., S. Grimaldi, J.P. Finet, P. Tordo, M. Fontanille and Y. Gnanou. 1998. Controlled/living free-radical polymerization of styrene and n-butyl acrylate in the presence of a novel asymmetric nitroxyl radical. *ACS Symposium Series* 685: 225-235. DOI: 10.1021/bk-1998-0685.ch014.

Bian, K. and M.F. Cunningham. 2005. Nitroxide-mediated living radical polymerization of 2-hydroxyethyl acrylate and the synthesis of amphiphilic block copolymers. *Macromolecules* 38(3): 695-701.

Bon, S.A., A.G. Steward and D.M. Haddleton. 2000. Modification of the ω-bromo end group of poly (methacrylate) s prepared by copper (I)-mediated living radical polymerization. *Journal of Polymer Science Part A: Polymer Chemistry* 38(15): 2678-2686.

Boulikas, T. and I. Tsogas. 2008. Microtubule-targeted antitumor drugs: Chemistry, mechanisms and nanoparticle formulations. *Gene TherMolBiol.* 12: 343-387.

Bruns, N., W. Bannwarth and J.C. Tiller. 2008. Amphiphilic conetworks as activating carriers for the enhancement of enzymatic activity in supercritical CO_2. *Biotechnology and Bioengineering* 101: 19-26.

Calderón, M., M.A. Quadir, S.K. Sharma and R. Haag. 2010. Dendritic polyglycerols for biomedical applications. *Advanced Materials* 22: 190-218.

Caminade, A.M., D. Yan and D.K. Smith. 2015. Dendrimers and hyperbranched polymers. *Chem. Soc. Rev.* 44: 3870-3873.

Chambon, P., L. Chen, S. Furzeland, D. Atkins, J.V.M. Weaver and D.J. Adams. 2011. Poly(N-isopropylacrylamide) branched polymer nanoparticles. *Polymer Chemistry* 2: 941.

Chaudhary, S.K. and O. Hernandez. 1979. 4-dimethylaminopyridine: an efficient and selective catalyst for the silylation of alcohols. *Tetrahedron Letters* 20(2): 99-102.

Chen, S., X.-Z. Zhang, S.X. Cheng, R.X. Zhuo and Z.W. Gu. 2008. Functionalized amphiphilic hyperbranched polymers for targeted drug delivery. *Biomacromolecules* 9: 2578-2585.

Chen, J., J. Ouyang, J. Kong, W. Zhong and M.M. Xing. 2013. Photo-cross-linked and pH-sensitive biodegradable micelles for doxorubicin delivery. *ACS Applied Materials & Interfaces* 5: 3108-3117.

Chisholm, M., N. Hudson, N. Kirtley, F. Vilela and D.C. Sherrington. 2009. Application of the "Strathclyde Route" to branched vinyl polymers in suspension polymerization: Architectural, thermal, and rheological characterization of the derived branched products. *Macromolecules* 42(20): 7745-7752.

Couturier, J.L., C. Henriet-Bernard, C. Le Mercier, P. Tordo and J.F. Lutz. 2003. U.S. Patent No. 6,569,967. Washington, DC: U.S. Patent and Trademark Office.

Dai, Y., P. Yu, X. Zhang and R. Zhuo. 2016. Gold nanoparticles stabilized by amphiphilic hyperbranched polymers for catalytic reduction of 4-nitrophenol. *Journal of Catalysis* 337: 65-71.

Das, T., S. Sengupta, U.K. Ghorai, A. Dey and A. Bandyopadhyay. 2015. Sequential amphiphilic and pH responsive hyperbranched copolymer: Influence of hyper branching/pendant groups on reversible self-assembling from polymersomes to aggregates and usefulness in waste water treatment. *RSC Advances* 124: 102932-102941.

Dong, H., Li, Y., Cai, S., Zhuo, R., Zhang, X. and Liu, L. 2008. A facile one-pot construction of supramolecular polymer micelles from α-cyclodextrin and poly (ε-caprolactone). *Angewandte Chemie International Edition* 47: 5573-5576.

Flory, P.J. 1941. Molecular size distribution in three dimensional polymers. I: Gelation 1. *J. Am. Chem. Soc.* 63(11): 3083-3090.

Frechet, J.M., M. Henmi, I. Gitsov, S. Aoshima, M.R. Leduc and R.B. Grubbs. 1995. Self-condensing vinyl polymerization: An approach to dendritic materials. *Science* 269: 1080-1083.

Forrest, M.L., J.A. Yáñez, C.M. Remsberg, Y. Ohgami, G.S. Kwon and N.M. Davies. 2008. Paclitaxel prodrugs with sustained release and high solubility in poly (ethylene glycol)-b-poly (ε-caprolactone) micelle nanocarriers: Pharmacokinetic disposition, tolerability and cytotoxicity. *Pharmaceutical Research*, 25: 194-206.

Gaber, N.N., Y. Darwis, K.K. Peh and Y.T.F. Tan. 2006. Characterization of polymeric micelles for pulmonary delivery of beclomethasonedipropionate. *Journal of Nanoscience and Nanotechnology* 6: 3095-3101.

Gao, C. and D. Yan. 2004. Hyperbranched polymers: From synthesis to applications. *Progress in Polymer Science* 29: 183-275.

Grayson, S.M. and W.T. Godbey. 2008. The role of macromolecular architecture in passively targeted polymeric carriers for drug and gene delivery. *Journal of Drug Targeting* 16: 329-356.

Grimaldi, S., F. Lemoigne, J.P. Finet, P. Tordo, P. Nicol, M. Plechot and Y. Gnanou. 2001. U.S. Patent No. 6,255,448. Washington, DC: U.S. Patent and Trademark Office.

Guan, Z., P.M. Cotts, E.F. McCord and S.J. McLain. 1999. Chain walking: A new strategy to control polymer topology. *Science* 283: 2059-2062.

Hawker, C.J. 1995. Architectural control in "living" free radical polymerizations: preparation of star and graft polymers. *Angewandte Chemie International Edition in English* 34(13-14): 1456-1459.

Hawker, C.J., J.M. Frechet, R.B. Grubbs and J. Dao. 1995. Preparation of hyperbranched and star polymers by a "living", self-condensing free radical polymerization. *Journal of the American Chemical Society* 117(43): 10763-10764.

Ho, C.H., M. Thiel, S. Celik, E.K. Odermatt, I. Berndt, R. Thomann and J.C. Tiller. 2012. Conventional and microwave-assisted synthesis of hyperbranched and highly branched polylysine towards amphiphilic core-shell nanocontainers for metal nanoparticles. *Polymer* 53: 4623-4630.

Irfan, M. and M. Seiler. 2010. Encapsulation using hyperbranched polymers: From research and technologies to emerging applications. *Industrial & Engineering Chemistry Research*, 49(3): 1169-1196.

Isaure, F., P.A.G. Cormack and D.C. Sherrington. 2006. Facile synthesis of branched water-soluble poly(dimethylacrylamide)s in conventional and parallel reactors using free radical polymerization. *Reactive and Functional Polymers* 66: 65-79.

Jensen, L.B., J. Griger, B. Naeye, A.K. Varkouhi, K. Raemdonck, R. Schiffelers and H.M. Nielsen. 2012. Comparison of polymeric siRNA nanocarriers in a murine LPS-activated macrophage cell line: Gene silencing, toxicity and off-target gene expression. *Pharmaceutical Research* 29: 669-682.

Jeong, J.H. and T.G. Park. 2001. Novel polymer - DNA hybrid polymeric micelles composed of hydrophobic poly (D, L-lactic-co-glycolic acid) and hydrophilic oligonucleotides. *Bioconjugate Chemistry* 12: 917-923.

Johnson, L.K., C.M. Killian and M. Brookhart. 1995. New Pd (II)- and Ni (II)-based catalysts for polymerization of ethylene and alpha-olefins. *Journal of the American Chemical Society* 117: 6414-6415.

Kazmaier, P.M., K.A. Moffat, M.K. Georges, R.P. Veregin and G.K. Hamer. 1995. Free-radical polymerization for narrow-polydispersity resins. Semiempirical molecular orbital calculations as a criterion for selecting stable free-radical reversible terminators. *Macromolecules* 28(6): 1841-1846.

Kim, Y.H. and O.W. Webster. 1988. Hyperbranched polyphenylenes. *Polym Prepr.* 29(2): 310-311.

Kim, Y.H. and O.W. Webster. 1990. Water soluble hyperbranchedpolyphenylene: A unimolecular micelle. *Journal of the American Chemical Society* 112: 4592-4593.

Kim, Y.H. and O.W. Webster. 1992. Hyperbranchedpolyphenylenes. *Macromolecules* 25: 5561-5572.

Kitajyo, Y., T. Imai, Y. Sakai, M. Tamaki, H. Tani, K. Takahashi and T. Kakuchi. 2007. Encapsulation–release property of amphiphilichyperbranched d-glucan as a unimolecular reverse micelle. *Polymer* 48(5), 1237-1244.

Lee, S.C., K.M. Huh, J. Lee, Y.W. Cho, R.E. Galinsky and K. Park. 2007. Hydrotropic polymeric micelles for enhanced paclitaxel solubility: In vitro and in vivo characterization. *Biomacromolecules* 8: 202-208.

Liu, B., M. Yang, X. Li, X. Qian, Z. Shen, Y. Ding and L. Yu. 2008. Enhanced efficiency of thermally targeted taxanes delivery in a human xenograft model of gastric cancer. *Journal of Pharmaceutical Sciences* 97: 3170-3181.

Liu, C., Y.Y. Fei, H.L. Zhang, C.Y. Pan and C.Y. Hong. 2018. Effective construction of hyperbranched multicyclic polymer by combination of ATRP, UV-induced cyclization, and self-accelerating click reaction. *Macromolecules* 52: 176-184.

Ma, J., J.W. Bartels, Z. Li, K. Zhang, C. Cheng and K.L. Wooley. 2010. Synthesis and solution-state assembly or bulk state thiol-ene crosslinking of pyrrolidinone- and alkene-functionalized amphiphilic block fluorocopolymers: From functional nanoparticles to anti-fouling coatings. *Australian Journal of Chemistry* 63: 1159-1163.

Matyjaszewski, K., B.E. Woodworth, X. Zhang, S.G. Gaynor and Z. Metzner. 1998. Simple and efficient synthesis of various alkoxyamines for stable free radical polymerization. *Macromolecules* 31(17): 5955-5957.

Matyjaszewski, K. 2018. Advanced materials by atom transfer radical polymerization. *Advanced Materials* 30: 1706441.

Mecking, S., L.K. Johnson, L. Wang and M. Brookhart. 1998. Mechanistic studies of the palladium-catalyzed copolymerization of ethylene and α-olefins with methyl acrylate. *Journal of the American Chemical Society* 120: 888-899.

Ohta, Y., T. Kanou, A. Yokoyama and T. Yokozawa. 2013. Synthesis of well-defined, amphiphilicpoly(ethylene glycol)-b-hyperbranched polyamide. *Journal of Polymer Science Part A: Polymer Chemistry* 51: 3762-3766.

Ohta, Y., Y. Abe, K. Hoka, E. Baba, Y.P. Lee, C.A. Dai and T. Yokozawa. 2019. Synthesis of amphiphilic, Janus diblock hyperbranched copolyamides and their self-assembly in water. *Polymer Chemistry* 10: 4246-4251.

Paleos, C.M., D. Tsiourvas, Z. Sideratou and L.A. Tziveleka. 2010. Drug delivery using multifunctional dendrimers and hyperbranched polymers. *Expert Opinion on Drug Delivery* 7: 1387-1398.

Peleshanko, S., R. Gunawidjaja, S. Petrash and V.V. Tsukruk. 2006. Synthesis and interfacial behavior of amphiphilic hyperbranched polymers: Poly (ethylene oxide)-polystyrene hyperbranches. *Macromolecules* 39(14): 4756-4766. Copyright (2006) American Chemical Society.

Powell, K.T., C. Cheng and K.L. Wooley. 2007. Complex amphiphilic hyperbranched fluoropolymers by atom transfer radical self-condensing vinyl (co) polymerization. *Macromolecules* 40: 4509-4515.

Singh, D. and N.P.S. Chauhan. 2019. Amphiphilic hyperbranched polymers. *In*: Advanced Functional Polymers for Biomedical Applications. pp. 111-128. Elsevier, Netherlands.

Tobis, J., L. Boch, Y. Thomann and J.C. Tiller. 2011. Amphiphilic polymer conetworks as chiral separation membranes. *Journal of Membrane Science* 372: 219-227.

Trivedi, R. and U.B. Kompella. 2010. Nanomicellar formulations for sustained drug delivery: Strategies and underlying principles. *Nanomedicine* 5: 485-505.

Tsarevsky, N.V., J. Huang and K. Matyjaszewski. 2009. Synthesis of hyperbranched degradable polymers by atom transfer radical (Co)polymerization of inimers with ester or disulfide groups. *Journal of Polymer Science Part A: Polymer Chemistry* 47(24): 6839-6851.

Wan, D., J. Yuan and H. Pu. 2009. Macromolecular nanocapsule derived from hyperbranchedpolyethylenimine (HPEI): Mechanism of guest encapsulation versus molecular parameters. *Macromolecules* 42: 1533-1540.

Weaver, J.V.M., R.T. Williams, B.J.L. Royles, P.H. Findlay, A.I. Coopera and S.P. Rannard. 2008. pH-responsive branched polymer nanoparticles. *Soft Matter* 4: 985-992.

Weng, Z., Y. Zheng, A. Tang and C. Gao. 2014. Synthesis, dye encapsulation, and highly efficient colouring application of amphiphilic hyperbranched polymers. *Australian Journal of Chemistry* 67: 103-111.

Xiong, M.P., J.A. Yáñez, C.M. Remsberg, Y. Ohgami, G.S. Kwon, N.M. Davies and M.L. Forrest. 2008. Formulation of a geldanamycinprodrug in mPEG-b-PCL micelles greatly enhances tolerability and pharmacokinetics in rats. *Journal of Controlled Release* 129: 33-40.

Yoo, H.S. and T.G. Park. 2001. Biodegradable polymeric micelles composed of doxorubicin conjugated PLGA–PEG block copolymer. *Journal of Controlled Release* 70: 63-70.

Zeng, X., Y. Zhang, Z. Wu, P. Lundberg, M. Malkoch and A.M. Nyström. 2012. Hyperbranched copolymer micelles as delivery vehicles of doxorubicin in breast cancer cells. *Journal of Polymer Science Part A: Polymer Chemistry* 50: 280-288.

Zhang, X., Y. Dai and G. Dai. 2020. Advances in amphiphilic hyperbranched copolymers with an aliphatic hyperbranched 2,2-bis (methylol) propionic acid-based polyester core. *Polymer Chemistry A* 11(5): 964-973.

Zheng, Y., A. Tang, Z. Weng, S. Cai, Y. Jin, Z. Gao and C. Gao. 2016. Amphiphilic hyperbranched polymers: Synthesis and host-guest supermolecular coloring application. *Macromolecular Chemistry and Physics* 217: 380-389.

Zhou, S., D. Zhang, L. Bai, J. Zhao, Y. Wu, H. Zhao and X. Ba. 2016. The synthesis of backbone thermo and pH responsive hyperbranched poly (bis (N,N-propyl acryl amide))s by RAFT. *Polymers* 8: 135.

CHAPTER

4

Biodegradable Polymers

Sayan Deb Dutta and Ki-Taek Lim*
Department of Biosystems Engineering, Kangwon National University,
Chuncheon 24341, Republic of Korea

1. Introduction

Conventional non-biodegradable polymers such as polystyrene, polypropylene, and polyethylene have raised considerable concern because of their massive accumulation in the environment for many years after disposal. These polymer materials seem to be ill suited for applications in which plastics are used for a very short time and disposed of immediately after use (Dussud et al. 2018, Gross and Kalra 2002, Luyt and Malik 2019). Moreover, plastics are often buried underground after disposal, posing a danger to plants, soil inhabitants, and microbes. The physical recycling of these materials is often unfeasible and usually undesirable (Gross and Kalra 2002). To address these concerns, the development of environmentally safe and biodegradable polymers has become essential. Biodegradable polymers (BPs) are naturally derived polymers designed to degrade upon disposal by the action of living microorganisms, such as fungi, algae, bacteria, parasites, and other unicellular microorganisms. Microorganisms effectively break down those polymer chains by hydrolytic enzymes and by releasing simple monomers and oligomers that can be utilized by plants and microbes. Figure 1 represents an overview of plant-derived biodegradable polymer production and the utilization process. Biodegradable polymers are often derived from plants and plant byproducts (Haider et al. 2019, Zindani et al. 2019). There are three main types of BPs: synthetic, natural, and hybrid materials (Iqbal et al. 2019).

The most convenient applications of BPs include various packaging materials (food containers, carry bags, trash bags, loose-fill foam, film wrapping, laminated sheets), disposable materials (various fabrics, diaper sheets, cotton swabs), consumer goods (tableware, kitchenware, toys,

*Corresponding author: ktlim@kangeon.ac.kr

cosmetic cases), and agricultural goods (planters, mulch films) (Asghari et al. 2017, Gross and Kalra 2002, Kenry and Liu 2018, Scaffaro et al. 2019). The natural biodegradable polymer includes various polysaccharides (e.g., cellulose, nanocellulose, starch, alginate, chitin/chitosan, hyaluronic acid) or proteins (e.g., gelatin, collagen, fibrin, silk, soy) (Dutta et al. 2019a, Iqbal et al. 2019, Patel et al. 2019a, Swetha et al. 2010). However, the most frequently used synthetic BPs include poly(lactic acid) (PLA), poly(glycolic acid) (PGA), poly(lactic-co-glycolide) (PLGA), poly(ε-caprolactone) (PCL), poly(ethylene glycol) (PEG), and poly(p-dioxanone) (Asghari et al. 2017, Dutta et al. 2019b, Eatemadi et al. 2016, Gross and Kalra 2002, Guo and Ma 2014, Iqbal et al. 2019). The production and commercialization of BPs are, however, restricted because of extensive competition with regular plastics that are low-quality, non-biodegradable, cheap and familiar to the customer. Thus, the development and infrastructure for the disposal of these BPs in the bioactive environment must be expanded and will require capital funding (Asghari et al. 2017, Gross and Kalra 2002). This chapter briefly addresses potential sources of various BPs and their environmental benefits along with their production strategies.

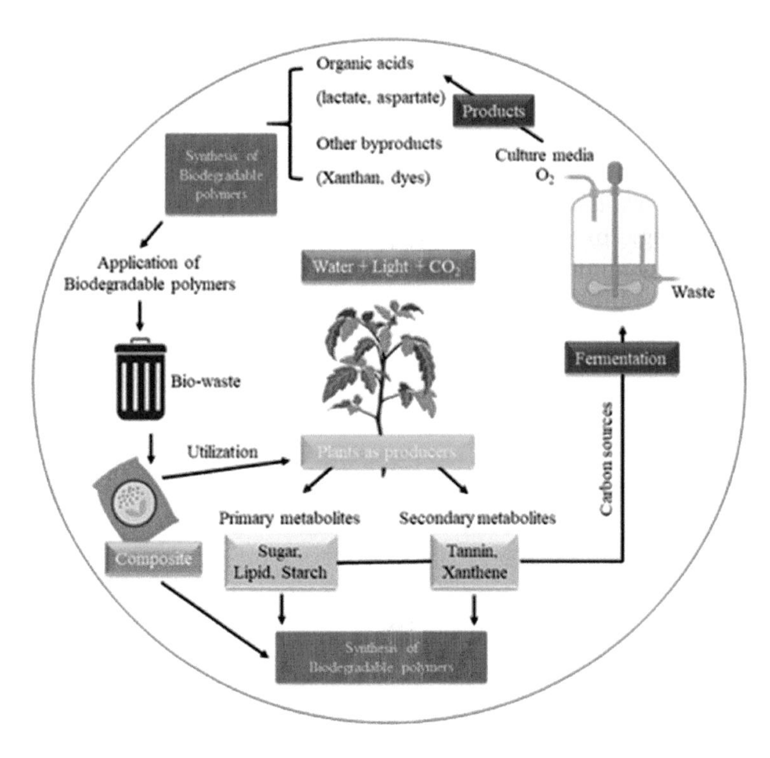

Fig. 1. Production of biodegradable polymers (BPs) from agricultural products and fermentation processes. After the disposal of bio-wastes, it is utilized by the microbes and anaerobes. Microbes degrade those wastes into natural substances, such as CO_2, water, pure monomer, and other humic matter.

2. Synthetic biodegradable polymers

2.1. Polyglycolic acid (PGA)

Polyglycolide or poly(glycolic acid) is a biodegradable and biocompatible aliphatic polyester, commonly used for biomedical applications (Fig. 2a). PGA can be obtained by ring-opening polymerization of glycolic acid monomers (Asghari et al. 2017, Ikada and Tsuji 2000). PGA is highly crystalline (44-55%) in nature and exhibits a high tensile modulus with low solubility in organic solvents. Because of its biocompatibility, PGA is often used to construct tissue engineering scaffolds. Kobayashi and co-workers reported that polyglycolide/hyaluronan (PGA/HA) nanocomposites were sufficient to reform vasculature within 5 d of implantation in the animal model (Kobayashi et al. 2013). Another study indicated that freeze-dried PGA/HA scaffold improved the chondrogenic potential of human mesenchymal stem cells (hMSCs)-based *in vivo* cartilage regeneration in a rabbit model (Patrascu et al. 2013). Outcomes from these experiments demonstrated that PGA is an excellent biodegradable biomaterial and can be used for tissue and blood vessel regeneration.

2.2. Polylactic acid (PLA)

Polylactide or polylactic acid is another biodegradable, bioadsorbable, thermoplastic aliphatic polyester derived from lactic acid. Lactide is a chiral molecule and exists in two isoforms: *l*-lactide and *d*-lactide. Polymerization of lactic acid results in the formation of a mixture of semi-crystalline (*d*, *l*)-lactic acid (Fig. 2b). *l*-lactic acid is naturally occurring and exhibits a high degree of crystallinity (≥37%). PLA and PLA-based nanocomposites have been used as semipermeable microcapsules containing enzymes, hormones, drugs, vaccines, and other bio-products. PLA is also used as medical implants in the form of a rod, screw, pin, or mesh (Asghari et al. 2017, dos Santos

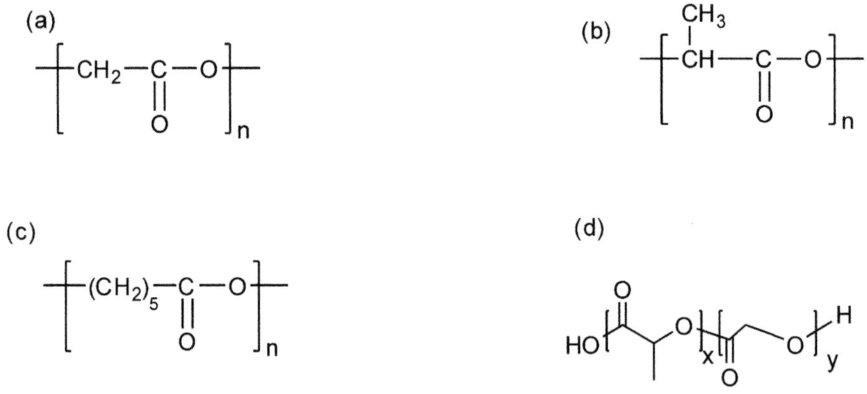

Fig. 2. Structure of some synthetic biodegradable biopolymers: (a) poly(glycolide), (b) poly(lactide), (c) poly(ε-caprolactone), and (d) poly(lactic-co-glycolic acid).

et al. 2017, Li et al. 2017). In one study, Lin and co-workers prepared chitosan (CH)-coated hydroxyapatite/polylactic acid (HA/PLA) nanofibers for bone tissue regeneration (C.-C. Lin et al. 2014).

2.3. Poly(ε-caprolactone) (PCL)

Polycaprolactone (PCL) is a type of semi-crystalline biodegradable biopolymer (Nair and Laurencin 2007). PCL can be prepared by ring-opening polymerization of ε-caprolactone using SnO_2 as catalyst under heating condition (Asghari et al. 2017, Middleton and Tipton 2000). PCL is highly crystalline with a low melting point (55-60°C) and soluble in a range of solvents (Fig. 2c). Owing to its unique structure and slow degradation property, PCL was often chosen as a long-term drug or vaccine delivery vehicle. PCL-coated levonorgestrel is used to prepare long-term contraceptive devices such as Capronor® (Nair and Laurencin 2005). In another study, gelatin/PCL electrospun fiber was reported to enhance the biocompatibility of chondrocytes and favored cartilage regeneration *in vitro* (Zheng et al. 2014). One of our recent studies highlighted the role of electrospun PCL/cellulose nanocrystal nanofibers in bone regeneration (Dutta et al. 2019b). Many other studies indicate that PCL is a biocompatible scaffold to be used in cartilage and bone tissue regeneration (N. Fu et al. 2016, S. Fu et al. 2012, Ren et al. 2017, Yeo and Kim 2012). An overview of PCL and PCL-based nanocomposites used in tissue engineering application is presented in Table 1.

2.4. Poly(lactide-co-glycolide) (PLGA)

Next to polycaprolactone, PLGA is the most extensively investigated polyester so far. PLGA is also a biodegradable biopolymer with numerous medical applications (Asghari et al. 2017). PLGA is synthesized from two different monomers of lactic acid and glycolic acid by ring-opening polymerization (Middleton and Tipton 2000). Both *l*- and *d*-lactic acids can be used for the synthesis of PLGA (Fig. 2d). However, different ratios of lactic acid and glycolic acid have been used to prepare PLGA for a wide range of biomedical applications. PLGA has been known for a long time as an ideal material for cell adhesion and proliferation, making it suitable for tissue engineering applications. Interestingly, PLGA mixed with nano-hydroxyapatite (nHA) showed improved proliferation, differentiation of MC3T3-E1 osteoblast cells (Qian et al. 2014). This study confirmed that PLGA/nHA composite scaffolds were ideal for tissue engineering applications.

3. Natural biodegradable polymers

3.1. Hyaluronic acid (HA)

Hyaluronic acid (HA) is a naturally occurring biopolymer, chiefly made up of non-sulfated GAG that contains alternating units of D-glucuronic acid and D-*N*-acetyl glucosamine, linked by adjacent β-1, 4 and β-1, 3-glycosidic

Table 1. Examples of PCL and PCL-based nanocomposites used in tissue engineering applications

PCL composite	*Nature*	*Applications*	*References*
PCL/polyethylene glycol(PEG)/nano hydroxyapatite(nHA)	Injectable hydrogel	Bone regeneration	S. Fu et al. 2012
PCL/gelatin	Electrospun fiber	Bone regeneration	Ji et al. 2013, Ren et al. 2017
PCL/PEG/acellular bone matrix	Injectable hydrogel	Bone regeneration	Ni et al. 2014
PCL/cellulose nanocrystals (CNCs)	Electrospun fiber	Bone regeneration	Dutta et al. 2019b
PCL/CNCs/graphene oxide (GO)	Electrospun fiber	Bone regeneration	Patel et al. 2019b
PCL/Iron oxide (Fe_3O_4)	3D printed scaffold	Bone regeneration, drug delivery, hyperthermia	J. Zhang et al. 2014
PCL/polylactic-co-glycolide (PLGA)/β-tricalcium phosphate (β-TCP)	3D printed scaffold	Bone regeneration	Shim et al. 2014
PCL/hydroxyapatite (HA)	Microporous membrane	Bone regeneration	Basile et al. 2015
PCL/strontium-containing hydroxyapatite (SrHA)	3D printed scaffold	Bone regeneration	Liu et al. 2019
PCL/carbon nanotube (CNT)/ Fe_3O_4	Membrane	Bone regeneration	Świętek et al. 2019
PCL/polytetrahydrofuran urethane (PTHFU)	Electrospun fiber	Cartilage regeneration	Jiang et al. 2019
PCL/collagen	3D printed scaffold	Cartilage regeneration	Theodoridis et al. 2019

linkages (Fig. 3a) (Shi et al. 2016). Naturally occurring HA contains a straight-chain anionic polysaccharide structure and is extensively dispersed throughout the human body. Under mild conditions, HA can be physically crosslinked by freeze-thawing, without chemical crosslinking (H. Zhang et al. 2013). HA hydrogels can easily be prepared by chemical crosslinking, enzyme crosslinking, or photo-crosslinking. An example of chemically crosslinked HA hydrogel includes the use of poly(ethylene glycol) diacrylate in a humidified atmosphere (Young and Engler 2011). Horseradish peroxidase (HRP) combined with hydrogen peroxide (H_2O_2) is used to prepare tyramine-HA conjugated hydrogels. It is an excellent example of enzymatic conversion of HA hydrogels, where HRP plays a role as a biological crosslinker. Based on its biochemical properties, HA can be used as an ECM material for wound healing applications (Born 2006, Highley et al. 2016).

3.2. Cellulose

Cellulose is the naturally occurring biodegradable biopolymer (Dutta et al. 2019a). It is composed of a linear chain of glucose units, interconnected by β-1, 4-glycosidic bonds (Mohite and Patil 2014, Ullah et al. 2016). Cellulose can be extracted from plants through different chemical treatments or by microbial action. Bacterial cellulose (BC) fibers have a high aspect ratio, along with a diameter of 20 to 100 μm. Besides, researchers are focusing on plant-derived cellulose because of its cost-effective production and high yield value compared to BC. Cellulose is insoluble in pure water and, thus, it is challenging to make cellulose-based hydrogels and nanocomposites (Jung et al. 2005). Many solvent systems have been developed to solubilize cellulose completely. Examples of such solvents include LiCl/dimethylacetamide (DMAc) (Yoshino et al. 1996), *N*-methyl morpholine-*N*-oxide (NMMO) (Kim et al. 2007), and ionic liquids as well as alkali/urea (or thiourea) (Dutta et al. 2019a) aqueous systems. A wide range of microbes produce cellulose, such as *Acetobacter*, *Agrobacterium*, *Achromobacter*, *Aerobacter*, *Azotobacter*, *Salmonella*, *Escherichia*, and *Rhizobium*, *Aspergillus*, *Trichoderma*, *Penicillium*, and *Humicola* spp. (Peppas et al. 2000). BC fibers exhibit a high degree of polymerization compared to plant-derived cellulose (Dutta et al. 2019a).

3.3. Chitosan

Chitosan is another naturally derived biodegradable biopolymer, principally made up of a linear chain of β-1, 4-linked D-glucosamine and *N*-acetyl-D-glucosamine units (Fig. 3c). Chitosan can be obtained from the extensive deacetylation of chitin polymers (Malafaya et al. 2007). Chitin and chitosan are the main polymers found in the exoskeleton of many insects. Since chitosan is a pH-sensitive material, it can be deposited electrically on electrodes (Shi et al. 2016). Their pH-sensitivity and gel-forming nature further suggest the use of chitosan-based hydrogels in the delivery of certain drugs, such as ibuprofen and clotrimazole (Shi et al. 2016). Chitosan hydrogels can be easily prepared by chemical crosslinking.

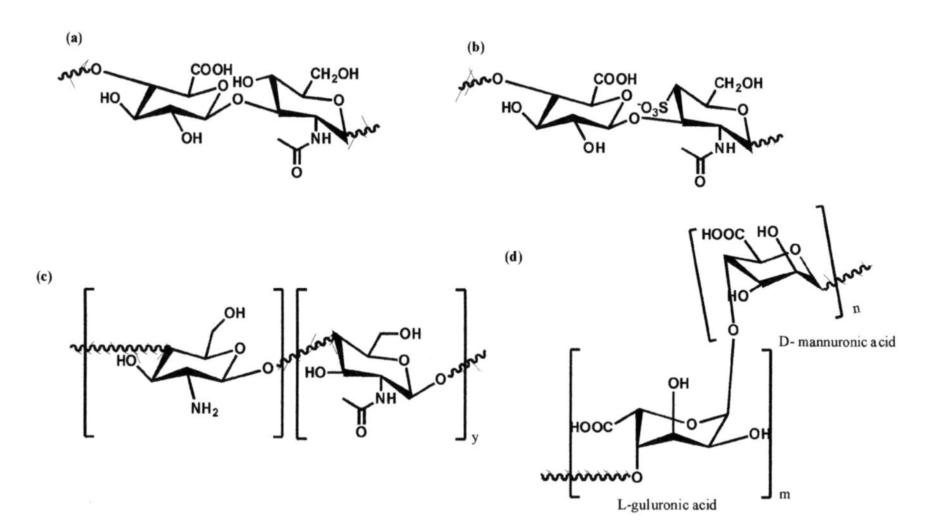

Fig. 3. Structure of some naturally derived biodegradable biopolymers: (a) hyaluronic acid, (b) chondroitin sulfate, (c) chitosan, and (d) alginate.

3.4. Gelatin

Gelatin is derived from collagen biopolymer by hydrolysis of the collagen chain. Gelatin is a solid substance, water-soluble, biodegradable, thermo-responsive, and biocompatible. Gelatin forms a colloidal structure in water and behaves like a gel upon freezing (Asghari et al. 2017, Echave et al. 2019, Gross and Kalra 2002). Owing to their cell-stimulatory properties, gelatin and gelatin-based biopolymers are extensively used for 3D cell culture applications (D. Olsen et al. 2003). Gelatin has different states, such as Type A or net positive and Type B or net negative, depending on its extraction methods, which allows the binding of oppositely charged proteins while maintaining bioactivity. Because of this variable electrochemical property, gelatin is the primary material for the fabrication of microcapsules and microspheres for drug delivery applications (Karim and Bhat 2009, B.R. Olsen et al. 2000, D. Olsen et al. 2003).

3.5. Alginate

Alginate or alginic acid is commonly found within the cell walls and intercellular spaces of some seaweeds and brown algae (*Laminaria* sp.) that play a vital role in the structural integrity of those marine plants. Alginate is reported as non-toxic and, hence, extensively used in additive industries for making salads, cookies, and ice creams. Alginate is a non-branched binary polymer of β-D-mannuronic acid and α-L-glucouronic acid monomers that are connected by β-1,4 glycoside linkage (Fig. 3d) (Kaundal et al. 2015, Nair and Laurencin 2007). Because of its biocompatibility, alginate is also used in the medical field, for bone fracture treatment, wound dressings, and blood detoxification (Zhao et al. 2010).

There are many more naturally derived biodegradable polysaccharides and proteins isolated from plants and animal sources. Some are listed in Table 2.

4. Hybrid biopolymers

4.1. Poly(*N*-isopropyl acrylamide)

Poly(*N*-isopropyl acrylamide) or PNIPAAm is a type of biodegradable and thermo-responsive biopolymer. PNIPAAm is usually synthesized from *N*-isopropyl acrylamide monomers by free-radical polymerization in the presence of initiator (Asghari et al. 2017, Schild 1992). Because of its unique thermo-responsive properties, PNIPAAm is commonly used as biosensors, drug delivery agents, and even in tissue engineering applications.

4.2. Poly(caprolactone/lactide) copolymer

PCL/PLA hybrid polymer is often used as a biomaterial for tissue engineering and drug delivery applications (Nair and Laurencin 2007). PCL/PLA copolymer is synthesized by ring-opening polymerization and is biodegradable, biocompatible, and bioadsorbable. One study showed that PCL/PLA nanocomposite fiber loaded with thymol (herbal drug) can improve its wound healing property (Karami et al. 2013).

4.3. Poly((DL-lactic acid-co-glycolide)-*g*-ethylene glycol) (PLGA-*g*-PEG)

PLGA-*g*-PEG is another copolymer composed of poly(DL-lactic acid-co-glycolic acid) and polyethylene glycol. It is a hybrid biodegradable polymer (Nair and Laurencin 2007, Sidney et al. 2015). PLGA-*g*-PEG was reported to be an excellent drug carrier and often used as material for bone tissue engineering. A study investigated that PLGA-*g*-PEG scaffolds loaded with diclofenac sodium showed localized delivery in acute bone inflammation (Sidney et al. 2015). This result suggested that PLGA-*g*-PEG-based biodegradable polymer can be used as a superior drug delivery vehicle in the future.

5. Advantages and uses of biodegradable polymers

Most of the biodegradable polymers are water-soluble. Thus, they can be used as detergent coatings, scale inhibitors, flocculants, emulsifiers, packing agents, and many other products. These polymers can also be used as coating agents in various products, such as shampoo, toothpaste, food containers, skin lotions, and textiles. However, some of the synthetic biodegradable plastics made up of acrylic acid, polyvinyl alcohol, maleic anhydride, methacrylic acid, and similar components persist in the water for a long time. Organic plastic products, chiefly made up of cellulose and starch, could be used

Table 2. List of some naturally derived biodegradable polymers used in biomedical applications

Biopolymers	*Nature*	*Composition*	*Application*	*References*
Polysaccharide	Chondroitin sulfate	N-acetyl galactosamine (GalNAc)	Wound healing, cartilage regeneration	Chan et al. 2005, Kosir et al. 2000
Human and bovine-derived proteins	Collagen	A mixture of 33% glycine, 25% proline, and 25% hydroxyproline, 10-15% lysine	Wound dressing, tissue regeneration	Ferreira et al. 2012, K. Lin et al. 2019
	Cyanophycin	α-amino-α-carboxyl-linked L-aspartic acid residues	Biomedical	Nair and Laurencin 2007
	Poly-ε-L-lysine	Lysine monomers	Biomedical	Nair and Laurencin 2007
	Poly-γ-glutamic acid	Glutamate monomers	Microbial fermentation, drug delivery	Nair and Laurencin 2005, 2007
	Poly(aspartic acid)	Aspartate residues	Smart drug delivery	Matsumura et al. 2004, Nair and Laurencin 2007
	Elastin	Major constituents of vascular and lung tissue; heptapeptide or pentapeptide	Biomedical, hydrogel scaffolds preparation, dermal tissue regeneration, shape-memory hydrogels	Long et al. 2020, Wen et al. 2019, Y. Zhang et al. 2020
	Albumin	Found in serum and plasma of blood	Intravenous drug and gene delivery, cardiovascular regeneration	Uchida et al. 2005
	Gelatin	Hydrolysis product of collagen	Drug delivery	Nair and Laurencin 2007
Silkworm-derived proteins	Silk fibroin (SF)	Mixture of various proteins	Tissue regeneration, drug delivery	Melke et al. 2016, Pritchard and Kaplan 2011
	Silk sericin	Mixture of various proteins	Tissue regeneration	Vepari and Kaplan 2007

to avoid such accumulations (Nair and Laurencin 2007). Organic plastics composed of carboxymethyl cellulose and hydroxyethyl cellulose are most suitable for biodegradation after disposal (Gross and Kalra 2002, Yamamoto et al. 2005). Besides, poly(amino acids) with free carboxylic groups, such as poly(aspartic acid) and poly(glutamic acid), are ideal candidates for use as water-soluble polymers. However, these polymers are hydrolytically very unstable.

6. Perspectives and conclusions

Most of the biodegradable materials available in the market are based on natural polymers, such as cellulose, starch, collagen, gelatin, and other polymers. The production of biodegradable monomers and polymers from microbes, plants, or plant-derived byproducts represents sustainable and green chemistry. Breakthroughs in molecular cloning have yielded microbes and plants with enhanced synthesis of various natural polymers. Notably, this is one of the emerging techniques of the future. The combustion of non-biodegradable plastics causes emissions containing CO_2, H_2S, and other chemicals that are extremely harmful to livestock. Recent advances in organic chemistry, biochemistry, and novel bioprocesses are allowing the development of a wide range of novel and bio-based degradable polymer materials for biomedical as well as commercial applications. The favorable outcomes of biodegradable polymers lie in our ability to custom design or modify the existing materials to acquire appropriate biosafety, degradation, and physical properties to elicit suitable biological responses.

Acknowledgements

This work was supported by the Basic Science Research Program through the National Research Foundation of Korea (NRF) funded by the Ministry of Education (NRF-2018R1A6A1A03025582) and the National Research Foundation of Korea (NRF-2019R1D1A3A03103828).

References

Asghari, F., M. Samiei, K. Adibkia, A. Akbarzadeh and S. Davaran. 2017. Biodegradable and biocompatible polymers for tissue engineering application: A review. *Artificial Cells, Nanomedicine, and Biotechnology* 45(2): 185-192.

Basile, M.A., G.G. d'Ayala, M. Malinconico, P. Laurienzo, J. Coudane, B. Nottelet, F.D. Ragione, A. Oliva. 2015. Functionalized PCL/HA nanocomposites as microporous membranes for bone regeneration. *Materials Science and Engineering C* 48, 457-468.

Born, T. 2006. Hyaluronic acids. *Clinics in Plastic Surgery* 33(4): 525-538.

Chan, P., J. Caron, G. Rosa and M. Orth. 2005. Glucosamine and chondroitin sulfate regulate gene expression and synthesis of nitric oxide and prostaglandin E2 in articular cartilage explants. *Osteoarthritis and Cartilage* 13(5): 387-394.

dos Santos, V., R.N. Brandalise and M. Savaris. 2017. Biomaterials: Characteristics and properties. *Engineering of Biomaterials* 5-15. Springer.

Dussud, C., C. Hudec, M. George, P. Fabre, P. Higgs, S. Bruzaud, A.M. Delort, B. Eyheraguibel, A.L. Meistertzheim, J. Jacquin and J. Cheng. 2018. Colonization of non-biodegradable and biodegradable plastics by marine microorganisms. *Frontiers in Microbiology* 9: 1571.

Dutta, S.D., D.K. Patel and K.-T. Lim. 2019a. Functional cellulose-based hydrogels as extracellular matrices for tissue engineering. *Journal of Biological Engineering* 13(1): 55.

Dutta, S.D., D.K. Patel, Y.-R. Seo, C.-W. Park, S.-H. Lee, J.-W. Kim, J. Kim, H. Seonwoo and K.-T. Lim. 2019b. In vitro biocompatibility of electrospun poly (ε-caprolactone)/cellulose nanocrystals-nanofibers for tissue engineering. *Journal of Nanomaterials* 2019: 1-12.

Eatemadi, A., H. Daraee, N. Zarghami, H. Melat Yar and A. Akbarzadeh. 2016. Nanofiber: Synthesis and biomedical applications. *Artificial Cells, Nanomedicine, and Biotechnology* 44(1): 111-121.

Echave, M.C., R. Hernáez-Moya, L. Iturriaga, J.L. Pedraz, R. Lakshminarayanan, A. Dolatshahi-Pirouz, . . . G. Orive. 2019. Recent advances in gelatin-based therapeutics. *Expert Opinion on Biological Therapy* 19(8): 773-779.

Ferreira, A.M., P. Gentile, V. Chiono and G. Ciardelli. 2012. Collagen for bone tissue regeneration. *Acta Biomaterialia* 8(9): 3191-3200.

Fu, N., J. Liao, S. Lin, K. Sun, T. Tian, B. Zhu and Y. Lin. 2016. PCL-PEG-PCL film promotes cartilage regeneration in vivo. *Cell Proliferation* 49(6): 729-739.

Fu, S., P. Ni, B. Wang, B. Chu, L. Zheng, F. Luo, J. Luo and Z. Qian. 2012. Injectable and thermo-sensitive PEG-PCL-PEG copolymer/collagen/n-HA hydrogel composite for guided bone regeneration. *Biomaterials* 33(19): 4801-4809.

Gross, R.A. and B. Kalra. 2002. Biodegradable polymers for the environment. *Science* 297(5582): 803-807.

Guo, B. and P.X. Ma. 2014. Synthetic biodegradable functional polymers for tissue engineering: A brief review. *Science China Chemistry* 57(4): 490-500.

Haider, T.P., C. Völker, J. Kramm, K. Landfester and F.R. Wurm. 2019. Plastics of the future? The impact of biodegradable polymers on the environment and on society. *Angewandte Chemie International Edition* 58(1): 50-62.

Highley, C.B., G.D. Prestwich and J.A. Burdick. 2016. Recent advances in hyaluronic acid hydrogels for biomedical applications. *Current Opinion in Biotechnology* 40: 35-40.

Ikada, Y. and H. Tsuji. 2000. Biodegradable polyesters for medical and ecological applications. *Macromolecular Rapid Communications* 21(3): 117-132.

Iqbal, N., A.S. Khan, A. Asif, M. Yar, J.W. Haycock and I.U. Rehman. 2019. Recent concepts in biodegradable polymers for tissue engineering paradigms: A critical review. *International Materials Reviews* 64(2): 91-126.

Ji, W., F. Yang, J. Ma, M.J. Bouma, O.C. Boerman, Z. Chen, J.J. van den Beucken and J.A. Jansen. 2013. Incorporation of stromal cell-derived factor-1α in PCL/gelatin electrospun membranes for guided bone regeneration. *Biomaterials* 34(3): 735-745.

Jiang, T., S. Heng, X. Huang, L. Zheng, D. Kai, X.J. Loh and J. Zhao. 2019. Biomimetic Poly (poly (ε-caprolactone)-polytetrahydrofuran urethane) based nanofibers enhanced chondrogenic differentiation and cartilage regeneration. *Journal of Biomedical Nanotechnology* 15(5): 1005-1017.

Jung, J.Y., J.K. Park and H.N. Chang 2005. Bacterial cellulose production by Gluconacetobacter hansenii in an agitated culture without living non-cellulose producing cells. *Enzyme and Microbial Technology* 37(3): 347-354.

Karami, Z., I. Rezaeian, P. Zahedi and M. Abdollahi. 2013. Preparation and performance evaluations of electrospun poly (ε-caprolactone), poly (lactic acid), and their hybrid (50/50) nanofibrous mats containing thymol as an herbal drug for effective wound healing. *Journal of Applied Polymer Science* 129(2): 756-766.

Karim, A. and R. Bhat. 2009. Fish gelatin: Properties, challenges, and prospects as an alternative to mammalian gelatins. *Food Hydrocolloids* 23(3): 563-576.

Kaundal, A., P. Kumar and A. Chaudhary. 2015. A review on mucoadhesive buccal tablets prepared using natural and synthetic polymers. *World Journal of Pharmacy and Pharmaceutical Sciences* 4(7): 475-500.

Kenry and B. Liu. 2018. Recent advances in biodegradable conducting polymers and their biomedical applications. *Biomacromolecules* 19(6): 1783-1803.

Kim, Y.-J., J.-N. Kim, Y.-J. Wee, D.-H. Park and H.-W. Ryu. 2007. Bacterial cellulose production by Gluconacetobacter sp. PKY5 in a rotary biofilm contactor. *Applied Biochemistry and Biotechnology* 137(1-12): 529.

Kobayashi, H., D. Terada, Y. Yokoyama, D.W. Moon, Y. Yasuda, H. Koyama and T. Takato. 2013. Vascular-inducing poly (glycolic acid)-collagen nanocomposite-fiber scaffold. *Journal of Biomedical Nanotechnology* 9(8): 1318-1326.

Kosir, M.A., C.C. Quinn, W. Wang and G. Tromp. 2000. Matrix glycosaminoglycans in the growth phase of fibroblasts: More of the story in wound healing. *Journal of Surgical Research* 92(1): 45-52.

Li, J., J. Ding, T. Liu, J.F. Liu, L. Yan and X. Chen 2017. Poly (lactic acid) controlled drug delivery. *Industrial Applications of Poly (lactic acid)* 109-138. Springer.

Lin, C.-C., S.-J. Fu, Y.-C. Lin, I.-K. Yang and Y. Gu. 2014. Chitosan-coated electrospun PLA fibers for rapid mineralization of calcium phosphate. *International Journal of Biological Macromolecules* 68: 39-47.

Lin, K., D. Zhang, M.H. Macedo, W. Cui, B. Sarmento and G. Shen. 2019. Advanced collagen-based biomaterials for regenerative biomedicine. *Advanced Functional Materials* 29(3): 1804943.

Liu, D., W. Nie, D. Li, W. Wang, L. Zheng, J. Zhang, J. Zhang, C. Peng, X. Mo and C. He. 2019. 3D printed PCL/SrHA scaffold for enhanced bone regeneration. *Chemical Engineering Journal* 362: 269-279.

Long, X., Z. Wu, T. Deng, H. Guo and J. Guan. 2020. Hydrogels from silk fibroin and multiarmed hydrolyzed elastin peptide. *Surface Innovations* 1-8.

Luyt, A.S. and S.S. Malik. 2019. Can biodegradable plastics solve plastic solid waste accumulation? *Plastics to Energy* 403-423. Elsevier.

Malafaya, P.B., G.A. Silva and R.L. Reis. 2007. Natural-origin polymers as carriers and scaffolds for biomolecules and cell delivery in tissue engineering applications. *Advanced Drug Delivery Reviews* 59(4-5): 207-233.

Matsumura, Y., T. Hamaguchi, T. Ura, K. Muro, Y. Yamada, Y. Shimada, K. Shirao, T. Okusaka, H. Ueno, M. Ikeda and N. Watanabe. 2004. Phase I clinical trial and pharmacokinetic evaluation of NK911, a micelle-encapsulated doxorubicin. *British Journal of Cancer* 91(10): 1775-1781.

Melke, J., S. Midha, S. Ghosh, K. Ito and S. Hofmann. 2016. Silk fibroin as biomaterial for bone tissue engineering. *Acta Biomaterialia* 31: 1-16.

Middleton, J.C. and A.J. Tipton. 2000. Synthetic biodegradable polymers as orthopedic devices. *Biomaterials* 21(23): 2335-2346.

Mohite, B.V. and S.V. Patil. 2014. A novel biomaterial: Bacterial cellulose and its new era applications. *Biotechnology and Applied Biochemistry* 61(2): 101-110.

Nair, L.S. and C.T. Laurencin. 2005. Polymers as biomaterials for tissue engineering and controlled drug delivery. *Tissue Engineering I* 47-90. Springer.

Nair, L.S. and C.T. Laurencin. 2007. Biodegradable polymers as biomaterials. *Progress in Polymer Science* 32(8-9): 762-798.

Ni, P., Q. Ding, M. Fan, J. Liao, Z. Qian, J. Luo, X. Li, F. Luo, Z. Yang and Y. Wei. 2014. Injectable thermosensitive PEG-PCL-PEG hydrogel/acellular bone matrix composite for bone regeneration in cranial defects. *Biomaterials* 35(1): 236-248.

Olsen, B.R., A.M. Reginato and W. Wang. 2000. Bone development. *Annual Review of Cell and Developmental Biology* 16(1): 191-220.

Olsen, D., C. Yang, M. Bodo, R. Chang, S. Leigh, J. Baez, D. Carmichael, M. Perälä, E.R. Hämäläinen, M. Jarvinen and J. Polarek. 2003. Recombinant collagen and gelatin for drug delivery. *Advanced Drug Delivery Reviews* 55(12): 1547-1567.

Patel, D.K., S.D. Dutta and K.-T. Lim. 2019a. Nanocellulose-based polymer hybrids and their emerging applications in biomedical engineering and water purification. *RSC Advances* 9(33): 19143-19162.

Patel, D.K., Y.-R. Seo, S.D. Dutta and K.-T. Lim. 2019b. Enhanced osteogenesis of mesenchymal stem cells on electrospun cellulose nanocrystals/poly (ε-caprolactone) nanofibers on graphene oxide substrates. *RSC Advances* 9(62): 36040-36049.

Patrascu, J.M., J.P. Krüger, H.G. Böss, A.K. Ketzmar, U. Freymann, M. Sittinger, . . . C. Kaps. 2013. Polyglycolic acid-hyaluronan scaffolds loaded with bone marrow-derived mesenchymal stem cells show chondrogenic differentiation in vitro and cartilage repair in the rabbit model. *Journal of Biomedical Materials Research Part B: Applied Biomaterials* 101(7): 1310-1320.

Peppas, N., Y. Huang, M. Torres-Lugo, J. Ward and J. Zhang. 2000. Physicochemical foundations and structural design of hydrogels in medicine and biology. *Annual Review of Biomedical Engineering* 2(1): 9-29.

Pritchard, E.M. and D.L. Kaplan. 2011. Silk fibroin biomaterials for controlled release drug delivery. *Expert Opinion on Drug Delivery* 8(6): 797-811.

Qian, J., W. Xu, X. Yong, X. Jin and W. Zhang. 2014. Fabrication and in vitro biocompatibility of biomorphic PLGA/nHA composite scaffolds for bone tissue engineering. *Materials Science and Engineering: C* 36: 95-101.

Ren, K., Y. Wang, T. Sun, W. Yue and H. Zhang. 2017. Electrospun PCL/gelatin composite nanofiber structures for effective guided bone regeneration membranes. *Materials Science and Engineering: C* 78: 324-332.

Scaffaro, R., A. Maio, F. Sutera, E.F. Gulino and M. Morreale. 2019. Degradation and recycling of films based on biodegradable polymers: A short review. *Polymers* 11(4): 651.

Schild, H.G. 1992. Poly (N-isopropylacrylamide): Experiment, theory and application. *Progress in Polymer Science* 17(2): 163-249.

Shi, Z., X. Gao, M.W. Ullah, S. Li, Q. Wang and G. Yang. 2016. Electroconductive natural polymer-based hydrogels. *Biomaterials* 111: 40-54.

Shim, J.-H., M.-C. Yoon, C.-M. Jeong, J. Jang, S.-I. Jeong, D.-W. Cho and J.-B. Huh. 2014. Efficacy of rhBMP-2 loaded PCL/PLGA/β-TCP guided bone regeneration membrane fabricated by 3D printing technology for reconstruction of calvaria defects in rabbit. *Biomedical Materials* 9(6): 065006.

Sidney, L.E., T.R. Heathman, E.R. Britchford, A. Abed, C.V. Rahman and L.D. Buttery. 2015. Investigation of localized delivery of diclofenac sodium from poly (D, L-lactic acid-co-glycolic acid)/poly (ethylene glycol) scaffolds using an in vitro osteoblast inflammation model. *Tissue Engineering Part A* 21(1-2): 362-373.

Swetha, M., K. Sahithi, A. Moorthi, N. Srinivasan, K. Ramasamy and N. Selvamurugan. 2010. Biocomposites containing natural polymers and hydroxyapatite for bone tissue engineering. *International Journal of Biological Macromolecules* 47(1): 1-4.

Świętek, M., A. Brož, J. Tarasiuk, S. Wroński, W. Tokarz, A. Kozieł, . . . L. Bačáková. 2019. Carbon nanotube/iron oxide hybrid particles and their PCL-based 3D composites for potential bone regeneration. *Materials Science and Engineering: C* 104: 109913.

Theodoridis, K., E. Aggelidou, M. Manthou, E. Demiri, A. Bakopoulou and A. Kritis. 2019. Assessment of cartilage regeneration on 3D collagen-polycaprolactone scaffolds: Evaluation of growth media in static and in perfusion bioreactor dynamic culture. *Colloids and Surfaces B: Biointerfaces* 183: 110403.

Uchida, M., A. Ito, K.S. Furukawa, K. Nakamura, Y. Onimura, A. Oyane, . . . T. Tateishi. 2005. Reduced platelet adhesion to titanium metal coated with apatite, albumin-apatite composite or laminin-apatite composite. *Biomaterials* 26(34): 6924-6931.

Ullah, M.W., M. Ul-Islam, S. Khan, Y. Kim, J.H. Jang and J.K. Park. 2016. In situ synthesis of a bio-cellulose/titanium dioxide nanocomposite by using a cell-free system. *RSC Advances* 6(27): 22424-22435.

Vepari, C. and D.L. Kaplan. 2007. Silk as a biomaterial. *Progress in Polymer Science* 32(8-9): 991-1007.

Wen, Q., S.M. Mithieux and A.S. Weiss. 2019. Elastin Biomaterials in Dermal Repair. *Trends in Biotechnology* (2020 Mar 1) 38(3): 280-291.

Yamamoto, M., U. Witt, G. Skupin, D. Beimborn and R.J. Müller. 2005. Biodegradable aliphatic-aromatic polyesters: "Ecoflex®". *Biopolymers Online: Biology • Chemistry • Biotechnology • Applications* 4.

Yeo, M.G. and G.H. Kim 2012. Preparation and characterization of 3D composite scaffolds based on rapid-prototyped PCL/β-TCP struts and electrospun PCL coated with collagen and HA for bone regeneration. *Chemistry of Materials* 24(5): 903-913.

Yoshino, T., T. Asakura and K. Toda. 1996. Cellulose production by Acetobacter pasteurianus on silicone membrane. *Journal of Fermentation and Bioengineering* 81(1): 32-36.

Young, J.L. and A.J. Engler. 2011. Hydrogels with time-dependent material properties enhance cardiomyocyte differentiation in vitro. *Biomaterials* 32(4): 1002-1009.

Zhang, H., F. Zhang and J. Wu. 2013. Physically crosslinked hydrogels from polysaccharides prepared by freeze-thaw technique. *Reactive and Functional Polymers* 73(7): 923-928.

Zhang, J., S. Zhao, M. Zhu, Y. Zhu, Y. Zhang, Z. Liu and C. Zhang. 2014. 3D-printed magnetic Fe_3O_4/MBG/PCL composite scaffolds with multifunctionality of bone regeneration, local anticancer drug delivery and hyperthermia. *Journal of Materials Chemistry B* 2(43): 7583-7595.

Zhang, Y., M.S. Desai, T. Wang and S.-W. Lee. 2020. Elastin-based thermo-responsive shape-memory hydrogels. *Biomacromolecules* (2020 Jan 22) 21(3): 1149-1156.

Zhao, L., M.D. Weir and H.H. Xu. 2010. An injectable calcium phosphate-alginate hydrogel-umbilical cord mesenchymal stem cell paste for bone tissue engineering. *Biomaterials* 31(25): 6502-6510.

Zheng, R., H. Duan, J. Xue, Y. Liu, B. Feng, S. Zhao, Y. Zhu, Y. Liu, A. He, W. Zhang and W. Liu. 2014. The influence of Gelatin/PCL ratio and 3-D construct shape of electrospun membranes on cartilage regeneration. *Biomaterials* 35(1): 152-164.

Zindani, D., K. Kumar and J.P. Davim. 2019. An insight into plant-based biodegradable composites. *Biodegradable Composites: Materials, Manufacturing and Engineering* 10.

CHAPTER

5

Functional Pseudo-Proteins

N. Zavradashvili[1], S. Kobauri[1], J. Puiggali[2] and R. Katsarava[1*]

[1] Institute of Chemistry and Molecular Engineering, Agricultural University of Georgia, Kakha Bendukidze University Campus, # 240 David Aghmashenebeli Alley, Tbilisi 0131, Georgia

[2] Departament d'Enginyeria Química, EEBE, Universitat Politècnica de Catalunya, Edifici I.2, C/Eduard Maristany, 10-14, 08019 Barcelona, Spain

1. Introduction

Polymers composed of α-amino acids (AAs) are considered to be among the most promising biomedical materials owing to the following features:

- They are biodegradable, which means they can "disappear" from the body after having served their function.
- They result in natural endogenous by-products or α-amino acids upon biodegradation, which offers a nutritive potential for cells.
- They have high affinity with tissues (tissue compatibility) owing to peptide-type hydrophilic NH-CO bonds.

Polymers made of AAs, whether naturally occurring or synthetically made, may have a diverse molecular architecture. In protein molecules, AAs have "head-to-tail" orientation, as depicted in Fig. 1A (α-amino group is "head" and α-carboxyl groups are "tail"). This natural proteinaceous architecture of macromolecules containing only amide NH-CO (peptide) groups is easily recognizable by the immune system of the organism. This can cause immune incompatibility of the biomaterial with the body tissues. Furthermore, the content of only peptide bonds in protein molecules significantly limits material properties of proteins. Therefore, an AA-based macromolecular system—synthetic analogs of proteins (SAPs) that have non-proteinaceous architecture, i.e., orientations of AAs in polymeric backbones as depicted in Fig. 1B, and containing other chemical links along with NH-CO bonds—look more promising. In such case the biodegradable polymers,

*Corresponding author: r.katsarava@agruni.edu.ge

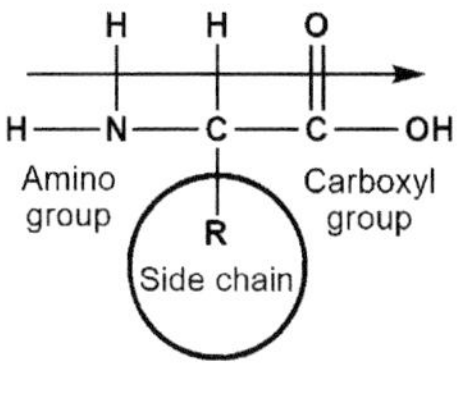

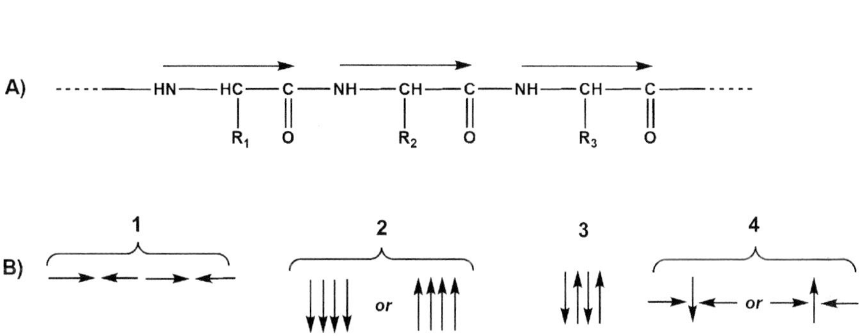

Fig. 1. Possible orientations of AAs in macromolecules. A, head-to-tail orientation (proteinaceous molecular architecture). B, other types of orientations (non-proteinaceous molecular architecture): 1, head-to-head and tail-to-tail; 2, parallel; 3, antiparallel; 4, mixed. Conditionally, the arrow is down if N-terminus of AA is in the polymer backbone, C-terminus in the lateral group. If vice versa, the arrow is up.

like proteins, will release AAs upon biodegradation and be less recognizable by the immune system of the organism owing to "incorrect" molecular architecture of macromolecules, thus providing immune compatibility. The incorporation in the polymeric backbones of other types of chemical links (such as ester, urea, urethane) along with amide NH-CO bonds is useful leverage to tune in a wide range of material properties of synthetic AA-based biodegradable polymers.

Various families of SAPs were designed and introduced in the field during the last two or three decades. These are poly(amino acid)s, pseudo-poly(amino acid)s, polydepsipeptides, and pseudo-proteins, the synthesis, material properties and applications of which are analyzed in previously reported reviews (Kobauri et al. 2019b, Zavradashvili et al. 2020).

Pseudo-proteins (PPs) are SAPs having head-to-head or tail-to-tail orientations of AAs. The PPs are among the most versatile SAPs and have a huge potential for practical applications as biodegradable biomaterials (Kobauri et al. 2019b, Zavradashvili et al. 2020). The PPs with the said orientations of AAs can be obtained using two types of dimerized AAs as basic building blocks, either diacyl-*bis*-α-amino acids (DABAs) or *bis*-(α-amino acid)-alkylene diesters (DADEs) (Fig. 2).

N,N′-diacyl-*bis*-α-amino acids are prepared by dimerization of AAs with *bis*-electrophiles such as diacyl-chlorides (Katsarava et al. 1985, Kharadze et al. 1994, Kharadze et al. 1999) or bis-(chloroformate)s (Kartvelishvili et al. 1996); bis-electrophiles of other classes (e.g., phosgene, diisocyanates) can

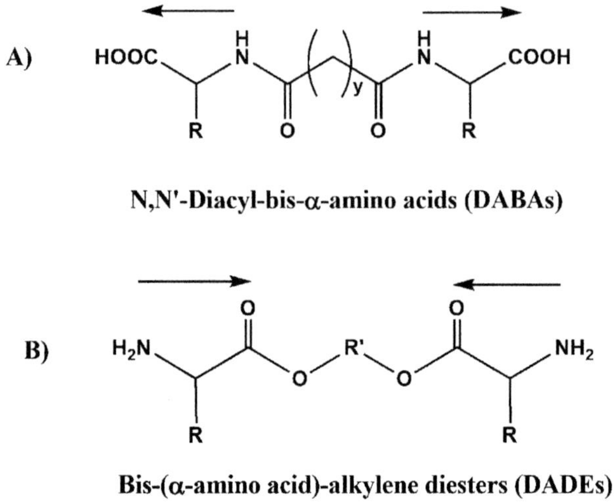

A)

N,N'-Diacyl-bis-α-amino acids (DABAs)

B)

Bis-(α-amino acid)-alkylene diesters (DADEs)

Fig. 2. The AA-based monomer with (A) tail-to-tail and (B) head-to-head orientations of AAs.

also be used for dimerizing AAs. The obtained diacids DABAs (diamide-diacids – Z = none) (Katsarava et al. 1985, Kharadze et al. 1994, Kharadze et al. 1998) or diurethane-diacids (Kartvelishvili et al. 1996) are not active enough to further interact with bis-nucleophiles. Therefore, they were activated by transforming in either activated diesters (Kharadze et al. 1994, Kharadze et al. 1998, Kartvelishvili et al. 1996) or bis-azlactones (Katsarava et al. 1985, Kharadze et al. 1994, Kharadze et al. 1999). This needs additional steps, chemicals and efforts that increase the cost of the desired polymers. Much more promising are PPs obtained on the basis of *bis*-(α-amino acid)-alkylene diesters (diamine-diester, DADEs) composed of cheap and vastly available components such as AAs, diols and *p*-toluenesulfonic acid. DADEs are synthesized as stable di-*p*-toluenesulfonic acid salts (TDADEs) in a very simple and cost-effective procedure: direct thermal condensation of two moles of α-amino acids with one mole of diols in the presence of two moles of *p*-toluenesulfonic acid, without using any condensing agent (Fig. 3). *p*-Toluenesulfonic acid in this reaction plays a dual role: it protects amino groups of an AA and catalyses the condensation reaction.

Various commercially available and inexpensive compounds—fatty diols both polymethylene (Arabuli et al. 1994, Katsarava and Gomurashvili 2011) and bicyclic diols obtained from renewable resources (Gomurashvili et al. 2000, Okada et al. 2001), numerous hydrophobic AAs (Arabuli et al. 1994, Katsarava and Gomurashvili 2011, Gomurashvili et al. 2000, Okada et al. 2001), and arginine (Katsarava and Gomurashvili 2011, Gomurashvili et al. 2000, Okada et al. 2001, Memanishvili et al. 2014)—were successfully used for synthesizing TDADEs by direct condensation of AAs with diols (Scheme 1). The latter easily proceeds in refluxed organic solvents that can remove water liberated after the condensation reaction azeotropically

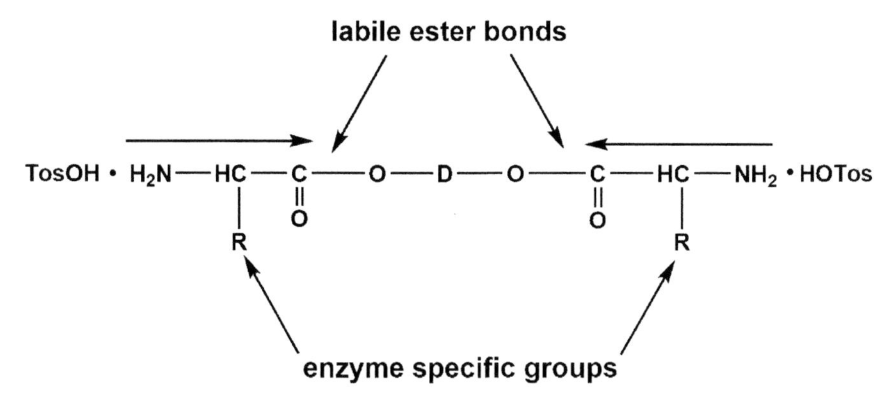

Fig. 3. The structure of TDADEs, key monomers for synthesizing PPs.

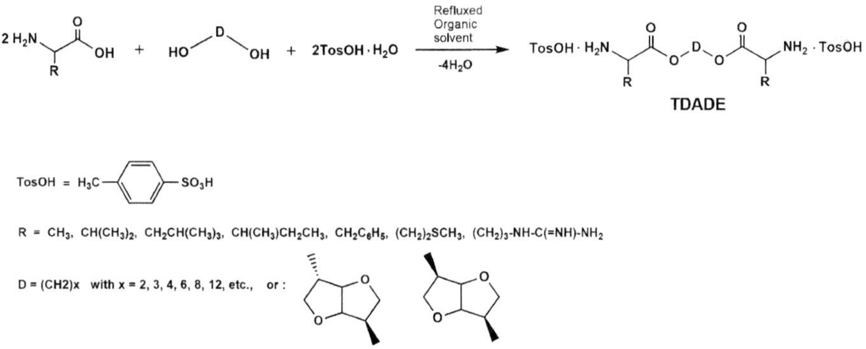

2 H_2N–CH(R)–COOH + HO–D–OH + 2TosOH · H_2O —(Refluxed Organic solvent, $-4H_2O$)→ TosOH · H_2N–CH(R)–CO–O–D–O–CO–CH(R)–NH_2 · TosOH

TDADE

TosOH = H_3C–C_6H_4–SO_3H

R = CH_3, $CH(CH_3)_2$, $CH_2CH(CH_3)_2$, $CH(CH_3)CH_2CH_3$, $CH_2C_6H_5$, $(CH_2)_2SCH_3$, $(CH_2)_3$-NH-C(=NH)-NH_2

D = (CH2)x with x = 2, 3, 4, 6, 8, 12, etc., or :

Scheme 1. Synthesis of diamine-diester monomers (TDADEs) by direct thermal condensation of AAs with fatty diols.

(benzene or toluene) (Arabuli et al. 1994, Katsarava and Gomurashvili 2011, Gomurashvili et al. 2000, Okada et al. 2001, Memanishvili et al. 2014). It was found that toxic and hazardous benzene and toluene could be replaced by much safer cyclohexane (Katsarava et al. 2016), which is important from a technological point of view when TDADEs are produced on a large scale.

The TDADEs are synthesized in high yields (up to 90-95%). The TDADEs made of hydrophobic AAs, which are of interest for industrial scale-up production of PPs, are purified by recrystallization from water. Additional merits of the TDADE monomers are stability upon storage, active terminal amino groups for successful chain growth along with the designed-in non-proteinaceous (head-to-head) orientation of amino acids, enzyme-specific lateral groups R, and hydrolysable ester bonds (Fig. 3) that goes later into macromolecules made of them.

The TDADEs are key bis-nucleophilic monomers for synthesizing a virtually unlimited number of PPs—representatives of a huge family of ester-polymers such as poly(ester amide)s (PEAs), poly(ester urethane)s (PEURs), and poly(ester urea)s (PEUs), and AA-BB type poly(depsipeptide)s (PDPs) having a wide range of material properties and reasonable biodegradation rates through the hydrolysis of ester moieties (Kobauri et al. 2019b, Zavradashvili et al. 2020, Katsarava and Gomurashvili 2011, Gomurashvili et al. 2000, Okada et al. 2001, Memanishvili et al. 2014, Katsarava et al. 2016, Ochkhikidze et al. 2020, Katsarava et al. 2016, Rodriguez-Galan et al. 2011a, Rodriguez-Galan et al. 2011b, Fonseca et al. 2014).

Numerous key bis-electrophilic monomers—counter-partners of TDADEs—were used for synthesizing the said classes of the PPs (Fig. 4). Various chlorides such as diacyl chlorides (DAC), diol bis-chloroformates (DBC), and phosgene (PG) or triphosgene (TPG) are suitable bis-electrophiles for synthesizing PPs via interfacial polycondensation (IP) (Kobauri et al. 2019b, Zavradashvili et al. 2020, Katsarava and Gomurashvili 2011, Morgan 1965). Activated diester monomers of various classes—activated diester of dicarboxylic acids (ADD), activated diol bis-carbonates (ADC), activated carbonates (AC)—are appropriate for synthesizing PPs via solution active polycondensation (SAcP) (Kobauri et al. 2019b, Zavradashvili et al. 2020, Katsarava and Gomurashvili 2011, Katsarava 1991, Katsarava 2003).

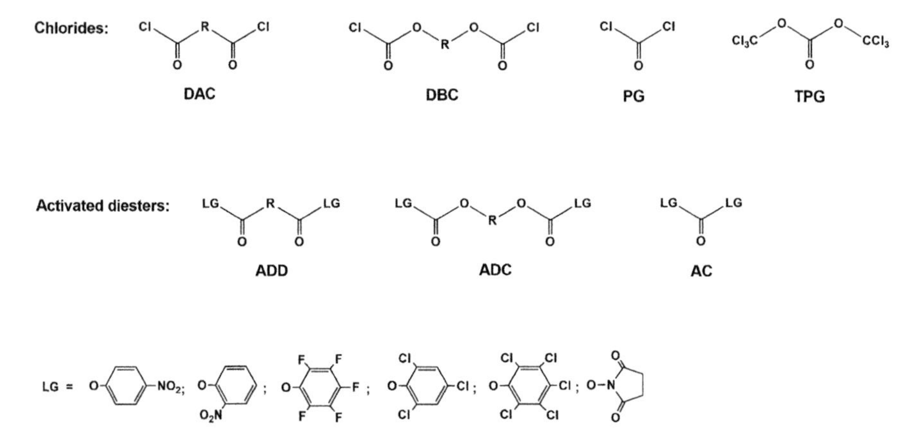

Fig. 4. Bis-electrophilic monomers, counter-partners of TDADEs in the synthesis of PPs.

DACs and ADDs are used for synthesizing the PEAs, DBCs and ADC for synthesizing the PEURs, PG (or TPG), and AC for synthesizing the PEUs (Kobauri et al. 2019b, Zavradashvili et al. 2020, Katsarava and Gomurashvili 2011, Katsarava 1991, Katsarava 2003), as depicted in Schemes 2-4.

The first reported PPs, which were synthesized via step-growth polymerization (by SAcP or IP) (Kobauri et al. 2019b, Zavradashvili et al. 2020, Katsarava and Gomurashvili 2011, Gomurashvili et al. 2000, Okada et al. 2001, Ochkhikidze et al. 2020, Katsarava et al. 2016, Rodriguez-Galan et al. 2011a, Rodriguez-Galan et al. 2011b), did not contain any pendant functional groups, except to end-groups, presumably one amino and one carboxylic (free or activated ester) end-group. These functional end-groups are very low in concentration particularly in high-molecular-weight PPs and, hence, they are

n TosOH · H2N ... NH2 · HOTos + n Cl–A–Cl or n LG–A–LG; pTSA and HCl acceptor; IP; SAcP; For LG see Fig. 4; PP-PEAs

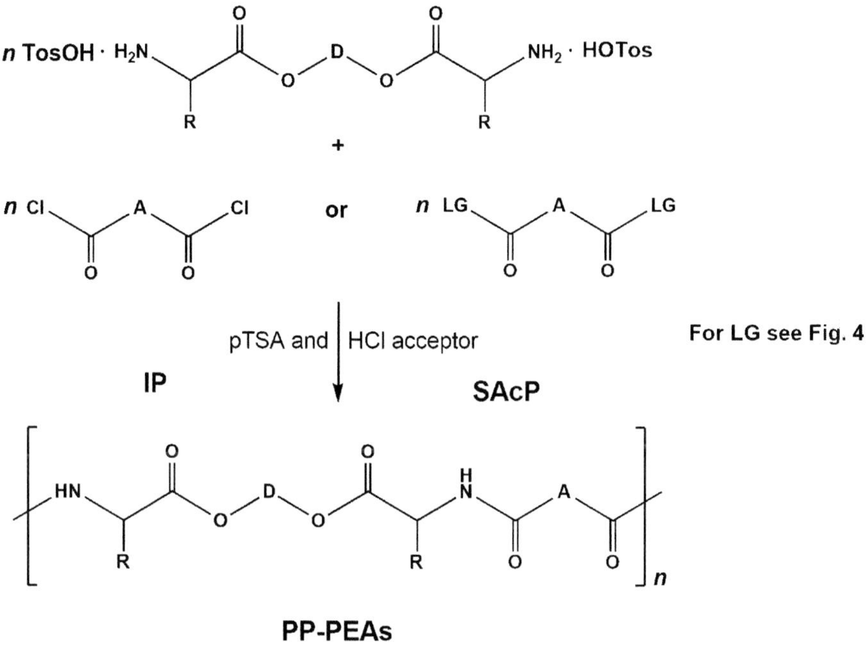

Scheme 2. The synthesis of PPs–PEAs via interfacial polycondensation (IP) and solution active polycondensation (SAcP).

n TosOH · H2N ... NH2 · HOTos + n Cl–O–D1–O–Cl or n LG–O–D1–O–LG; pTSA and HCl acceptor; IP; SAcP; For LG see Fig. 4; PP-PEURs

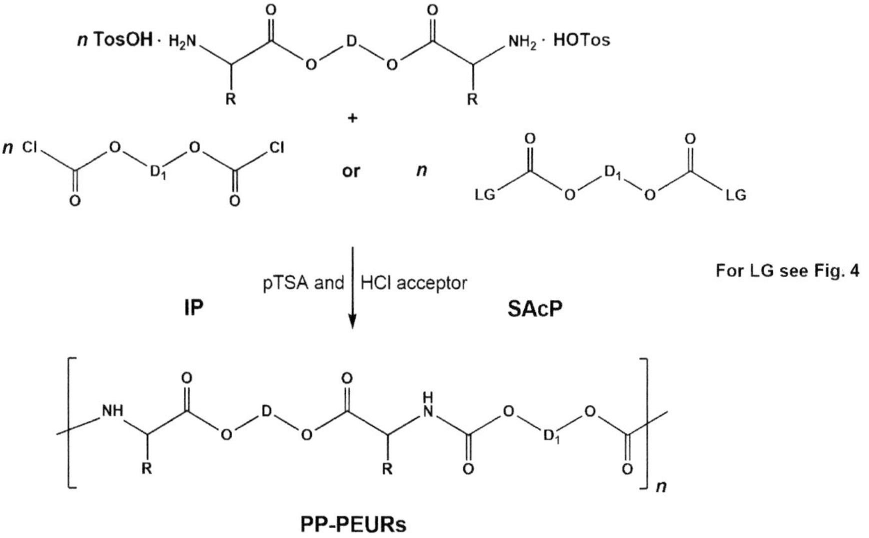

Scheme 3. The synthesis of PPs–PEURs via interfacial polycondensation (IP) and solution active polycondensation (SAcP).

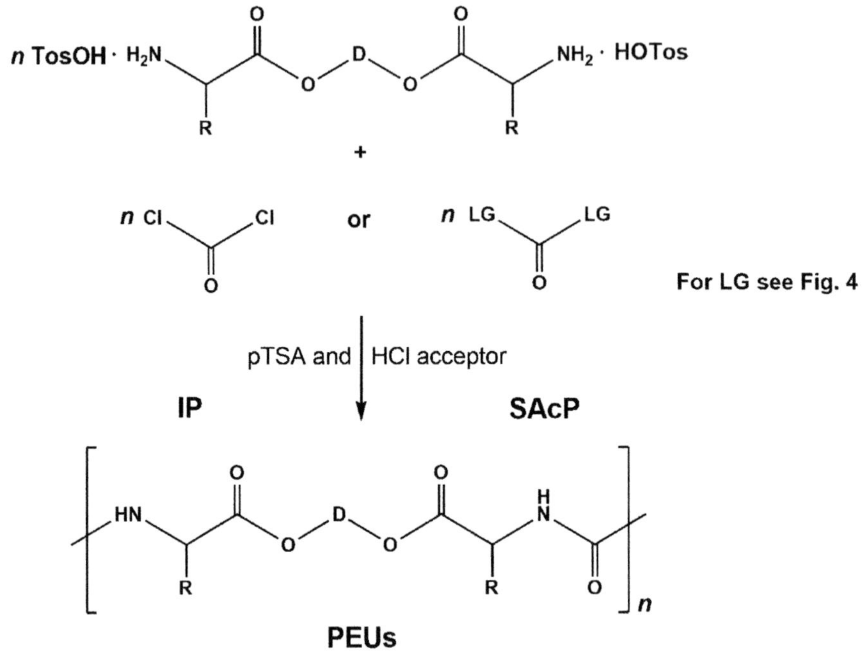

n TosOH · H_2N … NH_2 · HOTos

\+

n Cl … Cl or *n* LG … LG

For LG see Fig. 4

pTSA and HCl acceptor

IP SAcP

PEUs

Scheme 4. The synthesis of PPs–PEUs via interfacial polycondensation (IP) and solution active polycondensation (SAcP).

too low to use PPs as carriers via covalent linkage of any molecules including bioactive compounds and drugs. Besides, these rare functional groups cannot influence physical-chemical properties of the PPs. The concentration of terminal functional groups in the polycondensation could be even lower than we could predict theoretically, owing to the intramolecular interaction of terminal functional groups during polycondensation that results in cyclic polymers free of terminal functional groups (Kricheldorf 2010). The incorporation of lateral functional groups along the backbones of the PPs could provide a wide range of chemical/biochemical reactions to broaden the biomedical application of this family of biodegradable polymers. In addition, the lateral functional groups could also exhibit significant hydrophilicity and positive and/or negative charge characteristics, and such characteristics are believed to be able to significantly affect and regulate the physical, chemical and biological properties of polymers, for example, the ability to diffuse through cell membranes.

2. Synthesis of functional PPs

2.1. Functionalization of PPs using functionalized bis-nucleophilic co-monomers

The first representatives of functional PPs were co-PEAs synthesized using amino acid L-lysine as a diamine (co-monomer of TDADEs) (Jokhadze et al. 2007, Chu and Katsarava 2003). For incorporating L-lysine into the PP backbones as a diamine, i.e., via α and ε amino groups, it was transformed into C-protected benzyl ester (KBn) form with the purpose of subsequent catalytic hydrogenolysis of the obtained Bn-polymers.

L-lysine benzyl ester was synthesized as a stable di-*p*-toluenesulfonic acid salt (TKBn) by direct condensation of L-lysine monohydrochloride (KHCl) with benzyl alcohol (BnOH) in the presence of *p*-toluenesulfonic acid (TosOH) in refluxed benzene or toluene, as shown in Scheme 5.

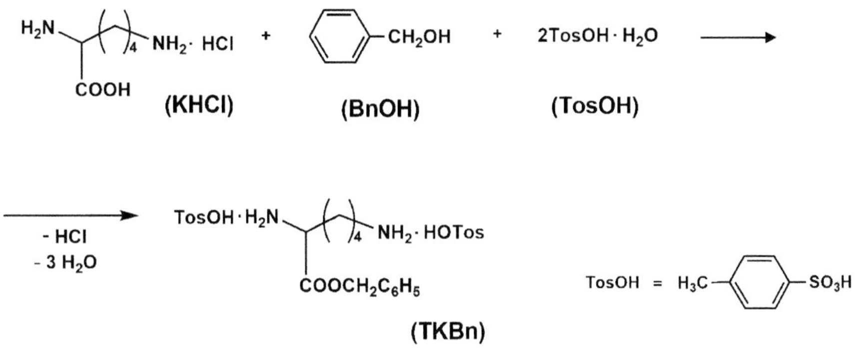

Scheme 5. Synthesis of lysine benzyl ester di-*p*-toluenesulfonic acid salt (TKBn).

The PPs as benzyl esters (PP-Bn) were synthesized by SAcP of TDADE/TKBn mixture with activated diesters of various classes: ADD or ADC (see Fig. 4). This synthetic strategy was developed for obtaining PP-Bn of two classes: co-PEAs (Jokhadze et al. 2007, Chu and Katsarava 2003) and co-PEURs (Chu and Katsarava 2003) using ADD and ADC, accordingly (Scheme 6).

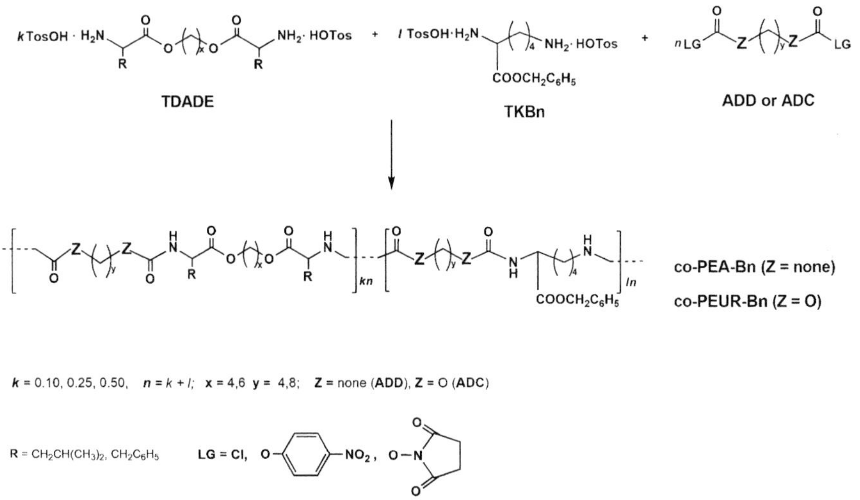

Scheme 6. Synthesis of lysine-based PP-Bn co-PEAs and co-PEURs via SAcP.

This strategy could also be applied to the third class of functional PPs, co-PEUs, using ACs (Katsarava and Gomurashvili 2011), which are a new type of monomer for synthesizing PEUs via SAcP (Katsarava et al. 1993, Kartvelishvili et al. 1997). However, a more rational way to synthesize the PP-Bn of co-PEU class was found to be IP using PG (or TPG) as a bis-electrophilic monomer (Katsarava et al. 2014), as depicted in Scheme 7.

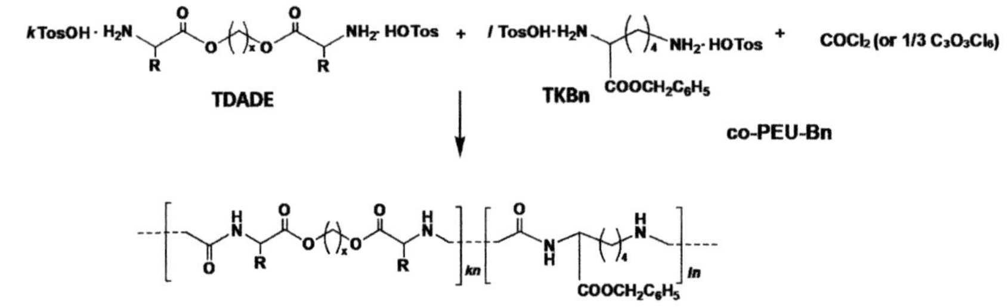

*k*TosOH · H_2N ... NH_2· HOTos + *l* TosOH·H_2N ... NH_2· HOTos + $COCl_2$ (or 1/3 $C_3O_3Cl_6$)

TDADE TKBn $COOCH_2C_6H_5$ co-PEU-Bn

Scheme 7. Synthesis of lysine-based PP-Bn co-PEUs via IP.

The co-polymeric benzyl esters were transformed into the PP-polyacids by catalytic hydrogenolysis using palladium black (Jokhadze et al. 2007, Chu and Katsarava 2003), as shown in Scheme 8 for co-PEA-Bn and co PEUR-Bn.

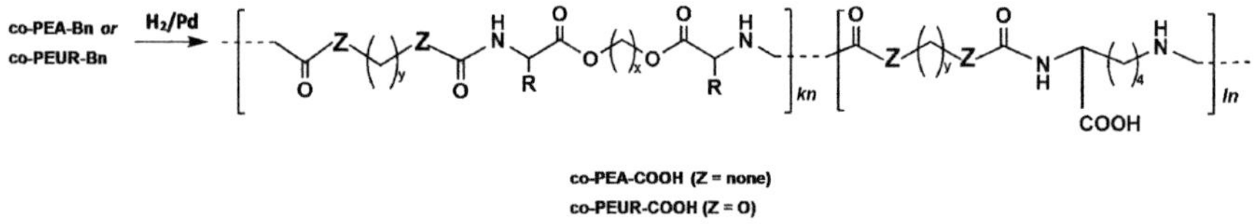

co-PEA-Bn or co-PEUR-Bn $\xrightarrow{H_2/Pd}$

co-PEA-COOH (Z = none)
co-PEUR-COOH (Z = O)

Scheme 8. Synthesis of lysine-based PP-polyacids co-PEA-COOH and co-PEUR-COOH by catalytic hydrogenolysis of corresponding PP-Bn.

A more rational synthesis of the lysine-based PP-polyacids consists in the use of salts of the amino acid lysine (K) as monomers in SAcP di-*p*-toluenesulfonate (TK) or bis-trifluoroacetate (TFAK), as shown in Fig. 5.

TK was found useful for synthesizing the PP-polyacids; however, its synthesis in crystalline form suitable for SAcP turned out to be rather problematic: the interaction of pure K (one mole) with *p*-toluenesulfonic acid's monohydrate (2 moles) in refluxed benzene or toluene (to remove water) resulted in two crystalline polymorphs with melting points 165 and 190°C,

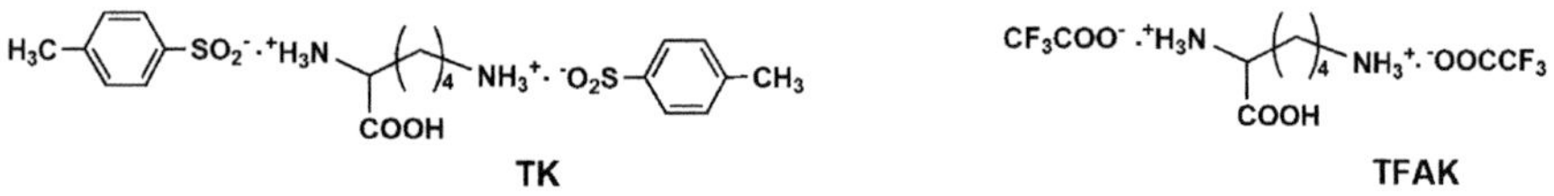

H_3C–C_6H_4–SO_3^-·^+H_3N ... NH_3^+·^-O_3S–C_6H_4–CH_3 (COOH) TK

CF_3COO^-·^+H_3N ... NH_3^+·$^-OOCCF_3$ (COOH) TFAK

Fig. 5. Salt forms of lysine used as monomers for synthesizing PP-polyacids di-*p*-toluenesulfonate (TK) or bis-trifluoroacetate (TFAK).

but in most cases this procedure led to uncrystallizable syrup (Katsarava and Mumladze, unpublished data). A better result was obtained with bis-trifluoroacetate TFAK, which was obtained reproducibly as a crystalline product by dissolving K in excess trifluoroacetic acid and refluxing 1.0 h (Ochkhikidze et al. 2020). The use of the TK and TFAK monomers makes it possible to synthesize PP-polyacids (Scheme 9) excluding additional steps of the synthesis of KBn and PP-Bn and to avoid the use of an expensive heavy metal (Pd) catalyst.

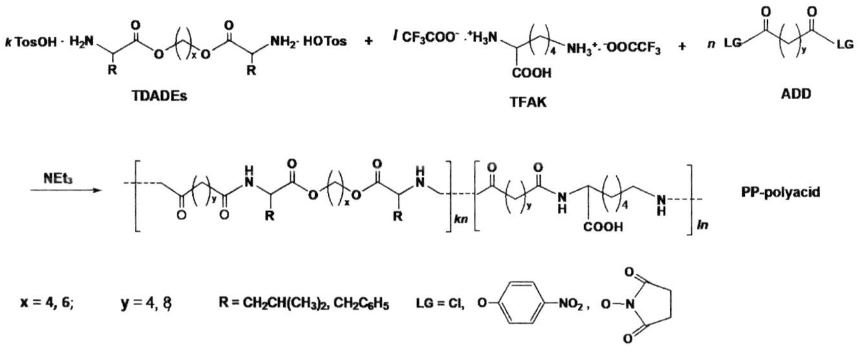

Scheme 9. Synthesis of PEAs-COOH on the basis of KTFA.

Another useful bis-nucleophilic co-monomer for functionalizing PPs is 1,3-Diamino-2-propanol (Ochkhikidze et al. 2020), which was used as handy di-*p*-toluenesulfonic acid salt TDAP (Fig. 6).

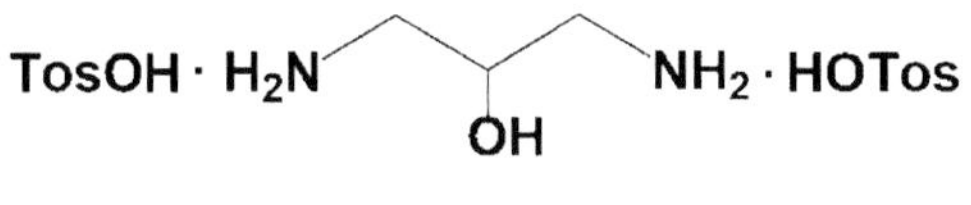

TDAP

Fig. 6. Hydroxyl-containing salt monomer for synthesizing PP-polyols 1,3-Diamino-2-propanol di-*p*-toluenesulfonate (TDAP).

PP-polyols (PEAs with lateral hydroxyl functional groups) were synthesized by SAcP of TDADEs/TDAP mixtures with ADDs (Ochkhikidze et al. 2020), as shown in Scheme 10.

It has to be noted that a high difference in reactivity of primary amino and hydroxyl groups towards ADDs made it possible to obtain linear polymers: PP-polyols soluble in organic solvents. Scheme 10 demonstrates the synthesis of PP-polyols of the PEA class, though it is applicable also to the synthesis of PP-polyols of other classes—PEURs and PEUs—using ADCs and ACs as bis-electrophilic monomers, accordingly.

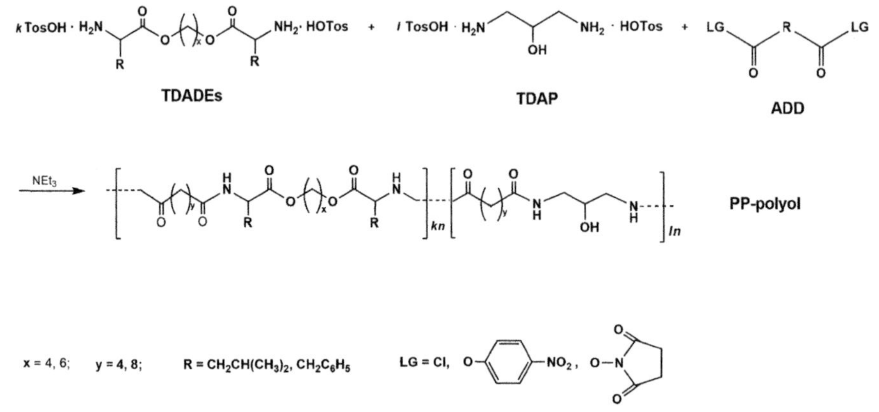

Scheme 10. Synthesis of PP-polyols on the basis of TDAP.

The PP-polyols could be subjected to chemical transformations via hydroxyl groups including acylation with unsaturated diacids and transformation into the polymers with unsaturations in lateral chains (see section 2.5 below).

2.2. Functionalization of PPs using N-carboxyanhydrides as co-monomers

The AA lysine can be incorporated into the PP backbone via α functional groups (α-NH_2 and α-COOH) by using ε-(benzyloxycarbonyl)-L-lysine *N*-carboxyanhydride (ZKNCA) as a modifying agent. For this, Deng et al. (2009) used a two-step process: in the first step the TDADE composed of L-phenylalanine and 1,6-hexadediol (F6) was interacted with ZKNCA at various mole (n/m) ratios, as depicted in Scheme 11.

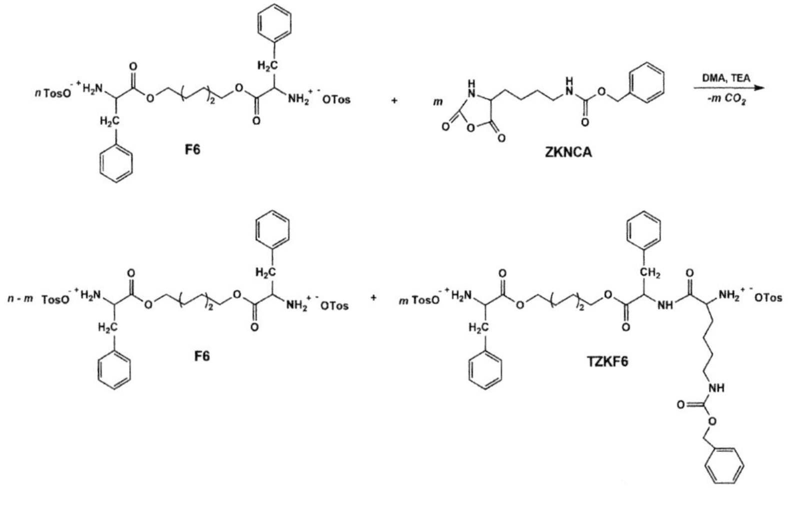

Scheme 11. TDADE monomer (F6) modified by interaction with ZKNCA.

This reaction (Scheme 11) at n>m resulted in mixtures of F6 and modified F6-TZKF6, which was polycondensed under the conditions of SAcP with ADD-di-p-nitrophenyl sebacate (NSe) resulting in high-molecular-weight Z-protected PP-coPEAs-8ZKF6 (Scheme 12, A).

After deprotecting 8ZKF6 using a mixture of trifluoroacetic acid, methanesulfonic acid and anisole (Scheme 12, B), functional PP-polyamine $8NH_2KF6$ in a salt form was obtained. For removing the acids, the polymer was dissolved in DMA and neutralized with triethylamine, and the obtained PP-polyamine was precipitated into ethyl acetate.

Analogously, Cui et al. (2012) synthesized functional PP using γ-benzyl-L-glutamate N-carboxyanhydride (BGNCA) and TDADE composed of L-leucine and 1,6-hexadediol (L6) (Fig. 7) instead of ZKNCA and F6, accordingly.

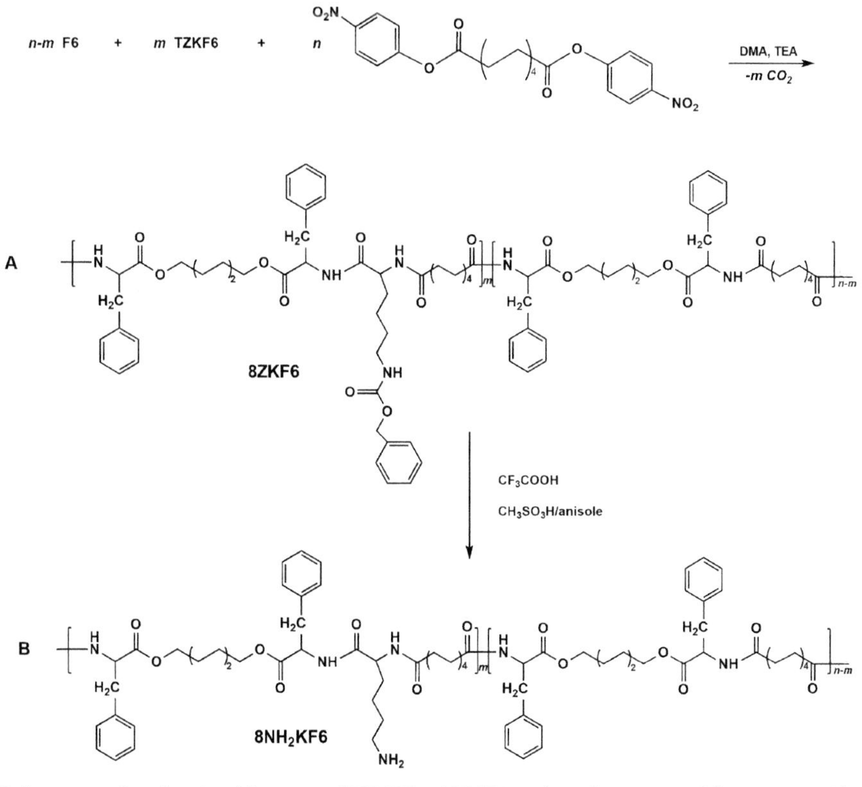

Scheme 12. Synthesis of functional PP-PEA 8ZKF6 and its deprotected form $8NH_2KF_6$.

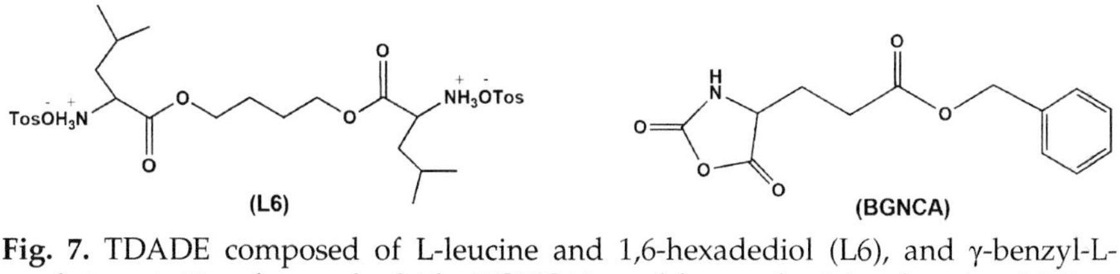

Fig. 7. TDADE composed of L-leucine and 1,6-hexadediol (L6), and γ-benzyl-L-glutamate N-carboxyanhydride (BGNCA) used for synthesizing functional PP.

The Bn-protected PP-coPEA (8GBnL6) was obtained under the conditions of SAcP according to Scheme 13, A. After removing the Bn-protecting group, 8GBnL6 was transformed into PP-polyacid 8GL6 (Scheme 13, B).

The PP-polyacid 8GL6 then was grafted with tetraaniline (TA) (Cui et al. 2012). The grafted PP 8GL6-TA exhibited good electroactivity, mechanical properties, and biodegradability and showed promise as bone repair scaffold material in tissue engineering (see Section 3).

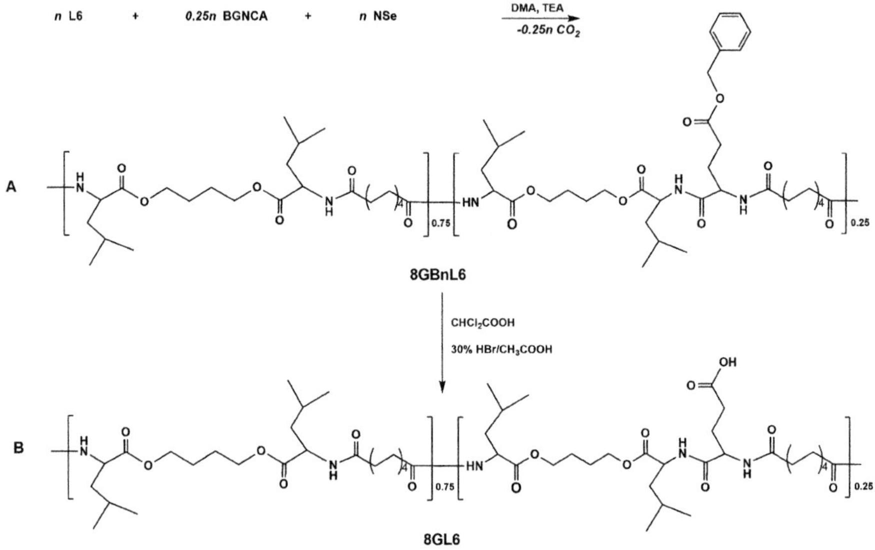

n L6 + *0.25n* BGNCA + *n* NSe — DMA, TEA; $-0.25n\ CO_2$

A

8GBnL6

$CHCl_2COOH$

30% HBr/CH_3COOH

B

8GL6

Scheme 13. Synthesis of functional PP-PEA 8GBnL6 and its deprotected form 8GL6.

2.3. Functionalization of PPs using functionalized DADEs

PP-polyacids of PEA classes were synthesized using a DADE made of orthogonal protected aspartic acid (Atkins et al. 2009, Knight et al. 2014). The DADE was synthesized via multistep techniques of the peptide chemistry as depicted in Scheme 14: in the first step, 2 moles of orthogonal protected aspartic acid (D) was condensed via α-carboxyl group with 1,4-butanediol in the presence of dicyclohexylcarbodiimide (DCC) as a coupling agent. After catalytic (Pd/C) hydrogenolysis of the intermediate di-benzyl ester, i.e., after removing the benzyl protection from the terminal α,α′-amino groups, the functional DADE composed of *tert*-butyl protected L-aspartic acid was obtained as a free base (DADE-D).

Afterwards, the DADE-D monomer, in either pure state or combination with TDADE monomers on the basis of hydrophobic amino acids (such as alanine or phenylalanine), was interacted with either ADDs or dicarboxylic acid dichloride (DDCs) using SAcP or IP techniques, accordingly. As a result, high-molecular-weight PP-PEAs with protected (as *tert*-butyl ester) lateral carboxyl groups PEA-*t*Bu were obtained, as depicted in Scheme 15, A. The protecting *tert*-butyl groups were removed under conditions that do not cause

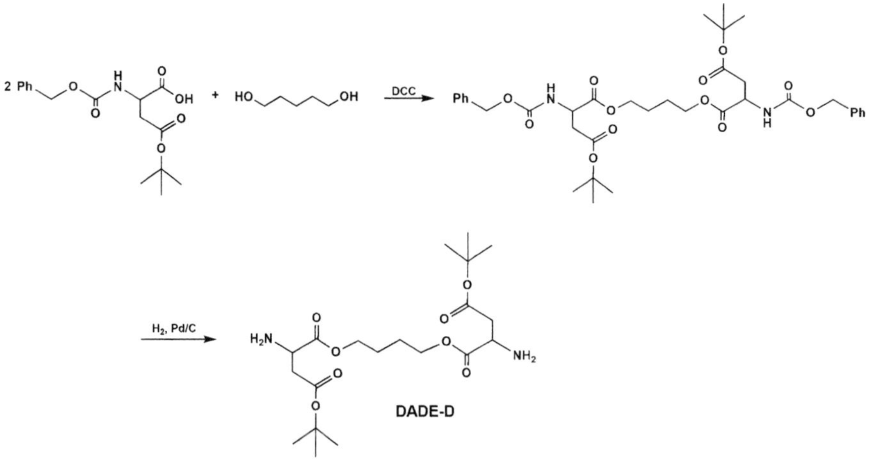

Scheme 14. Synthesis of DADE-D monomer composed of *tert*-butyl protected L-aspartic acid.

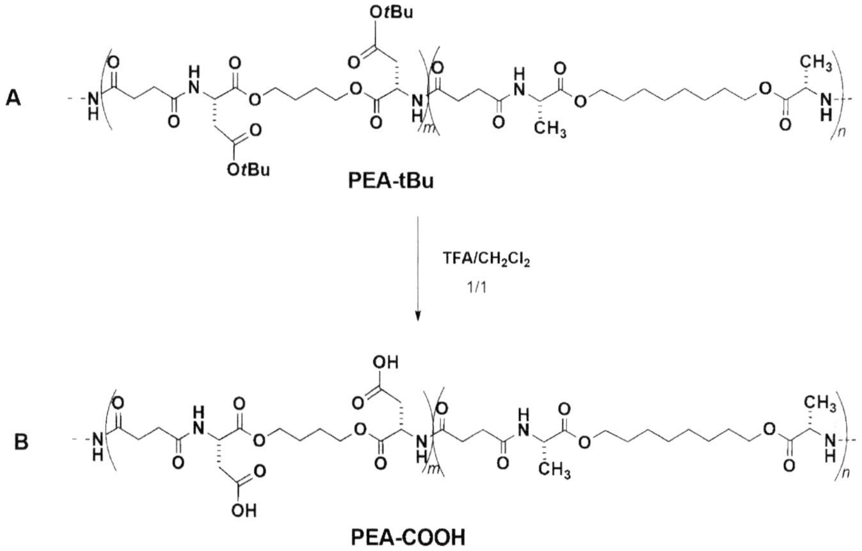

Scheme 15. Deprotection of *tert*-butyl protected PP PEA-*t*Bu.

polymer backbone degradation, by dissolving them in 1:1 trifluoroacetic acid (TFA):CH_2Cl_2 mixture (Scheme 15, B).

It was shown that free lateral carboxyl groups in the obtained PEA-COOH could readily be functionalized: they were converted into N-hydroxysuccinimidyl esters (Atkins et al. 2009), providing a useful template for further derivatization.

The same approach was used for synthesizing functional PPs (cationic PPs, cPPs) using orthogonal protected L-lysine (K) (Atkins et al. 2009, De Wit et al. 2008, Knight et al. 2011). In the first step, 2 moles of orthogonal protected L-lysine was condensed with fatty diols (1,4-butanediol or 1,8-octanediol) in the presence of DCC as a coupling agent. The functional L-lysine-based monomers with *tert*-butyloxycarbonyl (BOC)-protected side groups were obtained as free base (DADE-K) after catalytic (Pd/C) hydrogenolysis of the intermediate di-benzyl esters, i.e., removing benzyl protection from the terminal α,α′-amino groups, as depicted in Scheme 16.

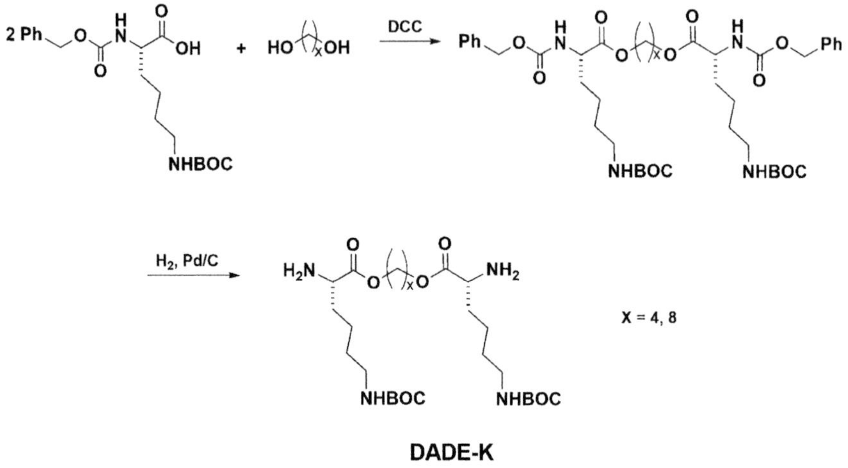

2 Ph O N(H) O OH + HO( )xOH —DCC→ Ph O N(H) O O( )xO O N(H) O Ph

NHBOC NHBOC NHBOC

—H_2, Pd/C→ H_2N O O( )xO O NH_2

X = 4, 8

NHBOC NHBOC

DADE-K

Scheme 16. Synthesis of DADE-K monomer composed of BOC-protected L-lysine.

The DADE-K monomer in either pure state or combination with TDADE monomers composed of hydrophobic amino acids (alanine, phenylalanine) was interacted with either ADD or DDC using SAcP or IP techniques, accordingly. As a result, high-molecular-weight PP-PEAs/coPEAs with BOC-protected lateral amino groups were obtained: PEA/coPEA-BOC, as depicted in Scheme 17, A (Atkins et al. 2009, De Wit et al. 2008, Knight et al. 2011). The protecting BOC groups were removed under conditions that do not cause polymer backbone degradation, by dissolving them in 1:1 TFA:CH_2Cl_2 mixture (Scheme 17, B).

The cPPs PEAs-NH_2·TFA contain "living" primary amino groups capable of subsequent transformations. Another important class of cPPs were obtained on the basis of amino acid L-arginine (R) in which a source of positive charge is chemically less active guanidine group. Therefore, this kind of cPPs are mostly used as polyelectrolytes applicable as both gene transfection and antimicrobial agents though, if needed, they can also be subjected to chemical transformations via acylation (Prasad et al. 1967) or alkylation (Powell et al. 2003) of guanidine lateral groups.

R-based cPPs are made of TDADEs (Memanishvili et al. 2014, Yamanouchi et al. 2008, Wu et al. 2011, Pang and Chu 2010), which represent tetra-*p*-toluenesulfonic acid (TosOH) salts: two molecules of TosOH are bound with

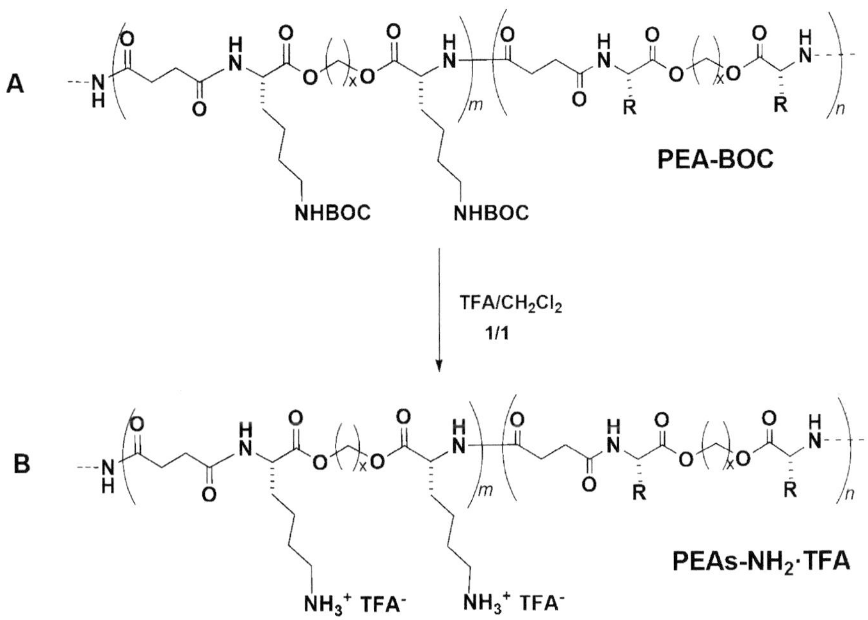

Scheme 17. Deprotection of BOC-protected PP PEA-BOC.

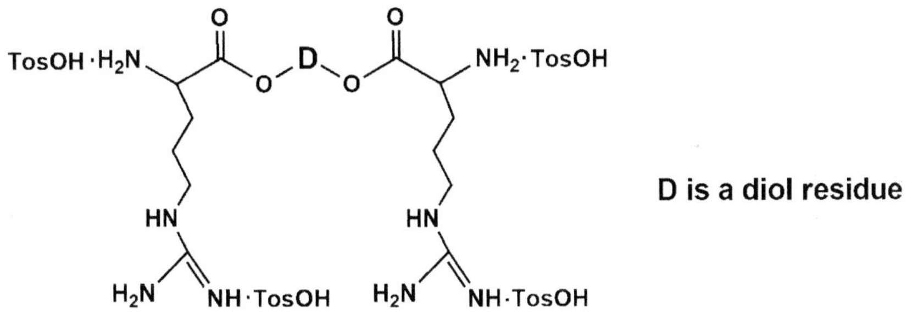

Fig. 8. The structure of R-based TDADEs.

α,α′-amino groups, as in all TDADEs, and another two are bound with lateral guanidine groups of R, as depicted in Fig. 8.

Various diols (HO-D-OH)—regular polymethylene diols (Memanishvili et al. 2014, Yamanouchi et al. 2008, Wu et al. 2011, Pang and Chu 2010) and oligo-ethylene glycols (Memanishvili et al. 2014, Pang and Chu 2010)—were used for synthesizing R-based TDADEs. The cPPs were synthesized via SAcP of R-based TDADEs with activated diesters of various classes.

Yamanouchi et al. (2008) and Wu et al. (2011) synthesized cPPs of PEA class by solution polycondensation of R-based TDADEs with ADDs. Memanishvili et al. (2014), via the interaction of activated diesters of different classes—ADDs, ADCs, and ACs—with various R-based TDADE under the conditions of SAcP, obtained cPPs of all three classes—PEAs, PEURs, and PEUs—which are depicted in Fig. 9.

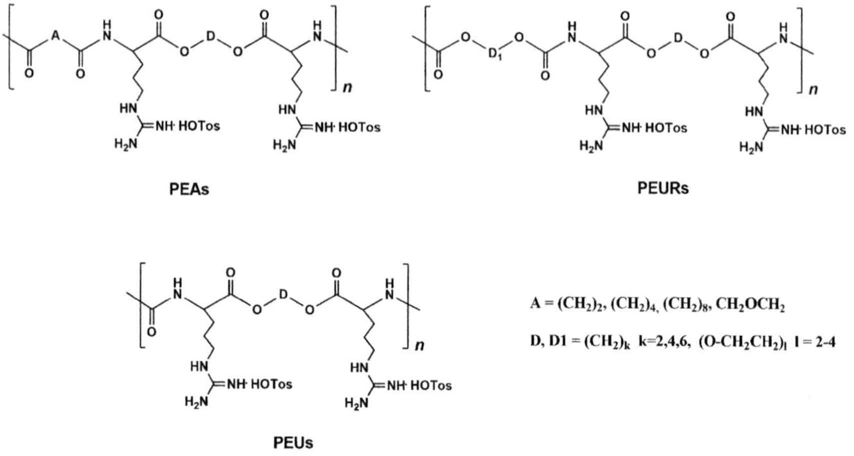

PEAs PEURs

PEUs

A = $(CH_2)_2$, $(CH_2)_4$, $(CH_2)_8$, CH_2OCH_2

D, D1 = $(CH_2)_k$ k=2,4,6, $(O\text{-}CH_2CH_2)_l$ l = 2-4

Fig. 9. The structures of R-based cationic PPs of various classes—PEAs, PEURs, and PEUs.

Along with the cPPs containing polymethylene moieties in the backbones, cPPs having PEG-like macromolecules were obtained by using bis-nucleophilic monomers (R-based TDADEs) and bis-electrophiles (ADDs and ADCs) containing ether linkages in molecules (Memanishvili et al. 2014). The cPPs with PEG-like backbones showed better solubility in water and higher cell compatibility as compared with polymethylene chains containing analogs.

Interesting TDADEs containing unsaturated lateral chains (Fig. 10) were obtained on the basis of so-called non-proteinogenic amino acids such as allylglycine (Pang and Chu 2010, Pang et al. 2010, Pang et al. 2014, Kantaria et al. 2018, Zavradashvili et al. 2019b) and propargylglycine (Kantaria et al. 2018, Zavradashvili et al. 2019b) using direct condensation of the amino acids with diols, as depicted in Scheme 1. The monomers were used for

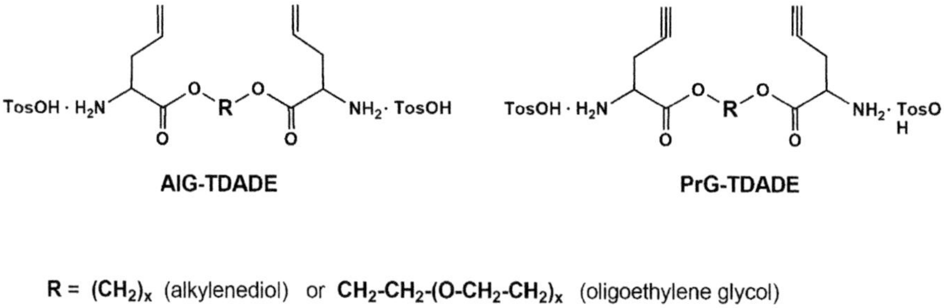

AlG-TDADE PrG-TDADE

R = $(CH_2)_x$ (alkylenediol) or $CH_2\text{-}CH_2\text{-}(O\text{-}CH_2\text{-}CH_2)_x$ (oligoethylene glycol)

Fig. 10. The structures of AlG- and PrG-based TDADEs

synthesizing unsaturated PPs (uPPs) containing unsaturations in the lateral chains.

Pang and Chu (2010) used the AlG-TDADEs as comonomers of regular TDADEs (e.g., composed of phenylalanine) for synthesizing uPP-PEA by copolymerization with ADDs under the conditions of SAcP according to Scheme 18.

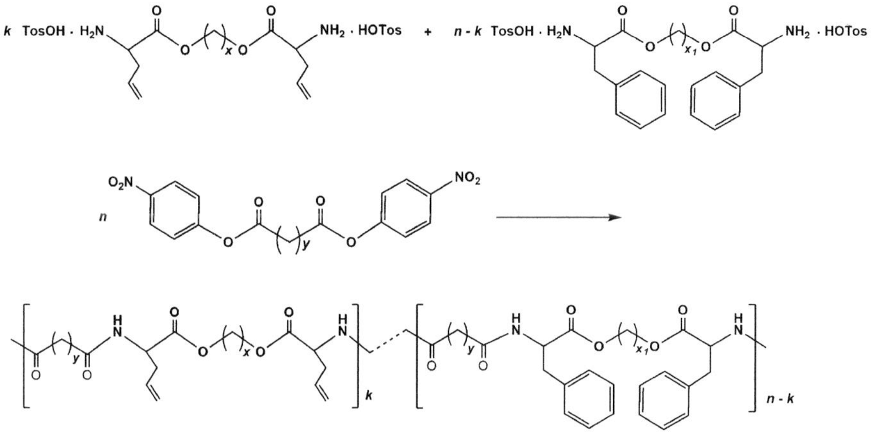

Scheme 18. Synthesis of unsaturated PP-PEA via SAcP.

AlG-TDADE and PrG-TDADE were also polycondensed with sebacoyl chloride under the conditions of IP (Scheme 19), resulting in uPP-PEAs (Kantaria et al. 2018).

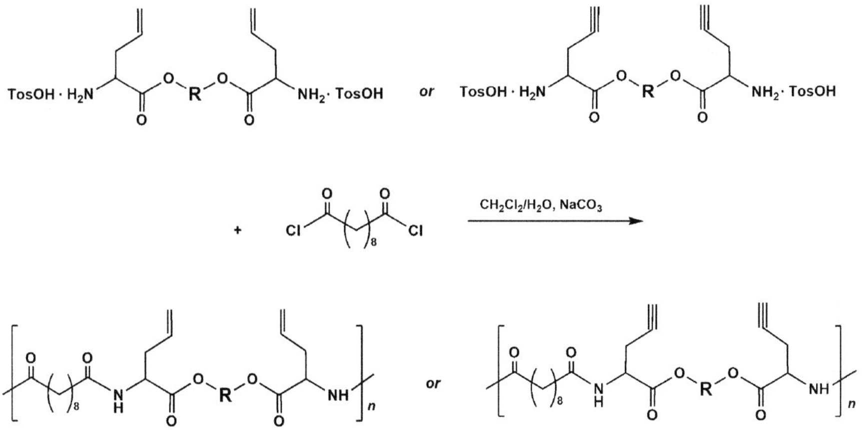

Scheme 19. Synthesis of unsaturated PP-PEA by IP using sebacoylchloride.

The uPPs composed of AlG-TDADEs containing lateral double bonds are of interest for fabricating hydrogels by UV photo-crosslinking with commercial poly(ethylene glycol) diacrylate (PEG-DA) or Pluronic diacrylate (Pluronic-DA) (Pang and Chu 2010). The uPPs composed of PrG-TDADEs could be of interest for subsequent transformation using azide-alkyne click reaction. Besides, uPPs composed of non-proteinogenic AAs and nanoparticles made of them are of interest also as physiologically active systems (Kantaria et al. 2018, Zavradashvili et al. 2019b).

A series of water-soluble cationic uPP-PEAs were synthesized by solution copolycondensation of R-based TDADEs (Fig. 8) and AlG-TDADEs (Fig. 10) with activated di-p-nitrophenyl esters of dicarboxylic acids (ADDs) (Pang et al. 2010, Pang et al. 2014, Kantaria et al. 2018, Zavradashvili et al. 2019b), as depicted in Scheme 20. The cationic uPPs showed low cytotoxicity (Pang et al. 2010) and formed biodegradable and biocompatible cationic hybrid hydrogels after photo-crosslinking with PEG-DA (Pang et al. 2014).

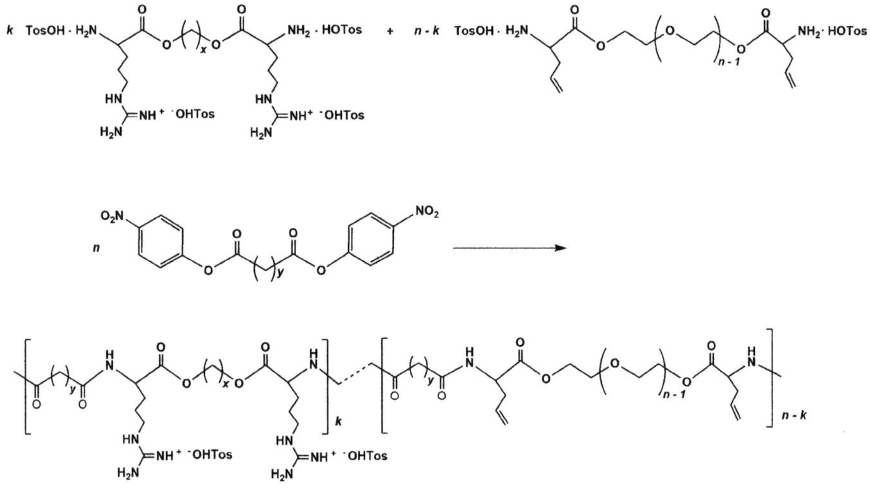

Scheme 20. Synthesis of cationic uPP-PEAs by SAcP of R-based TDADEs and AlG-TDADEs with ADDs.

2.4. Functionalization of PPs using functionalized bis-electrophilic monomers

Guo, Chkhaidze and co-workers (Guo et al. 2005, Chkhaidze et al. 2011) obtained uPP-PEAs with double bonds in the polymeric backbones using (1) unsaturated ADD such as di-p-nitrophenyl fumarate (NFu) and (2) unsaturated TDADE such as di-*p*-toluenesulfonic acid salt of bis-(L-phenylalanine)-2-butene-1,4-diester (PBe) (Fig. 11).

Three types of uPPs were synthesized by SAcP of TDADEs with NFu (Fig. 12, A), PBe with ADDs (Fig. 12, B), and NFu with PBe (Fig. 12, C).

The uPPs can be subjected to subsequent chemical transformations by interacting with various nucleophiles (Chkhaidze et al. 2011), or to transform them into biodegradable hydrogels by photo-crosslinking with PEG-DA (Guo and Chu 2005).

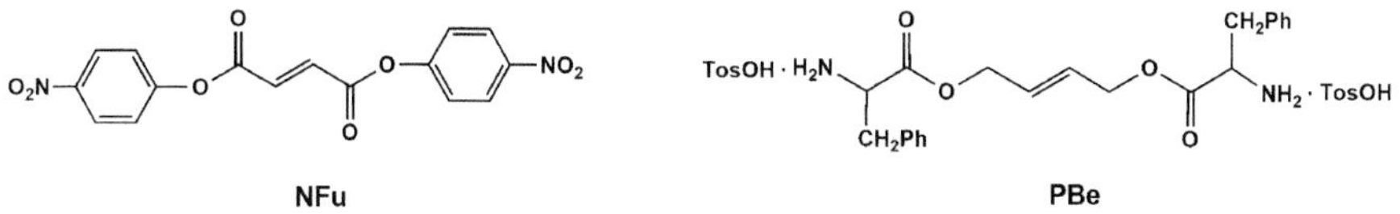

O
O_2N O O NO_2
O
NFu

O CH$_2$Ph
TosOH · H_2N O O NH_2 · TosOH
CH$_2$Ph O
PBe

Fig. 11. The unsaturated monomers for synthesizing UPPs NFu and PBe.

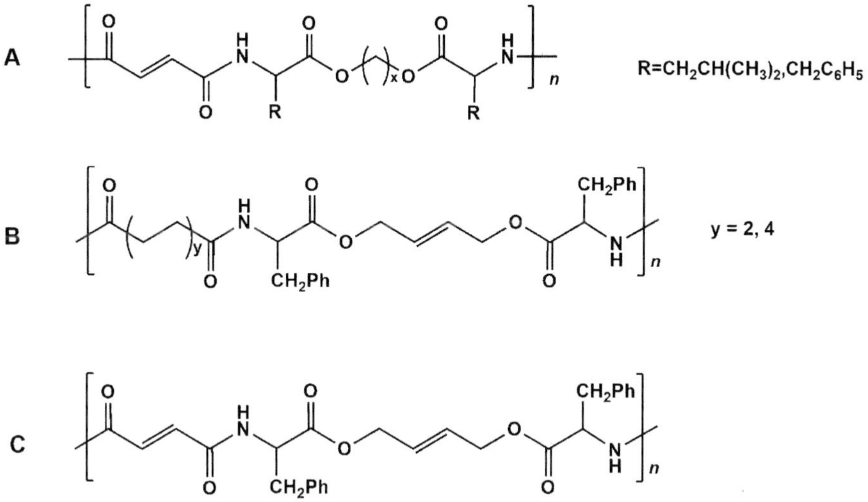

A
O O O
H N O ( )x O H N n
O R R
R=$CH_2CH(CH_3)_2$,$CH_2C_6H_5$

B
O O CH$_2$Ph
( )y H N O O N H n
O CH$_2$Ph O
y = 2, 4

C
O O CH$_2$Ph
H N O O N H n
O CH$_2$Ph O

Fig. 12. Three types of uPPs with unsaturations in the polymeric backbones.

Wu et al. (2012) obtained a family of biodegradable, biocompatible, water-soluble cationic uPP-PEAs via SAcP of NFu (Fig. 11) with R-based TDADEs (Fig. 8), as depicted in Scheme 21.

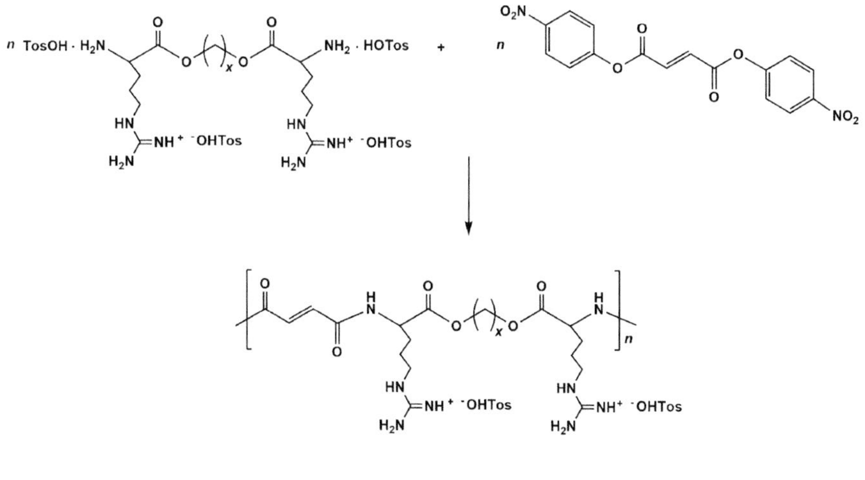

n TosOH · H_2N O O ( )x O O NH_2 · HOTos + *n*
HN =NH$^+$ $^-$OHTos H_2N HN =NH$^+$ $^-$OHTos H_2N
O_2N O O O O NO_2

O H N O O ( )x O O H N n
O
HN =NH$^+$ $^-$OHTos H_2N HN =NH$^+$ $^-$OHTos H_2N

Scheme 21. Synthesis of cationic uPP-PEAs by SAcP of NFu with R-based TDADEs.

The cationic fumaric acid-based uPP-PEAs depicted in Scheme 21, like AlG-based uPPs from Scheme 20, were transformed into biodegradable hybrid hydrogels after photo-crosslinking with Pluronic-DA (Wu et al. 2012).

Zavradashvili et al. (2013) first reported on the synthesis of PP-PEAs containing oxirane cycles in the backbones. The epoxy-PPs (ePPs) were obtained by SAcP of a functionalized ADD—di-p-nitrophenyl esters of epoxy (*cis* and *trans*) succinic acids with TDADEs. High-molecular-weight ePPs with desirable material properties were obtained on the basis of di-*p*-nitrophenyl-*trans*-epoxy-succinate, N*t*ES (Scheme 22).

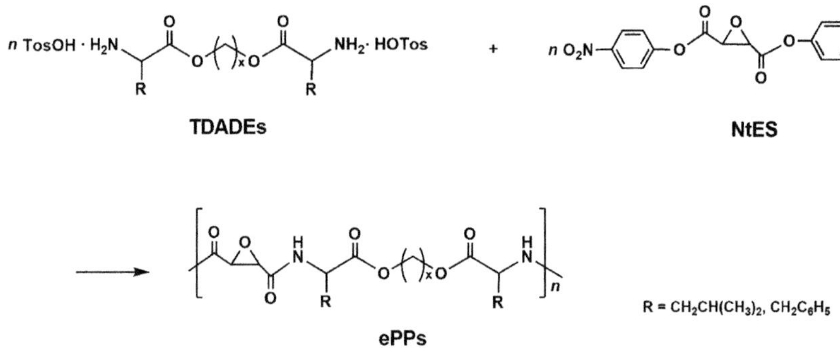

TDADEs NtES

ePPs

$R = CH_2CH(CH_3)_2, CH_2C_6H_5$

Scheme 22. Synthesis of ePP-PEAs.

The ePPs could chemically be modified by amine-functionalized reagents under mild conditions, or crosslinked with diamines (Zavradashvili et al. 2013). The use of hydrophilic diamines, e.g., PEG-diamines, could result in biodegradable hydrogels.

2.5. Functionalization of PPs by polymer-analogous transformations of various active pre-PPs

The PPs containing double bond moieties and oxirane rings in the backbones are convenient substrates for the covalent attachment of bioactive molecules, for numerous polymer-analogous reactions with heterobifunctional compounds that result in functionalized PPs, for transforming into insoluble biodegradable gels using various crosslinkers, and other purposes.

Double bonds in uPPs made of NFu (Fig. 12) are activated by two adjacent carbonyl groups and can be subjected to chemical transformations by interacting with various nucleophiles under mild conditions (Chkhaidze et al. 2011): thiols such as thioglycolic acid and mercaptoethanol, and amines such as β-alanine and n-butylamine, as depicted in Scheme 23. As noted, the uPPs were transformed into biodegradable hydrogels by photo-crosslinking with PEG-DA (Guo and Chu 2005).

The facile interaction of fumaric acid-based uPPs with thiols and aliphatic amines makes them suitable carriers for covalent conjugation of bio-active substances that contain SH and NH_2 groups. The uPPs made of

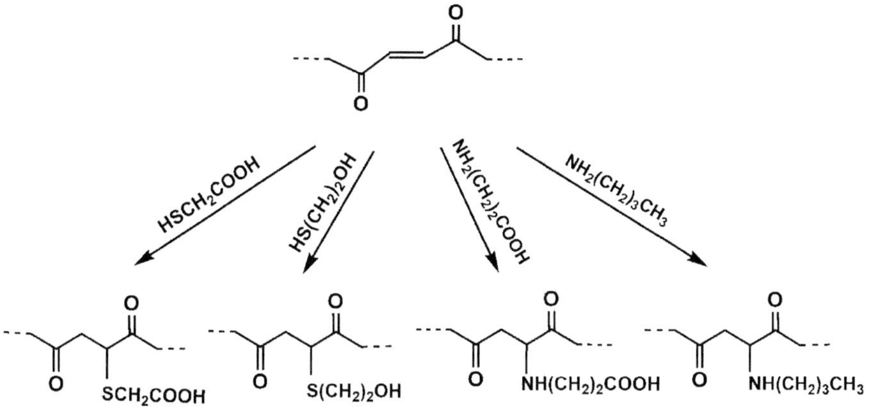

Scheme 23. Chemical transformations of PPs made of fumaric acid.

PBe and containing double bonds in a diol residue could be subjected to various transformations as well (Guo et al. 2005).

As explained in section 2.3, the double bond moieties capable of further chemical transformations could also be incorporated into the lateral chains of PPs using 2-allylglycine-based monomers AlG-TDADE (Pang and Chu 2010, Pang et al. 2010, Pang et al. 2014, Kantaria et al. 2018, Zavradashvili et al. 2019b). In an alternative approach (Zavradashvili et al. 2008), double bond moieties were incorporated into the lateral chains of PPs via a series of polymer-analogous transformations of lysine-based PP-polyacids from Scheme 8: initially, monoethanolamine was attached to the PP-polyacid using carbonyldiimidazole as a coupling agent, with subsequent acylation of the obtained polyol with chlorides of unsaturated mono acids such as methacrylic or cinnamic acids (Scheme 24).

It has to be noted that all the PP-polyacids discussed in the previous sections could be subjected to similar transformations. The PP-polyols made of TDAP according to Scheme 10 are also highly suitable substrates for such transformations and in one stage. All the unsaturated PPs can further be chemically modified as well as subjected to thermal and photochemical curing that substantially expands material properties and, hence, the scope of applications of PPs as absorbable surgical devices and drug carriers.

Like the fumaric acid-based uPPs, the ePPs from Scheme 25 could also be subjected to chemical modifications under mild conditions. Oxirane cycles of N*t*ES residues along the backbones interact with both nucleophililes (allylamine, dibutylamine, 4-amino-TEMPO, etc.) and electrophiles (acryloyl chloride, methacrylic anhydride) under mild conditions (Zavradashvili et al. 2013), resulting in a huge variety of functionalized PPs. A high reactivity of the ePP-substrates towards both nucleophilic and electrophilic reagents makes them promising ready-to-use carriers for the covalent attachment of drugs and a wide range of bioactive agents. The ePPs could be a good platform for synthesizing various functional polymers such as polyols,

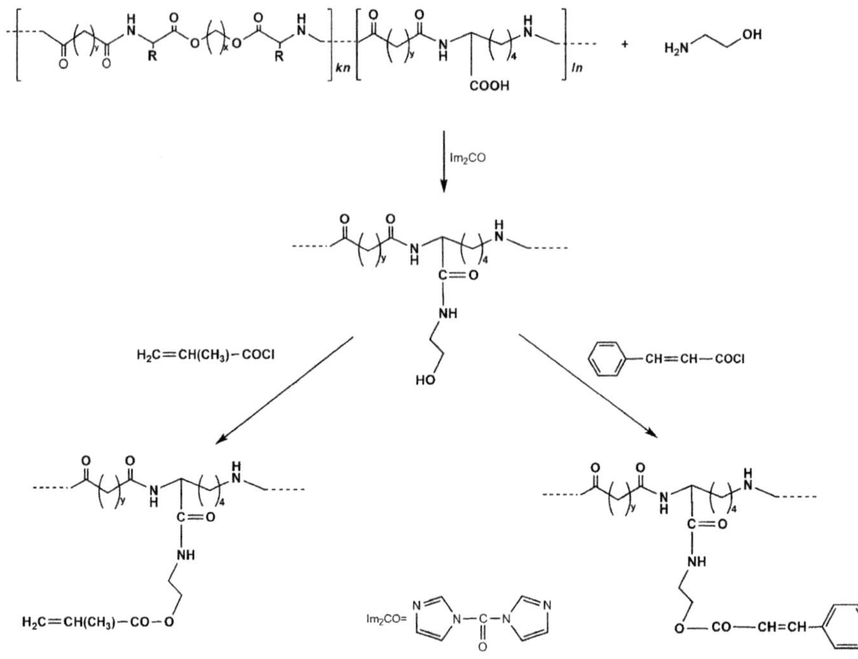

Scheme 24. Synthesis of uPPs with double bond moieties in the lateral chains.

polyelectrolytes and surfactants bearing negative or positive charges, as well as biodegradable hydrogels.

The covalent attachment of free iminoxyl radical 4-amino-TEMPO (TAM) to the ePP under mild conditions (Zavradashvili et al. 2013) is given as an example in Scheme 25.

Other examples of the applications of ePPs for constructing functional macromolecular systems are discussed in section 3. Thus, the functional PPs comprising double bonds and oxirane cycles in the backbones represent useful platforms for constructing a huge library of functional PPs.

3. Some practical applications of functional PPs

New biomimetic biomaterials are mostly based on functional and degradable polymers, which become appropriate for the immobilization of signaling molecules and growth factors. Different types of AA-containing BPs have been actively developed with the aim to obtain biomimicking functional biomaterials (Sun et al. 2011).

The built-in functional groups (e.g., amine, carboxyl, hydroxyl or thiol groups) of PPs we discussed above are ideal for the conjugation of active pharmacological agents with distinct activity. In addition, these materials constituted of metabolizable building blocks can show good biological, thermal and mechanical properties, and high degradability (both hydrolytic and enzymatic). All these factors justify the wide range of biomedical

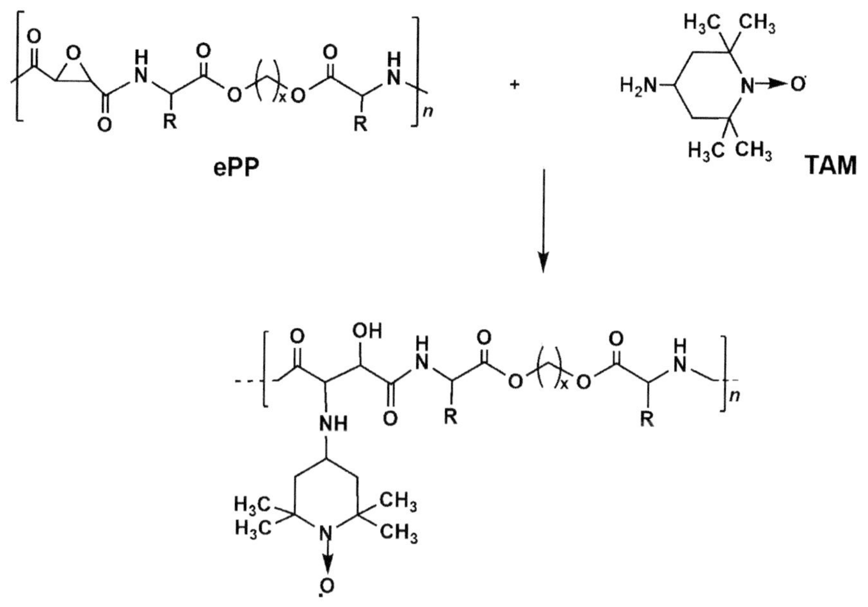

Scheme 25. Covalent attachment of TAM to ePP.

applications that are being developed and that mainly involve drug and gene delivery systems (Yamanouchi et al. 2008, Zilinskas et al. 2012), coatings for stents (Jokhadze et al. 2007, Chu and Katsarava 2003, Lee et al. 2002, Gomurashvili et al. 2008, DeFife et al. 2009), tissue engineering and medical imaging (De Wit et al. 2008, Knight et al. 2011, Sun et al. 2011, Horwitz et al. 2010).

All the PP-polyacids discussed in the present review are of interest as carriers for covalent attachment of various active molecules. Free iminoxyl radical TAM was attached to PP-polyacids, as shown in Scheme 8, using carbonyldiimidazole as a coupling agent (Chu and Katsarava 2003). An alternative way, free of coupling agents, via the interaction of TAM with ready-to-use matrix such as ePP (Zavradashvili et al. 2013) is shown in Scheme 25.

The TAM-attached PPs are promising as vascular stent coatings supressing uncontrollable cell proliferation and restenosis (Chu and Katsarava 2003). TAM mimics biological functionality of nitric oxide (NO•), which is a very small but highly reactive and unstable free radical with expanding known biological functions. NO• is extremely labile and short-lived (about 6–10 s); however, in a derivative form like TAM, the stability is improved significantly. NO• and its radical derivatives have been known to play a very important role in a host of expanding biological functions such as inflammation, neurotransmission, blood clotting, blood pressure, cardiovascular disorders, rheumatic and autoimmune diseases, antitumor activity with a high therapeutic index, antimicrobial property, sensitization or protection of cells and tissues against irradiation, oxidative stress, respiratory distress syndrome and cytoprotective property in reperfusion injury (Wang et al. 2005, Mayer and Bernd 2012). It was shown that TAM attached to a biodegradable polymer (polyglycolic acid) even via terminal functional groups exhibited the property of suppressing the proliferation of human smooth muscle cells *in vitro* (Lee et al. 1998), an indication that TAM-attached biomaterials may be able to suppress excess tissue growth in certain clinical conditions, like the restenosis problem of vascular stents. The PPs with controllable quantity of the attached TAM could be much more promising for numerous applications. It is important to know, however, that excessive introduction of NO• into the body may have adverse effects (Ou et al. 1997, Francoeur and Denis 1995). Having PP-based preparation with both controllable quantity of NO-radical and biodegradation rate makes it possible to tune these two parameters in a wide range. TAM-attached biomaterials could also be considered as scavengers of free radicals that could reduce tissue injury by neutralizing the generated toxic free radicals (Huang et al. 2006).

The PP-polyacid of PEA class composed of leucine and lysine (x = 6, y = 8, k = 0.75. l = 0.25 in Schemes 6 and 8) showed high elastic properties (elongation at break up to 800-1000%) and excellent adhesion to stainless steel, the characteristics indispensable for application as a vascular stent coating (Katsarava and Gomurashvili 2011, Jokhadze et al. 2007, Chu and Katsarava 2003). Biological studies confirmed an extremely high potential of these polymers for application as drug eluting vascular stent coating; the polymers showed excellent blood and tissue compatibility and are suitable for vascular stent-based local drug delivery targeting restenosis (Lee et al. 2002, Gomurashvili et al. 2008). It has to be noted that a high blood compatibility

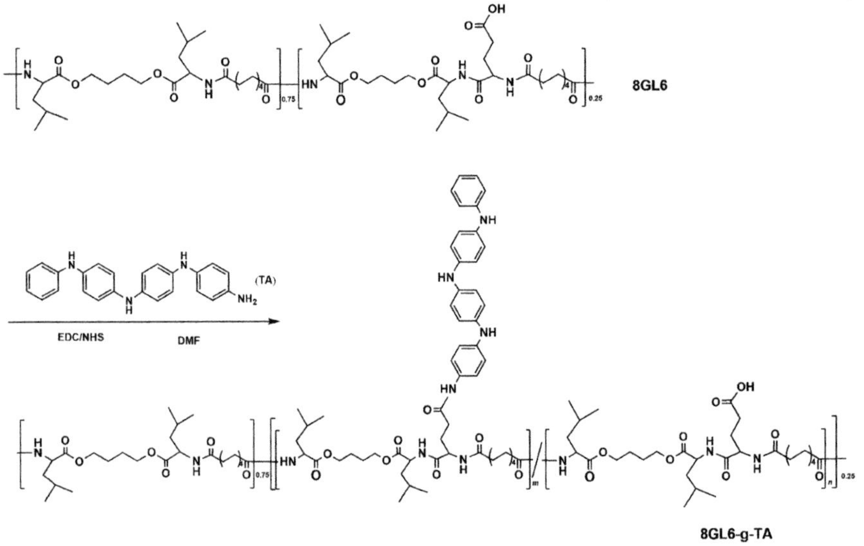

Scheme 26. Synthesis of 8GL6-g-TA.

of the PP-polyacids could be ascribed to the existence of carboxyl group (i.e., negative charge) at the surface of the stent coating. It was shown (Tang et al. 2008) that negatively charged surfaces absorb little to no fibrin compared to other surfaces, thus preventing clot formation. It was also demonstrated (DeFife et al. 2009) that these polymers may support a more natural healing response by attenuating the pro-inflammatory reaction to the implant and promoting growth of appropriate cells for repair of the tissue architecture.

Another sample of successful application of the PP-polyacid in constructing physiologically active polymers is grafting electroactive tetraaniline (TA) to the polyacid 8GL6 depicted in Scheme 13, B (Cui et al. 2012) using 1-ethyl-3-(3-dimethylaminopropyl) carbodiimide hydrochloride (EDC) and N- hydroxysuccinimide (NHS) as a coupling agent (Scheme 26).

The grafted PP 8GL6-g-TA exhibited good electroactivity, mechanical properties, and biodegradability. A systematic biocompatibility study of the grafted PP 8GL6-g-TA *in vitro* demonstrated that they were non-toxic and led to favorable adhesion and proliferation of mouse preosteoblastic MC3T3-E1 cells. Moreover, 8GL6-TA stimulated by pulsed electrical signal could serve to promote the di-erentiation of MC3T3-E1 with TCPs. Hence, the biodegradable and electroactive PP 8GL6-g-TA possessed the properties in favor of long-term potential application *in vivo* (electrical stimulation directly to the desired area) as bone repair sca-old materials engineering (Cui et al. 2012).

Functionalized PP-PEAs have been demonstrated as appropriate materials for fabrication/regeneration of vascular tissues. Specifically, PEAs containing L-aspartic acid can support adhesion and proliferation of vascular smooth muscle cells (VSMC) (Karimi et al. 2010). These polymers can also be employed to load and deliver the transforming growth factor-β1 (TGF-β1), which is essential to support and modulate smooth muscle cell proliferation and differentiation (Stegemann and Nerem 2003), and promote the production of both ECM (Mann et al. 2001) and VSMC phenotype (Lin et al. 2011). Immunofluorescence experiments demonstrated the efficient conjugation of TGF-β1 to the surface of PP-PEAs through the pendant carboxylic groups (Fig. 13).

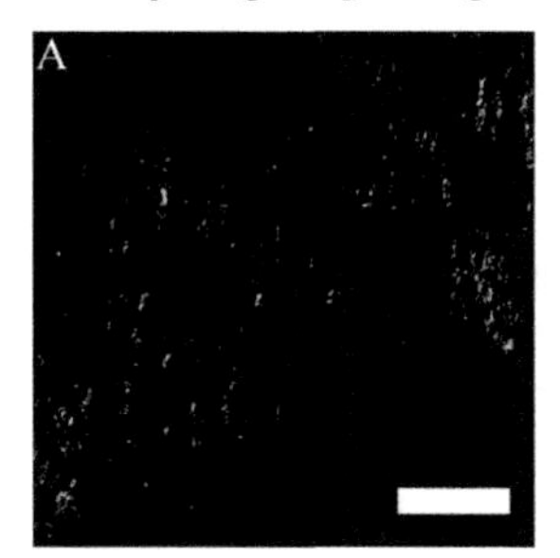

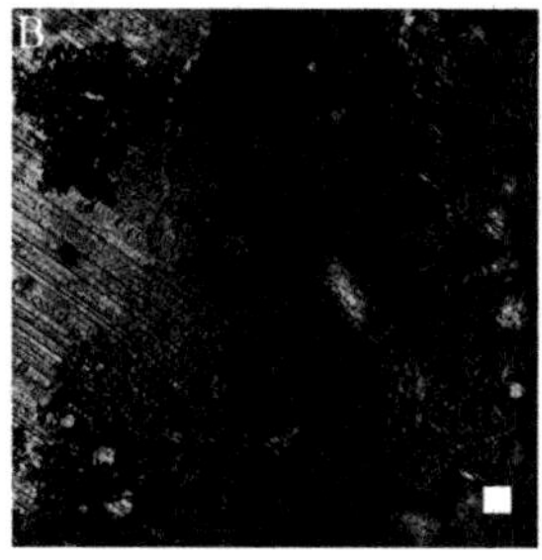

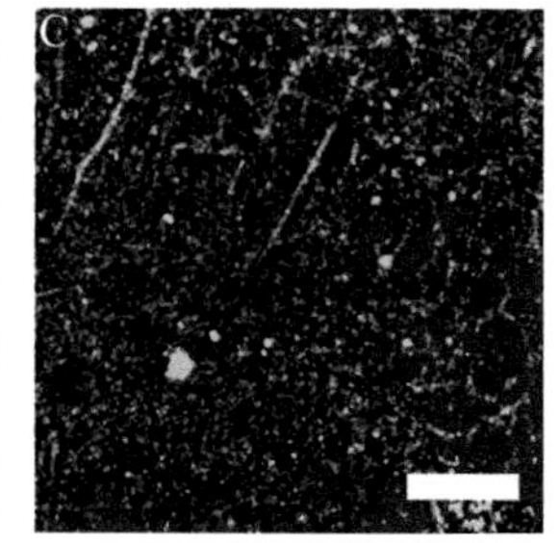

Fig. 13. Immunofluorescence images of an arginine-derived PEA alone (a), with TGF-β1 adsorbed to the surface (b) and TGF-β1 conjugated to the surface (c). Note that the last sample was rougher and exhibited a clearly higher fluorescence. Scale bar represents 50 μm. Reproduced with permission from Knight et al. (2014).

The PP-polyacids were obtained via multistep techniques of the peptide chemistry using DADE-D made of orthogonal protected aspartic acid (Schemes 14 and 15). For improving mechanical characteristics of the PP-polyacids, DADE-D was combined with TDADEs monomer composed of L-phenylalanine and 1,4-butanediol resulting in the PP-polyacids of co-PEAs class. Films made of these PP-polyacids were subjected to biological testing. It was found that human coronary artery smooth muscle cell (HCASMC) attachment, spreading and proliferation was observed on all PP-polyacid films tested (Knight et al. 2014). Vinculin expression at the cell periphery suggested that HCASMCs formed focal adhesions on the functional PPs, while the absence of smooth muscle α-actin (SMαA) expression implied the cells adopted a proliferative phenotype. The PP-polyacids were also electrospun to yield nanoscale three-dimensional (3D) scaffolds with average fiber diameters ranging from 130 nm to 294 nm. Immunoblotting studies suggested a potential increase in SMαA and calponin expression from HCASMCs cultured on 3D fibrous scaffolds when compared to 2D films. Immunofluorescence experiments demonstrated the efficient conjugation of TGF-β1 to the surface of PEAs through the pendant carboxylic groups (Fig. 13). This study demonstrated that PPs containing aspartic acid in the backbones are viable biomaterials for further investigation in vascular tissue engineering.

In sum, the PP-polyacids (anionic polymers) possess a high biomedical potential since they can form ionic complexes with cationic biomolecules including cationic drugs, basic peptides and blood proteins leading to several therapeutic applications (Samal et al. 2012). No doubt, for all of these applications anionic PPs are by far preferable.

Cationic PPs (cPPs) are also highly promising for numerous therapeutic applications. Cationic polymers (CPs) exhibit unique biological properties: they can form electrostatic complexes with anionic biomolecules including nucleic acids and proteins and are of interest as active biological compound carriers to be used in both gene therapy and biotechnology. In addition, they show stimuli-responsive, antimicrobial, antioxidant, antitumor, and anti-inflammatory bioactive properties. All these make CPs more promising for enhanced therapeutic potential (Samal et al. 2012, Moroson and Rotman 1975). CPs such as cPPs that are biodegradable and can be cleared from the body once they have served their function look especially valuable.

Various cPPs were designed and studied during the last decade. The cPPs with free primary amino groups (polyamines) were obtained via multistep techniques of the peptide chemistry using orthogonal protected lysine DADE-K (Schemes 16 and 17). NHS-fluorescein dye could easily be attached to the amine side groups of lysine-derived PEAs, which is an essential point for imaging applications. An *in vitro* cell culture study of the cPPs showed that they supported the proliferation of bovine aortic endothelial cell slightly better than gelatin-coated glass coverslips and may have potential applications for biomedical and pharmacological fields (Deng et al. 2009).

In addition, PP-PEAs having cationic lysine groups were found suitable to prepare ultrathin Langmuir-Blodgett films (Knight et al. 2012). Such films were ideal to study the adhesion of HCASMC by waveguide evanescent field fluorescence microscopy, which reveals the interfacial contact region between cells and substratum (Hassanzadeh et al. 2008). Results demonstrated that PP-PEAs were able to promote integrin signaling. This is highly important for cell survival, migration, and proliferation and in general for the development of scaffold-guided vascular tissue engineering. Previous studies on amine and carboxyl functionalized AA-PEAs pointed out that the former better supported endothelial cell adhesion, growth, and monolayer formation (Sun et al. 2011).

cPPs with free primary amino groups (polyamines) suitable for biomedical applications could be prepared also by polymer-analogous transformations of uPPs discussed above via C-S coupling using amino-thiol reagents such as cysteamine. The cPPs with free amino groups are of interest as chemically active ("living") polymers pendant amino group, which can be subjected to further chemical transformations under mild conditions. This was demonstrated by the coupling of model compounds-N-acetyl-L-valine and 2-[2-(2-methoxyethoxy) ethoxy] acetic acid to cPPs from Scheme 17 using various coupling agents (De Wit et al. 2008). These model compounds were selected with reactivity that mimics a drug molecule or protected peptide containing a free carboxyl group.

Another very important class of positively charged polymers are cPPs composed of amino acid L-arginine (R). Key monomers for synthesizing this kind of cPP are R-based TDADEs—tetra-*p*-toluenesulfonic acid salts of bis-(L-arginine)-alkylene diesters as depicted in Fig. 8. Chemical structures of various classes of cPPs made of these monomers are given in Fig. 9. The lateral guanidine group of arginine, contrary to primary amino groups, is chemically less active, which makes it possible to use directly R-based TDADEs for synthesizing the cPPs under the conditions of SAcP with soft acylating agents such as activated diesters (Fig. 4). For this reason the R-based cPPs are less suitable for polymer-analogous transformations and they are mostly used as polyelectrolytes applicable as both gene transfection and antimicrobial agents (though the guanidine group can, if necessary, be subjected to acylation (Prasad et al. 1967) or alkylation (Powell et al. 2003)).

Yamanouchi et al. (2008) and Wu et al. (2011) synthesized (PEAs) cPPs by SAcP of R-based TDADEs with ADDs. These polymers showed a high binding capacity toward plasmid DNA, and the binding activity was inversely correlated to the number of methylene groups in the diol segment of R-based cPPs. All the cPPs transfected smooth muscle cells with an efficiency that was comparable to the commercial transfection reagent Superfect. However, unlike Superfect, the cPPs, over a wide range of dosages, had minimal adverse effects on cell morphology, viability or apoptosis. It was demonstrated that the cPPs were able to deliver DNA into nearly 100% of cells under optimal polymer-to-DNA weight ratios, and that such a high level of delivery was achieved through an active endocytosis mechanism. A large portion of DNA delivered, however, was trapped in acidic endocytotic compartments, and subsequently was not expressed. The authors concluded that the cPPs should be further modified to enhance their endosome escape (Yamanouchi et al. 2008).

Memanishvili et al. (2014), using activated diesters of different classes—ADDs, ADCs, and ACs—(Fig. 4) in the SAcP with R-based TDADEs (Fig. 8), synthesized cPPs of various classes—PEAs, PEURs, and PEUs (Fig. 9). In addition to the regular polymers (containing polymethylene moieties in the backbones), this team synthesized also the same classes of the polymers modified with ether fragments—the cPPs with PEG-like backbones. For synthesizing these polymers both bis-nucleophilic (R-based TDADEs) and bis-electrophilic monomers (ADDs and ADCs) containing ether linkages were exploited (Memanishvili et al. 2014). The polymers with PEG-like backbones showed better solubility in water and higher cell compatibility as compared with the analogs containing only polymethylene moieties. These polymers have potential for application as gene transfection (Memanishvili et al. 2014) and antibacterial (Kharadze et al. 2015) agents.

Zavradashvili et al. (2019a) developed novel CPs composed of polyamines such as endogenous tetraamine spermine (Spm) and synthetically made triamine N-(2-aminoethyl)-1,3-propanediamine (Apd) by the SAcP of their salts with various activated diesters. The ADD made of *trans*-epoxy succinic acid—N*t*ES depicted in Scheme 22—was also used for preparing polyamine-based epoxy-CPs for subsequent covalent attachment of arginine methyl ester (used as hydrochloride RMeC) via one-pot/two-step procedure (Scheme 27) developed previously for the polyamide made of N*t*ES and ethylenediamine (Zavradashvili et al. 2014).

It was found that CPs depicted in Scheme 27 formed nano-sized polyplexes with pDNA at rather low CP/pDNA weight ratios and showed less cytotoxicity and higher transfection ability compared to widely used PEI as well as commercially available transfection agents such as DharmaFECT and Lipofectamine. Furthermore, new CPs showed selective transfection activity toward certain cell lines (4T1, HeLa, NIH3T3, and CCD 27SK), which is important for their potential applications in gene therapy (Zavradashvili et al. 2019a). These results encouraged us to synthesize new "hybrid" CPs that represent the combination of polyamine-based systems and R-based cPPs (Fig. 14). Studies of biological properties of "hybrid" CPs depicted in Fig. 14 are in progress now (Zavradashvili et al. unpublished data).

As noted, CPs exhibit unique biological properties (Samal et al. 2012, Moroson and Rotman 1975). For the R-based cPPs it was already found they have potential for application as gene transfection (Memanishvili et al. 2014) and antibacterial (Kharadze et al. 2015) agents. More recently, an unexpected new activity of the R-based cPPs was demonstrated: they inhibited activity of nicotinic acetylcholine receptors (nAChRs) (Lebedev et al. 2019). This activity of cPPs may be beneficial if the task is the inhibition of the α7 nicotinic receptors in the cells in cases of lung cancer: targeting α7nAChR may result in novel therapeutic methods (Wang and Hu 2018).

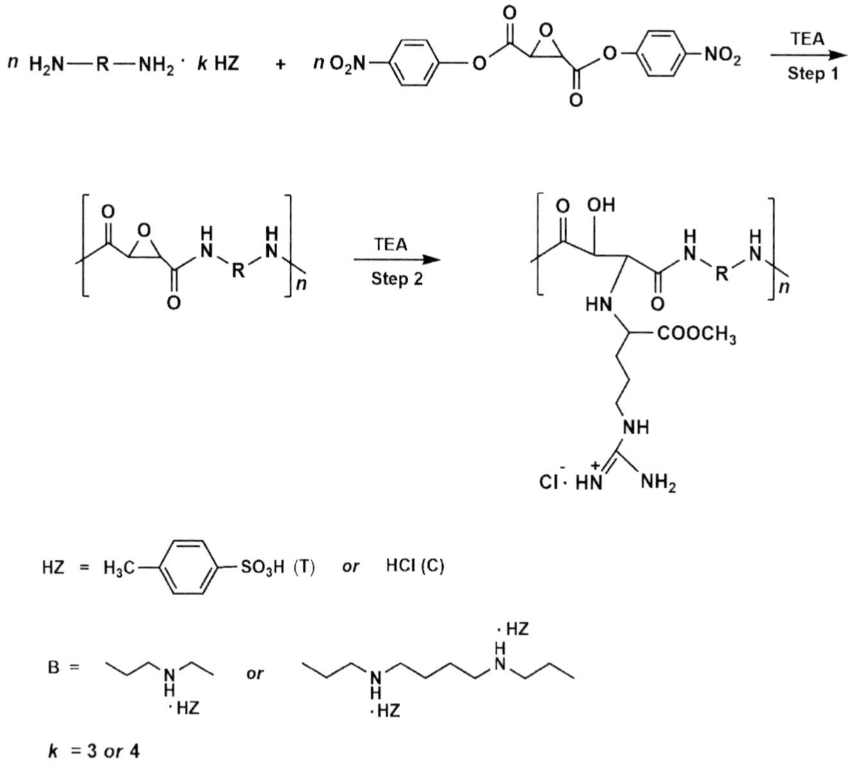

Scheme 27. Synthesis of R-attached CPs composed of polyamines via one-pot/two-step procedure.

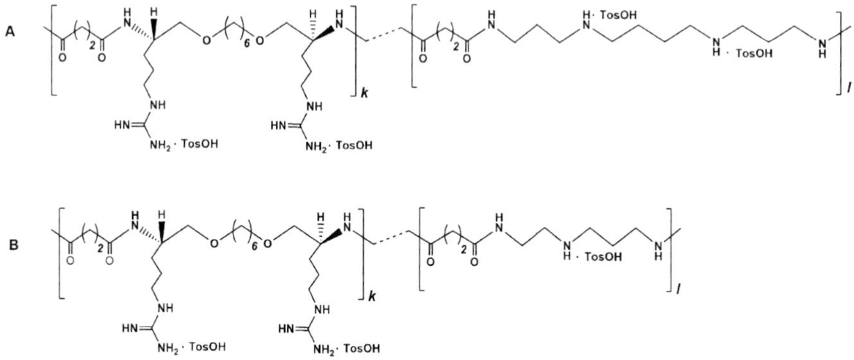

Fig. 14. Hybrid CPs composed of R-based TDADEs and Spm (A) and APD (B).

Along with linear and water-soluble R-based PPs, a series of biodegradable and biocompatible cationic hybrid hydrogels were obtained (Yamanouchi et al. 2008, Wu et al. 2011, Pang and Chu 2010, Pang et al. 2010, Pang et al. 2014) from water-soluble uPP precursors made of R-based TDADEs and

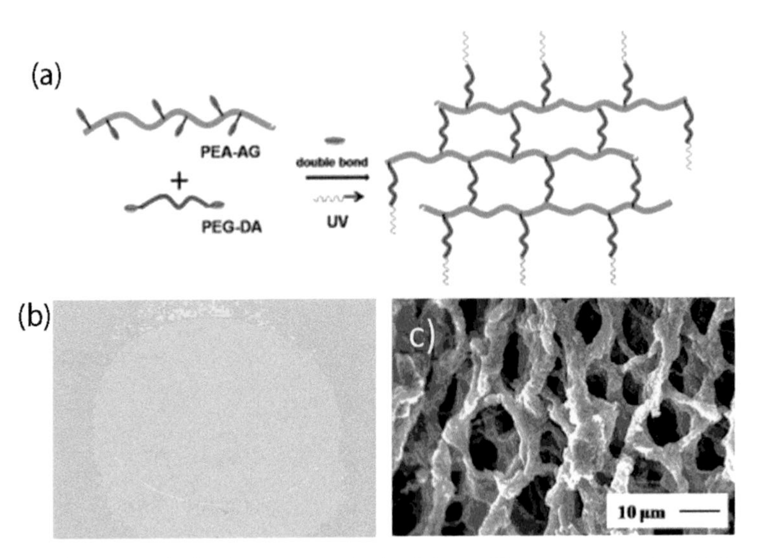

Fig. 15. (a) Photo-crosslinked structures derived from PEAs incorporating AlG units. (b) Optical image of a representative hybrid hydrogel. (c) SEM image showing the inner porous structure of the polymer shown in (b). Reproduced with permission from Pang and Chu (2010).

AlG-TDADEs (according to Scheme 20) using PEG-DA as a photo-crosslinker, as depicted in Fig. 15.

The hybrid hydrogels supported cell attachment and were non-toxic to the cells. Another type of cationic hybrid hydrogels was obtained from water-soluble fumaric acid-based unsaturated cPPs (according to Scheme 21) using Pluronic diacrylate as a photo-crosslinker (Wu et al. 2012). When compared with a pure Pluronic hydrogel, the cationic hybrid hydrogels greatly improved the attachment and proliferation of human fibroblasts on hydrogel surfaces. A bovine aortic endothelial cells viability test in the interior of the hydrogels has shown that the positively charged hybrid hydrogels can significantly improve the viability of the encapsulated endothelial cells compared to a pure Pluronic hydrogel.

Biodegradable cationic hybrid hydrogels were developed from glycidyl methacrylate chitosan (GMA-chitosan) and R-based uPPs (derived from arginine, butanediol, ethylene glycol and fumaryl chloride). Hydrogels were non-cytotoxic toward porcine aortic valve smooth muscle cells and showed good adhesion and proliferation of fibroblasts. Furthermore, hydrogels were advantageous in terms of balanced NO production and arginase activity, justifying potential applications as wound healing accelerators (He et al. 2014, Alapure et al. 2014).

The PPs are useful polymers for constructing nanocontainers for drug delivery purposes. It was found that the best PP tested for constructing the stable nanoparticles (NPs) was the PP-PEA composed of sebacic acid (8), L-leucine (L), and 1,6-hexanediol (6) labeled as 8L6 (Kantaria et al. 2016).

The zeta potential of NPs made of blends of 8L6 with R-based PP-PEA 8R6 could reach values as high as +28.0 mV, which is appropriated to stabilize the nanoparticle suspension, to use them for intracellular gene delivery, as well as to favor surface adhesion and cell penetration of NPs.

For practical applications of the NPs as drug delivery vehicles it is important to protect them from attack by the immune system of the organism. It is known that when nanosystems are in a physiological environment, they rapidly adsorb biomolecules such as proteins and lipids on their surface, forming a protein "corona" (Rahman et al. 2013, Sahoo et al. 2007). Some of the proteins (called opsonins) in the corona enhance the recognition of nanoparticle by the immune system. One of the best methods to protect the NPs from opsonin adsorption is their surface PEGylation. Surface PEGylation decreases the affinity of opsonins for adsorption on NPs: long chains of PEG form a random cloud around the NPs, thereby preventing absorption of opsonins and thus suppressing phagocytosis. The broadest synthetic/structural capabilities of PPs open a possibility to match macromolecular architectures of PEGylating and NPs forming polymers, thus providing a high affinity to them.

Three different approaches to the synthesis of PEGylating agents on the basis of functional PPs were developed. According to one, effective PEGylating agents were developed on the basis of ePPs (Scheme 22) oxirane cycles, which interact with various amines under mild conditions (Schemes 25 and 27). This reaction was applied for constructing a family of water-soluble PPs (labeled as PP-PEG) by interacting ePPs with amino-PEGs (Scheme 28). The amphiphilic PP-PEG suitable for the PEGylation of 8L6 was synthesized as follows: initially the ePP precursor was constructed as a copolymer (coPEA) constituted of regular hydrophobic fragments 8L6 (Fig. 16, block A), and epoxy-fragments (Fig. 16, block B). After interacting with amino-PEG-2000, water-soluble PP-PEG composed of hydrophobic block A and hydrophilic block B was obtained (Kantaria et al. 2019). The obtained amphiphilic PP-PEG contained in the backbones fragments similar to the backbone of the PP-PEA 8L6. This similarity provided a high affinity and, in turn, a firm anchoring of the PP-PEG with NPs made of 8L6, which was confirmed experimentally: the stable NPs were made on the basis of PP-PEA 8L6 using PP-PEG as both surfactant and PEGylated agent (Kantaria et al. 2019). A variety of new amphiphilic PPs with such dual function could be

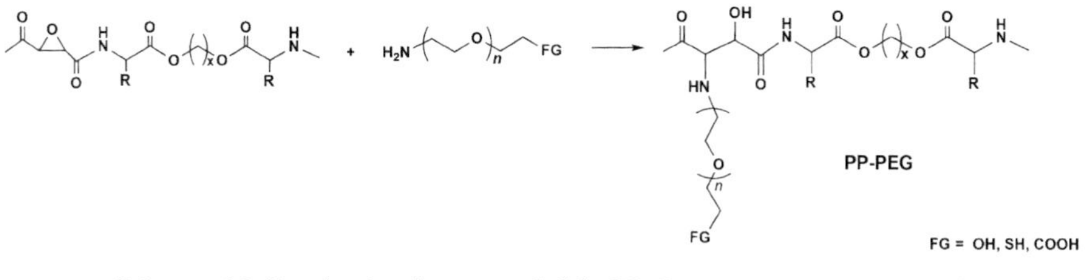

Scheme 28. Synthesis of water-soluble PPs by interacting ePPs with amino-PEG (PP-PEG).

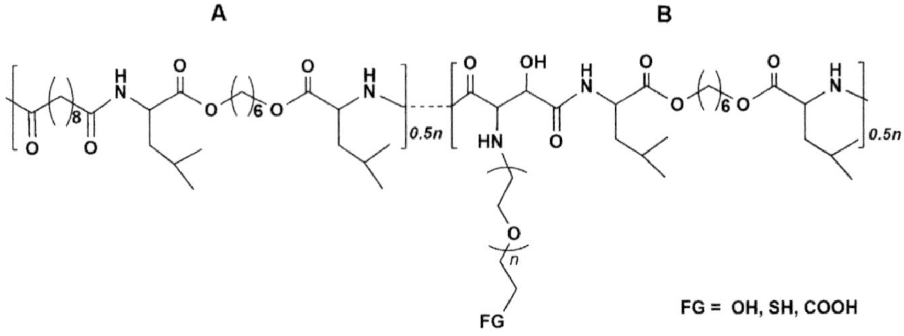

Fig. 16. Amphiphilic PP-PEG coPEA constituted of hydrophobic (A) and hydrophilic (B) fragments

made exploiting ePPs on the basis of other hydrophobic AAs, dicarboxylic acid and diols, and amino-PEGs of various MWs.

The second and very similar approach to the amphiphilic PP-PEGs with such dual function was applied to fumaric acid-based uPPs (Fig. 12). In this approach, for the covalent attachment of PEG-molecules to PPs, PEG-thiols are preferable to amino-PEGs (Scheme 29), though, according to Scheme 23, amino-PEGs could be used as well.

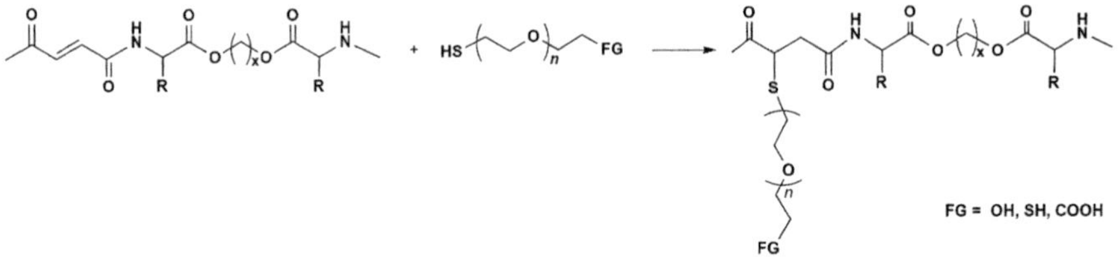

Scheme 29. Synthesis of water-soluble PPs by interacting uPPs with PEG-thiol.

The third approach to PP-based amphiphils seems the most simple and convenient (Kobauri et al. 2019a, Kobauri et al. 2013). According to this strategy, amphiphilic PPs as ABA-type triblock copolymers were obtained by a one-pot/two-step process, as demonstrated in Scheme 30 for 8L6-based PP as an example. In the first step, a telechelic PP (block B) with terminal activated ester groups was synthesized (Scheme 30, step 1); an average molecular weight (MW) of the B block can be tuned by varying the nucleophile/electrophile ($k/k+1$) ratio. In the second step (Scheme 30, step 2), the telechelic PP was interacted with amino-PEG-2000 (two moles per one mole of telechelic PP).

In this approach, as in the two approaches above, properties of amphiphils could be tuned using various AAs, diols, dicarboxylic acids, and amino-PEGs of different chain lengths.

In cases where heterobifunctional PEGs, containing OH, SH, COOH groups along with amino and thiol groups, are used in the reactions depicted in Schemes 25-27, functionalized PEGylating agents can be obtained.

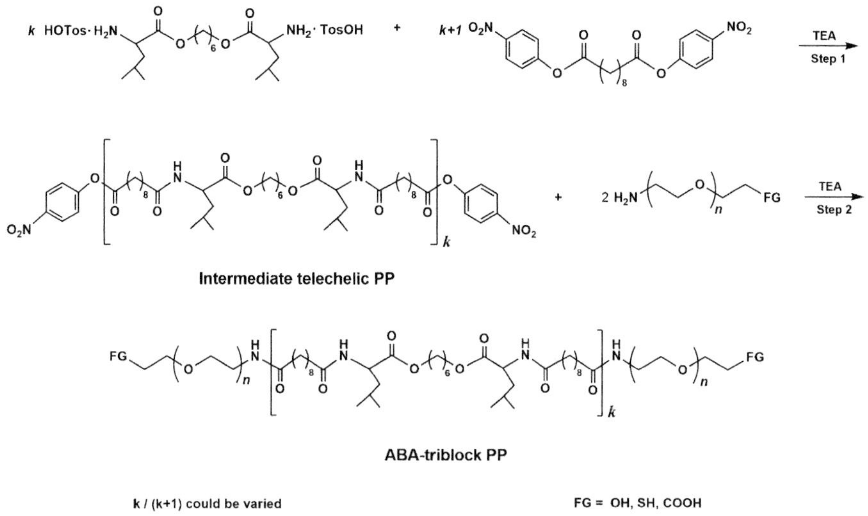

Intermediate telechelic PP

ABA-triblock PP

k / (k+1) could be varied

FG = OH, SH, COOH

Scheme 30. One-pot/two-step synthesis of water-soluble PP-based triblock copolymer: synthesis of intermediate telechelic PP (Step 1) and synthesis of ABA-triblock PP (Step 2).

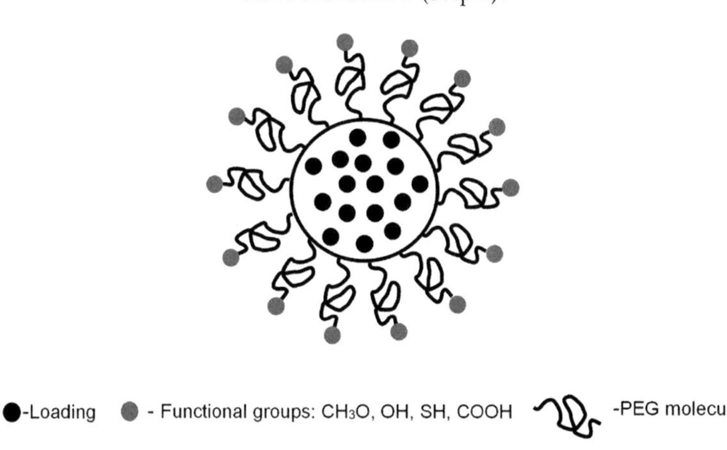

●-Loading ● - Functional groups: CH_3O, OH, SH, COOH -PEG molecule

Fig. 17. Schematic representation of the NP with functionalized PEG-cloud

The application of these agents when fabricating the NPs results in the particles with functionalized PEG-cloud (Fig. 17), which can be used for the conjugation of various markers, vectors, etc. to the surface of NPs.

PPs of non-ionic nature functionalized with hydroxyl groups are also of interest for constructing new biomaterials. For example, PP-PEAs based on serine units have been employed for the preparation of hydrogels via acrylation and photo-gelation (Deng et al. 2011). Hydrophilic/hydrophobic character and crosslinking degree were easily controlled through the ratio of protected (O-benzyl ester) and non-protected serine units in the final copolymer. Attachment and proliferation assays using bovine aortic endothelial cells and fibroblasts demonstrated that the interaction between polymer and cells were affected by the surface properties (i.e., hydrophobicity) of the final scaffold. OH-functionalized PPs, polyols promising for fabricating hydrogels, could also be prepared using functionalized bis-nucleophiles such as TDAP (Scheme 10) or via polymer-analogous transformations of active pre-PPs, as discussed above (see, e.g., Schemes 23, 24, 28 and 29).

PP-PEAs can also be designed to show a stimuli-responsive degradation. Such PPs belong to the so-called self-immolative polymer class (Peterson et al. 2012, Zhang et al. 2012) and can be designed making use of amino acid-based pendant spacers having a trigger moiety (Mejia and Gillies 2013). A nucleophilic amino acid side chain was generated after cleavage of the trigger group. The next step involved an intramolecular cyclization that formed a 5-membered ring and cleaved an ester linkage of the polymer backbone (Scheme 31). Satisfactory results were achieved by incorporating 2,4-diaminobutyric acid and homocysteine units, which are characterized by having stimuli-responsive triggers capping their pendant amine and thiol side chains, respectively (Mejia and Gillies 2013).

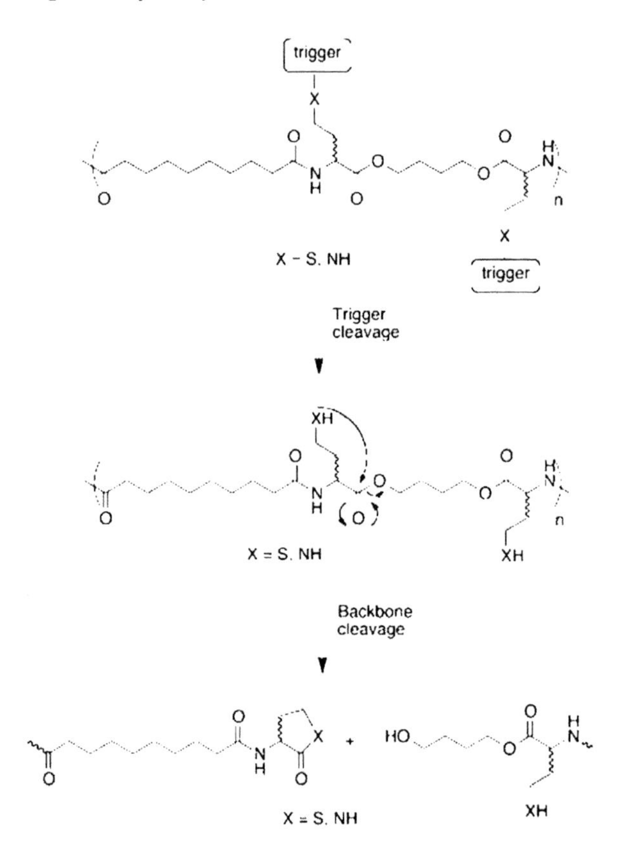

Scheme 31. Stimuli-responsive PEA having pendant groups susceptible to render a cyclization-induced backbone ester cleavage. Reproduced with permission from Mejia and Gillies (2013).

Cancer tissues are characterized by a redox imbalance caused by defective mitochondria and NADPH oxidase complexes. These two triggers lead to an excessive generation of both reactive oxygen and nitrogen species (ROS and RNS) and an increased consumption of cellular reducers (e.g., glutathione and NADH) that accelerate anabolism (Trachootham et al. 2009, Gius and Spitz 2006, Trachootham et al. 2008, Kroemer and Pouyssegur 2008, Bakalova et al. 2013). As a consequence of the derived processes, the redox balance in tumor cells changes according to the cancer progression from a reducing (early stage) to an oxidizing potential (terminal stage). In fact, the reducing potential in early stages can reach values of 2-3 orders of magnitude higher than that observed in healthy tissues (Cheng et al. 2011, Meng et al. 2009). Therefore, great interest is focused on the development of reduction-responsive polymeric nanocarriers since they could support an improved antitumor efficacy (Cui et al. 2013). For this purpose an original PP, SS-PEA, containing reductively cleavable disulfide bonds was designed. The SS-PEA was synthesized by SAcP of TDADE composed of L-phenylalanine and bis-(2-hydroxyethyl)disulfide with ADD (di-*p*-nitrophenyl adipate) in N,N-dimethylformamide (Sun et al. 2015). The SS-PEAs were synthesized with average molecular weights within 16.6-23.6 kDa. Polymers had a clear surface erosion (hydrolytic degradation) in α-chymotrypsin-containing medium and bulk degradation under a reductive environment containing dithiothreitol. A high cell compatibility was also demonstrated through adhesion and proliferation assays with L929 fibroblast cells. MTT assays revealed that doxorubicin (DOX)-loaded SS-PEA nanoparticles had a high antitumor activity approaching that of free DOX in drug-sensitive MCF-7 cells, and more than 10 times that of free DOX in drug-resistant MCF-7/ADR cells. These enzymatically and reductively degradable SS-PEAs (Fig. 18) have provided an appealing platform for biomedical technology, in particular controlled drug delivery applications.

In another approach, novel reductively degradable PP SS-PEA-graft-galactose (SS-PEA-Gal) copolymers were designed and developed to form smart nano-vehicles for active hepatoma-targeting DOX delivery. SS-PEA-Gal constructs were readily synthesized by SAcP of TDADE composed of L-phenylalanine and bis-(2-hydroxyethyl) disulfide and bis-vinyl sulfone functionalized cysteine hexanediol diester with ADD (di-*p*-nitrophenyl adipate), followed by conjugating with thiol-functionalized galactose (Gal-SH) via the Michael addition reaction (Lv et al. 2015). SS-PEA-Gal formed unimodal nanoparticles (PDI = 0.10–0.12) in water with average particle sizes within 9-138 nm. Notably, *in vitro* drug release studies showed that over 80% DOX was released from SS-PEA-Gal nanoparticles within 12 h under reductive conditions mimicking an intracellular environment, while low DOX release (<20%) was observed for reduction-insensitive PEA-Gal nanoparticles under otherwise the same conditions and SS-PEA-Gal nanoparticles under non-reductive conditions. The SS-PEA-Gal nanoparticles exhibited high specificity to asialoglycoprotein receptor (ASGP-R)-overexpressing HepG2 cells. MTT assays using HepG2 cells showed that DOX-loaded SS-PEA-Gal

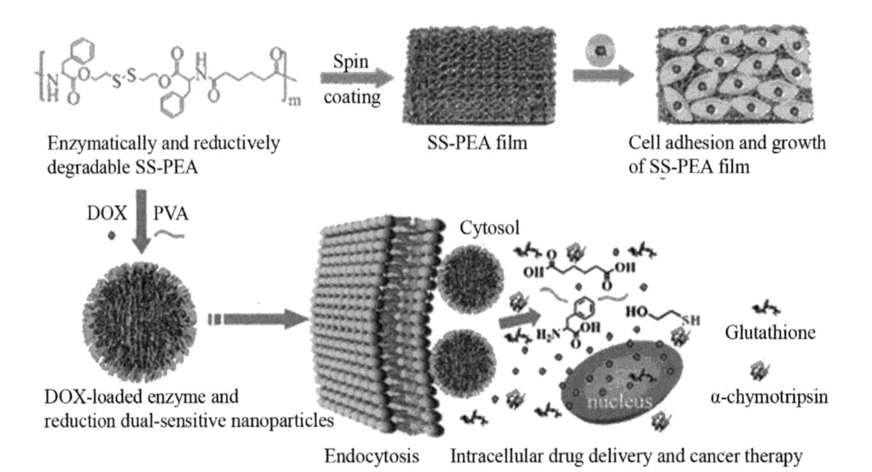

Fig. 18. Enzymatically and reductively degradable SS-PEA polymer based on phenylalanine units. Applications for cell culture and active intracellular anticancer drug delivery are shown. Reproduced with permission from Sun et al. (2015).

had a low half maximal inhibitory concentration (IC50) of 1.37 μg mL^{-1}, approaching that of free DOX. Flow cytometry and confocal laser scanning microscopy studies confirmed the efficient uptake of DOX-loaded SS-PEA-Gal nanoparticles by HepG2 cells as well as fast intracellular DOX release. Importantly, SS-PEA-Gal and PEA-Gal nanoparticles were non-cytotoxic to HepG2 and MCF-7 cells up to a tested concentration of 1.0 mg mL^{-1}. These tumor-targeting and reduction-responsive degradable nanoparticles have appeared as an interesting multi-functional platform for advanced drug delivery.

References

Alapure, Bhagwat V., Yan Lu, Mingyu He, Chih-Chang Chu, Hongying Peng, Filipe Muhale, Yue-Liang Brewerton, Bruce Bunnell and Song Hong. 2018. Accelerate healing of severe burn wounds by mouse bone marrow mesenchymal stem cell-seeded biodegradable hydrogel scaffold synthesized from arginine-based poly (ester amide) and chitosan. *Stem Cells and Development* 27(23): 1605-1620.

Arabuli, Natia, George Tsitlanadze, Lamara Edilashvili, Darejan Kharadze, Tsisana Goguadze, Vakhtang Beridze, Zaza Gomurashvili and Ramaz Katsarava. 1994. Heterochain polymers based on natural amino acids. Synthesis and enzymatic hydrolysis of regular poly (ester amide)s based on bis (L-phenylalanine) α, ω-alkylene diesters and adipic acid. *Macromolecular Chemistry and Physics* 195(6): 2279-2289.

Atkins, Katelyn M., David Lopez, Darryl K. Knight, Kibret Mequanint and Elizabeth R. Gillies. 2009. A versatile approach for the syntheses of poly (ester amide) s with pendant functional groups. *Journal of Polymer Science Part A: Polymer Chemistry* 47(15): 3757-3772.

Bakalova, Rumiana, Zhivko Zhelev, Ichio Aoki and Tsuneo Saga. 2013. Tissue redox activity as a hallmark of carcinogenesis: From early to terminal stages of cancer. *Clinical Cancer Research* 19(9): 2503-2517.

Cheng, Ru, Fang Feng, Fenghua Meng, Chao Deng, Jan Feijen and Zhiyuan Zhong. 2011. Glutathione-responsive nano-vehicles as a promising platform for targeted intracellular drug and gene delivery. *Journal of Controlled Release* 152(1): 2-12.

Chkhaidze, Ekaterine, David Tugushi, Darejan Kharadze, Zaza Gomurashvili, Chih-Chang Chu and Ramaz Katsarava. 2011. New unsaturated biodegradable poly (ester amide)s composed of fumaric acid, L-leucine and α, ω-alkylene diols. *Journal of Macromolecular Science, Part A* 48(7): 544-555.

Chu, Chih-Chang and Ramaz Katsarava. 2003. Elastomeric functional biodegradable copolyester amides and copolyester urethanes. U.S. Patent # 6,503,538.

Cui, Can, Ya-Nan Xue, Ming Wu, Yang Zhang, Ping Yu, Lei Liu, Ren-Xi Zhuo and Shi-Wen Huang. 2013. Cellular uptake, intracellular trafficking, and antitumor efficacy of doxorubicin-loaded reduction-sensitive micelles. *Biomaterials* 34(15): 3858-3869.

Cui, Haitao, Yadong Liu, Mingxiao Deng, Xuan Pang, Peibiao Zhang, Xianhong Wang, Xuesi Chen and Yen Wei. 2012. Synthesis of biodegradable and electroactive tetraaniline grafted poly (ester amide) copolymers for bone tissue engineering. *Biomacromolecules* 13(9): 2881-2889.

De Wit, Matthew A., Zixi Wang, Katelyn M. Atkins, Kibret Mequanint and Elizabeth R. Gillies. 2008. Syntheses, characterization, and functionalization of poly (ester amide)s with pendant amine functional groups. *Journal of Polymer Science Part A: Polymer Chemistry* 46(19): 6376-6392.

DeFife, Kristin M., Kathy Grako, Gina Cruz-Aranda, Sharon Price, Ron Chantung, Kassie Macpherson, Ramina Khoshabeh, Sindhu Gopalan and William G. Turnell. 2009. Poly (ester amide) co-polymers promote blood and tissue compatibility. *Journal of Biomaterials Science, Polymer Edition* 20(11): 1495-1511.

Deng, Mingxiao, Jun Wu, Cynthia A. Reinhart-King and Chih-Chang Chu. 2009. Synthesis and characterization of biodegradable poly (ester amide)s with pendant amine functional groups and in vitro cellular response. *Biomacromolecules* 10(11): 3037-3047.

Deng, Mingxiao, Jun Wu, Cynthia A. Reinhart-King and Chih-Chang Chu. 2011. Biodegradable functional poly (ester amide)s with pendant hydroxyl functional groups: Synthesis, characterization, fabrication and in vitro cellular response. *Acta Biomaterialia* 7(4): 1504-1515.

Fonseca, Ana C., Maria H. Gil and Pedro N. Simoes. 2014. Biodegradable poly(ester amide)s – A remarkable opportunity for the biomedical area: Review on the synthesis, characterization and applications. *Progress in Polymer Science* 39(7): 1291-1311.

Francoeur, Caroline and Michel Denis. 1995. Nitric oxide and interleukin-8 as inflammatory components of cystic fibrosis. *Inflammation* 19(5): 587-598.

Gius, David and Douglas R. Spitz. 2006. Redox signaling in cancer biology. *Antioxidants & Redox Signaling* 8(7-8): 1249-1252.

Gomurashvili, Zaza, Hans R. Kricheldorf and Ramaz Katsarava. 2000. Amino acid based bioanalogous polymers. Synthesis and study of new poly (ester amide) s composed of hydrophobic α-amino acids and dianhydrohexitoles. *Journal of Macromolecular Science: Pure and Applied Chemistry* 37: 215-227.

Gomurashvili, Zaza, Huashi Zhang, Jane Da, Turner D. Jenkins, Jonathan Hughes, Mark Wu, Leanne Lambert, Kathryn A. Grako, Kristin M. DeFife, Kassandra

MacPherson, Vassil Vassilev, Ramaz Katsarava and William G. Turnell. 2008. From drug-eluting stents to biopharmaceuticals: Poly (ester amide) a versatile new bioabsorbable biopolymer. pp. 10-26. *In*: Mahapatro, A. and Kulshrestha, A.S. (eds.). ACS Symposium Series 977: Polymers for Biomedical Applications. Oxford University Press, Oxford.

Guo, Kai and Chih-Chang Chu. 2005. Synthesis and characterization of novel biodegradable unsaturated poly (ester amide)/poly (ethylene glycol) diacrylate hydrogels. *Journal of Polymer Science Part A: Polymer Chemistry* 43(17): 3932-3944.

Guo, Kai, Chih-Chang Chu, Ekaterine Chkhaidze and Ramaz Katsarava. 2005. Synthesis and characterization of novel biodegradable unsaturated poly(ester amide)s . *Journal of Polymer Science Part A: Polymer Chemistry* 43(7): 1463-1477.

Hassanzadeh, Abdollah, Michael Nitsche, Silvia Mittler, Souzan Armstrong, Jeff Dixon and Uwe Langbein. 2008. Waveguide evanescent field fluorescence microscopy: Thin film fluorescence intensities and its application in cell biology. *Applied Physics Letters* 92(23): 233503.

He, Mingyu, Alicia Potuck, Yi Zhang and Chih-Chang Chu. 2014. Arginine-based polyester amide/polysaccharide hydrogels and their biological response. *Acta Biomaterialia* 10(6): 2482-2494.

Horwitz, Joshua A., Katrina M. Shum, Josephine C. Bodle, Ming Xiao Deng, Chih-Chang Chu and Cynthia A. Reinhart-King. 2010. Biological performance of biodegradable amino acid-based poly (ester amide)s: Endothelial cell adhesion and inflammation in vitro. *Journal of Biomedical Materials Research Part A* 95(2): 371-380.

Huang, Yanming, Lan Wang, Shengqiao Li, Xiaoshun Liu, Kwangdeok Lee, Eric Verbeken, Frans Van de Werf and Ivan de Scheerder. 2006. Stent-based tempamine delivery on neointimal formation in a porcine coronary model. *Acute Cardiac Care* 8(4): 210-216.

Jokhadze, Giuli, Marika Machaidze, Henry Panosyan, Chih-Chang Chu and Ramaz Katsarava. 2007. Synthesis and characterization of functional elastomeric poly (ester amide) co-polymers. *Journal of Biomaterials Science, Polymer Edition* 18(4): 411-438.

Kantaria, Temur, Nino Kupatadze, Giuli Otinashvili, Tengiz Kantaria, David Tugushi, Ramaz Katsarava, Ashot Saghyan, Anna Mkrtchyan and Sergey Poghosyan. 2018. Synthesis of new biodegradable poly(ester amide)s composed of non-proteinogenic α-amino acids. 1st International Scientific Conference "Current State of Pharmacy and Prospects of its Development" dedicated to 100th anniversary of Yerevan State University and the 75th anniversary of the NAS RA, 2018, Yerevan, Armenia.

Kantaria, Temur, Tengiz Kantaria, Sophio Kobauri, Mariam Ksovreli, Tinatin Kachlishvili, Nina Kulikova, David Tugushi and Ramaz Katsarava. 2016. Biodegradable nanoparticles made of amino-acid-based ester polymers: Preparation, characterization, and in vitro biocompatibility study. *Applied Sciences* 6(12): 444.

Kantaria, Temur, Tengiz Kantaria, Sophio Kobauri, Mariam Ksovreli, Tinatin Kachlishvili, Nina Kulikova, David Tugushi and Ramaz Katsarava. 2019. A new generation of biocompatible nanoparticles made of resorbable poly(ester amide) s. *Annals of Agrarian Science* 17: 49-58.

Karimi, Pooneh, Amin S. Rizkalla and Kibret Mequanint. 2010. Versatile biodegradable poly (ester amide)s derived from α-amino acids for vascular tissue engineering. *Materials* 3(4): 2346-2368.

Kartvelishvili, Tamara, Akaki Kvintradze and Ramaz Katsarava. 1996. Amino acid based bioanalogous polymers. Synthesis of novel poly (urethane amide)s based on N, N'-(trimethylenedioxydicarbonyl) bis (phenylalanine). *Macromolecular Chemistry and Physics* 197(1): 249-257.

Kartvelishvili, Tamara, George Tsitlanadze, Lamara Edilashvili, Nona Japaridze and Ramaz Katsarava. 1997. Amino acid based bioanalogous polymers. Novel regular poly (ester urethane)s and poly (ester urea)s based on bis (L-phenylalanine) α, ω-alkylene diesters. *Macromolecular Chemistry and Physics* 198(6): 1921-1932.

Katsarava Ramaz, David Tugushi and Zaza Gomurashvili. 2014. Poly (ester urea) Polymers and Methods of Use. U.S. Patent # 8,765,164.

Katsarava, Ramaz D. 1991. Progress and problems in activated polycondensation. *Russian Chemical Reviews* 60(7): 722-737.

Katsarava, Ramaz D., Darejan P. Kharadze, Lamara M. Avalishvili and Malkhaz M. Zaalishvili. 1985. Heterochain polymers based on natural α-amino acids. Bis-oxazolinone method of the synthesis of polyamides containing enzymatically cleavable bonds in the main chains. *Acta Polymerica* 36: 29-38.

Katsarava, Ramaz D., Tamara M. Kartvelishvili, Nona N. Japaridze, Tsisana A. Goguadze, Tamara A. Khosruashvili, Roald P. Tiger and Pjotr A. Berlin. 1993. Synthesis of polyureas by polycondensation of diamines with active derivatives of carbonic acid. *Die Makromolekulare Chemie: Macromolecular Chemistry and Physics* 194(12): 3209-3228.

Katsarava, Ramaz and Zaza Gomurashvili. 2011. Biodegradable polymers composed of naturally occurring α-amino acids. *In*: Lendlein, A., Sisson, A. (eds.). Handbook of Biodegradable Polymers-Isolation, Synthesis, Characterization and Applications. Wiley-VCH, Verlag GmbH & Co KGaA.

Katsarava, Ramaz, David Tugushi, Vakhtang Beridze and Nancy Tawil. 2016. Composition comprising a polymer and a bioactive agent and method of preparing thereof. U.S. Patent # 15/188,783.

Katsarava, Ramaz, Nina Kulikova and Jordi Puiggalí. 2016. Amino acid based biodegradable polymers – Promising materials for the applications in regenerative medicine. *Jacobs Journal of Regenerative Medicine* 1: 012.

Katsarava, Ramaz, Nino Mumladze (unpublished data)

Katsarava, Ramaz. 2003. Active polycondensation: From peptide chemistry to amino acid based biodegradable polymers. *Macromolecular Symposia* 199(1): 419-430. Weinheim: WILEY-VCH Verlag.

Kharadze, Darejan P., Tina N. Omiadze, George V. Tsitlanadze, Tsisana A. Goguadze, Natia M. Arabuli, Zaza D. Gomurashvili and Ramaz D. Katsarava. 1994. New biodegradable polymers derived from [N, N']-diacyl-bisphenylalanine. *Polymer Science* 36(9): 1214-1218.

Kharadze, Darejan, Larisa Kirmelashvili, Nino Medzmariashvili, Vakhtang Beridze, George Tsitlanadze, David Tugushi, Chih-Chang Chu and Ramaz Katsarava. 1999. Synthesis and α-chymotrypsinolysis of regular poly (ester amides) based on phenylalanine, diols, and terephthalic acid. *Polymer Science, Series A* 41(9): 883-890.

Kharadze, Darejan, Tamar Memanishvili, Ketevan Mamulashvili, Tina Omiadze, Larisa Kirmelashvili, Zaur Lomtatidze and Ramaz Katsarava. 2015. In vitro antimicrobial activity study of some new arginine-based biodegradable poly (ester urethane)s and poly (ester urea)s. *Journal of Chemistry and Chemical Engineering* 9: 524-532.

Knight, Darryl K., Elizabeth R. Gillies and Kibret Mequanint. 2011. Strategies in functional poly (ester amide) syntheses to study human coronary artery smooth muscle cell interactions. *Biomacromolecules* 12(7): 2475-2487.
Knight, Darryl K., Elizabeth R. Gillies and Kibret Mequanint. 2011. Strategies in functional poly (ester amide) syntheses to study human coronary artery smooth muscle cell interactions. *Biomacromolecules* 12(7): 2475-2487.
Knight, Darryl K., Rebecca Stutchbury, Daniel Imruck, Christopher Halfpap, Shigang Lin, Uwe Langbein, Elizabeth R. Gillies, Silvia Mittler and Kibret Mequanint. 2012. Focal contact formation of vascular smooth muscle cells on Langmuir-Blodgett and solvent-cast films of biodegradable poly (ester amide)s. *ACS Applied Materials & Interfaces* 4(3): 1303-1312.

Knight, Darryl K., Elizabeth R. Gillies and Kibret Mequanint. 2014. Biomimetic l-aspartic acid-derived functional poly (ester amide)s for vascular tissue engineering. *Acta Biomaterialia* 10(8): 3484-3496.

Kobauri, Sophio, David Tugushi, Vladimir P. Torchilin and Ramaz Katsarava. 2019a. Micelles Made of Pseudo-proteins for Solubilization of Hydrophobic Pharmaceuticals. International Conference in Advances in Medical Biotechnology, Jun 11-12, 2019, Barcelona, Spain.

Kobauri, Sophio, Temur Kantaria, Nino Kupatadze, Nazi Kutsiava, David Tugushi and Ramaz Katsarava. 2019b. Pseudo-proteins: A new family of biodegradable polymers for sophisticated biomedical applications. *Nano Technol & Nano Sci J* 2: 109. *Nano Technol & Nano Sci J* 1(1): 37-42.

Kobauri, Sophio, Vladimir P. Torchilin, David Tugushi and Ramaz Katsarava. 2013. PEG-PEA-PEG Triblock-copolymeric micelles as potential biodegradable nanocarriers for pharmaceuticals. *Contemporary Issues on Chemical Engineering* 41-45.

Kricheldorf, Hans R. 2010. Cyclic polymers: Synthetic strategies and physical properties. *Journal of Polymer Science Part A: Polymer Chemistry* 48(2): 251-284.

Kroemer, Guido and Jacques Pouyssegur. 2008. Tumor cell metabolism: Cancer's Achilles' heel. *Cancer Cell* 13(6): 472-482.

Lebedev, Dmitry S., Elena V. Kryukova, Igor A. Ivanov, Natalia S. Egorova, Nikita D. Timofeev, Ekaterina N. Spirova, Elizaveta Tufanova, Andrei Siniavin, Denis Kudryavtsev, Igor Kasheverov, Marios Zouridakis, Ramaz Katsarava, Nino Zavradashvili, Ia Iagorshvili, Socrates Tzartos and Victor Tsetlin. 2019. Oligoarginine Peptides: A new family of nicotinic acetylcholine receptor inhibitors. *Molecular Pharmacology* 96(5): 664-673.

Lee, Keun-Ho, Chih-Chang Chu, Fred Quimby and Suzanne Klaessig. 1998. Molecular design of biologically active biodegradable polymers for biomedical applications. *Macromolecular Symposia*, 130(1): 71-80. Basel: Hüthig & Wepf Verlag.

Lee, Seung H., Istvan Szinai, Kenneth Carpenter, Ramaz Katsarava, Giuli Jokhadze, Chih-Chang Chu, Yanming Huang, Eric Verbeken, Orville Bramwell, Ivan De Scheerder, Mun K. Hong. 2002. In-vivo biocompatibility evaluation of stents coated with a new biodegradable elastomeric and functional polymer. *Coronary Artery Disease* 13(4): 237-241.

Lin, Shigang, Martin Sandig and Kibret Mequanint. 2011. Three-dimensional topography of synthetic scaffolds induces elastin synthesis by human coronary artery smooth muscle cells. *Tissue Engineering Part A* 17(11-12): 1561-1571.

Lv, Jiaolong, Huanli Sun, Yan Zou, Fenghua Meng, Aylvin A. Dias, Marc Hendriks, Jan Feijen and Zhiyuan Zhong. 2015. Reductively degradable α-amino acid-based poly (ester amide)-graft-galactose copolymers: Facile synthesis, self-assembly, and hepatoma-targeting doxorubicin delivery. *Biomaterials Science* 3(7): 1134-1146.

Mann, Brenda K., Rachael H. Schmedlen and Jennifer L. West. 2001. Tethered-TGF-β increases extracellular matrix production of vascular smooth muscle cells. *Biomaterials* 22(5): 439-444.

Mayer, Bernd (ed.). 2012. *Nitric oxide*. Vol. 143. Springer Science & Business Media.

Mejia, José S. and Elizabeth R. Gillies. 2013. Triggered degradation of poly (ester amide)s via cyclization of pendant functional groups of amino acid monomers. *Polymer Chemistry* 4(6): 1969-1982.

Memanishvili, Tamar, Nino Zavradashvili, Nino Kupatadze, David Tugushi, Marekh Gverdtsiteli, Vladimir P. Torchilin, Christine Wandrey, Lucia Baldi, Sagar S. Manoli and Ramaz Katsarava. 2014. Arginine-based biodegradable ether-ester polymers with low cytotoxicity as potential genearriers. *Biomacromolecules* 15(8): 2839-2848.

Meng, Fenghua, Wim E. Hennink and Zhiyuan Zhong. 2009. Reduction-sensitive polymers and bioconjugates for biomedical applications. *Biomaterials* 30(12): 2180-2198.

Morgan, Paul W. 1965. Condensation Polymers: By Interfacial and Solution Methods. Vol. 10. Interscience Publishers.

Moroson, Harold and Marvin Rotman. 1975. Biomedical applications of polycations. *Polyelectrolytes and their Applications* 187-195. Springer, Dordrecht.

Ochkhikidze, Natia, Giorgi Titvinidze, Marekh Gverdtsiteli, Giuli Otinashvili, David Tugushi and Ramaz Katsarava. 2020. New polymer synthesis via O,O′-diacyl-bis-glycolic acids: AABB-polydepsi-peptides, poly(ester amide)s and functional polymers. *Journal of Macromolecular Science, Part A* (in Press).

Okada, Masahiko, Masashi Yamada, Makito Yokoe and Keigo Aoi. 2001. Biodegradable polymers based on renewable resources. V. Synthesis and biodegradation behavior of poly (ester amide)s composed of 1,4:3,6-dianhydro-d-glucitol, α-amino acid, and aliphatic dicarboxylic acid units. *Journal of Applied Polymer Science* 81(11): 2721-2734.

Ou, Junhai, Timothy M. Carlos, Simon C. Watkins, Joseph E. Saavedra, Larry K. Keefer, Young-Myeong Kim, Brian G. Harbrecht and Timothy R. Billiar. 1997. Differential effects of nonselective nitric oxide synthase (NOS) and selective inducible NOS inhibition on hepatic necrosis, apoptosis, ICAM-1 expression, and neutrophil accumulation during endotoxemia. *Nitric Oxide* 1(5): 404-416.

Pang, Xuan and Chih-Chang Chu. 2010. Synthesis, characterization and biodegradation of poly (ester amide)s based hydrogels. *Polymer* 51(18): 4200-4210.

Pang, Xuan, Jun Wu, Chih-Chang Chu and Xuesi Chen. 2014. Development of an arginine-based cationic hydrogel platform: Synthesis, characterization and biomedical applications. *Acta Biomaterialia* 10(7): 3098-3107.

Pang, Xuan, Jun Wu, Cynthia Reinhart-King and Chih-Chang Chu. 2010. Synthesis and characterization of functionalized water soluble cationic poly (ester amide)s. *Journal of Polymer Science Part A: Polymer Chemistry* 48(17): 3758-3766.

Peterson, Gregory I., Michael B. Larsen and Andrew J. Boydston. 2012. Controlled depolymerization: Stimuli-responsive self-immolative polymers. *Macromolecules* 45(18): 7317-7328.

Powell, David A., Philip D. Ramsden and Robert A. Batey. 2003. Phase-transfer-catalyzed alkylation of guanidines by alkyl halides under biphasic conditions: A convenient protocol for the synthesis of highly functionalized guanidines. *The Journal of Organic Chemistry* 68(6): 2300-2309.

Prasad, Raj Nandan and A.F. McKay. 1967. Acylation of guanidines and guanylhydrazones. *Canadian Journal of Chemistry* 45(19): 2247-2252.

Rahman, Masoud, Sophie Laurent, Nancy Tawil, L'Hocine Yahia and Morteza Mahmoudi. 2013. Protein corona: Applications and challenges. pp. 45-63. *In*: Boris Martinac (ed.). Protein-Nanoparticle Interactions. Springer, Berlin, Heidelberg.

Rodriguez-Galan, Alfonso, Lourdes Franco and Jordi Puiggali. 2011a. Degradable poly (ester amide)s for biomedical applications. *Polymers* 3(1): 65-99.

Rodriguez-Galan, Alfonso, Lourdes Franco and Jordi Puiggali. 2011b. Biodegradable poly(ester amide)s: synthesis and applications. pp. 207-272. *In*: Felton (ed.). Biodegradable Polymers: Processing, Degradation. Hauppauge. GP NOVA Science Publisher, New York.

Sahoo, Bankanidhi, Mithun Goswami, Suman Nag and Sudipta Maiti. 2007. Spontaneous formation of a protein corona prevents the loss of quantum dot fluorescence in physiological buffers. *Chemical Physics Letters* 445(4-6): 217-220.

Samal, Sangram Keshari, Mamoni Dash, Sandra Van Vlierberghe, David L. Kaplan, Emo Chiellini, Clemens Van Blitterswijk, Lorenzo Moroni and Peter Dubruel. 2012. Cationic polymers and their therapeutic potential. *Chemical Society Reviews* 41(21): 7147-7194.

Stegemann, Jan P. and Robert M. Nerem. 2003. Phenotype modulation in vascular tissue engineering using biochemical and mechanical stimulation. *Annals of Biomedical Engineering* 31(4): 391-402.

Sun, Huanli, Fenghua Meng, Aylvin A. Dias, Marc Hendriks, Jan Feijen and Zhiyuan Zhong. 2011. α-amino acid containing degradable polymers as functional biomaterials: Rational design, synthetic pathway, and biomedical applications. *Biomacromolecules* 12(6): 1937-1955.

Sun, Huanli, Ru Cheng, Chao Deng, Fenghua Meng, Aylvin A. Dias, Marc Hendriks, Jan Feijen and Zhiyuan Zhong. 2015. Enzymatically and reductively degradable α-amino acid-based poly (ester amide)s: Synthesis, cell compatibility, and intracellular anticancer drug delivery. *Biomacromolecules* 16(2): 597-605.

Tang, Liping, Paul Thevenot and Wenjing Hu. 2008. Surface chemistry influences implant biocompatibility. *Current Topics in Medicinal Chemistry* 8(4): 270-280.

Trachootham, Dunyaporn, Weiqin Lu, Marcia A. Ogasawara, Nilsa Rivera-Del Valle and Peng Huang. 2008. Redox regulation of cell survival. *Antioxidants & redox signaling* 10(8): 1343-1374.

Trachootham, Dunyaporn, Jerome Alexandre and Peng Huang. 2009. Targeting cancer cells by ROS-mediated mechanisms: A radical therapeutic approach? *Nature Reviews Drug Discovery* 8(7): 579-591.

Wang, Peng George, Tingwei Bill Cai and Naoyuki Taniguchi (eds.). 2005. Nitric Oxide Donors: For Pharmaceutical and Biological Applications. John Wiley & Sons.

Wang, Shengchao and Yue Hu. 2018. α7 nicotinic acetylcholine receptors in lung cancer. *Oncology Letters* 16(2): 1375-1382.

Wu, Jun, Martha A.Mutschler and Chih-Chang Chu. 2011. Synthesis and characterization of ionic charged water soluble arginine-based poly (ester amide). *Journal of Materials Science: Materials in Medicine* 22(3): 469-479.

Wu, Jun, Dequn Wu, Martha A. Mutschler and Chih-Chang Chu. 2012. Cationic hybrid hydrogels from amino-acid-based poly(ester amide): Fabrication, characterization, and biological properties. *Advanced Functional Materials* 22(18): 3815-3823.

Yamanouchi, Dai, Jun Wu, Andrew N. Lazar, K. Craig Kent, Chih-Chang Chu and Bo Liu. 2008. Biodegradable arginine-based poly (ester-amide)s as non-viral gene delivery reagents. *Biomaterials* 29(22): 3269-3277.

Zavradashvili, Nino, Giuli Jokhadze, Tamar Kviria and Ramaz Katsarava. 2008. Thermally- and photo-chemically curable biodegradable poly (ester amide)s with double bond moieties in lateral chains. pp. 173-179. *In*: Lekishvili, N., Zaikov, G.E. (eds.). Chemistry of Advanced Compounds and Materials. NOVA Science Publishers, Inc.

Zavradashvili, Nino, Giuli Jokhadze, Marekh Gverdtsiteli, Giuli Otinashvili, Nino Kupatadze, Zaza Gomurashvili, David Tugushi and Ramaz Katsarava. 2013. Amino acid based epoxy-poly (ester amide)s – A new class of functional biodegradable polymers: Synthesis and chemical transformations. *Journal of Macromolecular Science, Part A* 50(5): 449-465.

Zavradashvili, Nino, Tamar Memanishvili, Nino Kupatadze, Lucia Baldi, Xiao Shen, David Tugushi, Christine Wandrey and Ramaz Katsarava. 2014. Cell compatible arginine containing cationic polymer: One-pot synthesis and preliminary biological assessment. *Infectious Diseases and Nanomedicine I* 59-73. Springer, New Delhi.

Zavradashvili, Nino, Can Sarisozen, Giorgi Titvinidze, Giuli Otinashvili, Tengiz Kantaria, Davit Tugushi, Jordi Puiggali, Vladimir P. Torchilin and Ramaz Katsarava. 2019a. Library of cationic polymers composed of polyamines and arginine as gene transfection agents. *ACS Omega* 4(1): 2090-2101.

Zavradashvili, Nino, Giuli Otinashvili, Temur Kantaria, David Tugushi, Ashot Saghyan, Anna Mkrtchyan, Sergey Poghosyan and Ramaz Katsarava. 2019b. New cationic polymers composed of non-proteinogenic α-amino acids. 6th International Caucasian Symposium on Polymers & Advanced Materials, July 17-20, 2019, Batumi, Georgia.

Zavradashvili, Nino, Jordi Puiggali, Ramaz Katsarava. 2020. Artificial polymers made of α-amino acids – Poly(amino acid)s, pseudo-poly(amino acid)s, poly(depsipeptide)s, and pseudo-proteins. *Current Pharmaceutical Design* 26(5): 566-593.

Zavradashvili, Nino, Jordi Puiggali, L.J. del Valle and Ramaz Katsarava (unpublished data).

Zhang, Hua, Kimy Yeung, Jessica S. Robbins, Ryan A. Pavlick, Meng Wu, Ran Liu, Ayusman Sen and Scott T. Phillips. 2012. Self-powered microscale pumps based on analyte-initiated depolymerization reactions. *Angewandte Chemie International Edition* 51(10): 2400-2404.

Zilinskas, Gregory J., Abdolrasoul Soleimani and Elizabeth R. Gillies. 2012. Poly (ester amide)-poly (ethylene oxide) graft copolymers: Towards micellar drug delivery vehicles. *International Journal of Polymer Science*. Article ID 564348: 11, doi: 10.1155/2012/564348

CHAPTER

6

Functional Proteins

Keya Ganguly and Ki-Taek Lim*
Department of Biosystems Engineering, Kangwon National University, Chuncheon 24341, Republic of Korea

1. Introduction

The development of biomaterials with desirable traits is essential for the treatment of the ever-growing number of complicated human diseases (Courtney et al. 1993, Ivanova et al. 2014). Proteins and peptides serve as therapeutic sources for medical applications (Seo et al. 2012). However, their stability, bioavailability, and immunogenic responses are some of the significant challenges in biomedicine (Zinzalla and Thurston 2009). A widely adopted technique to overcome this challenge is the development of functionalized proteins, mainly by conjugation with a suitable polymer. Proteins and peptides have also been used for the functionalization of polymers for biomedical applications (Cobo et al. 2015). Functionalized protein-polymer complexes have been in use since the beginning of the 21st century as polymer therapeutics. These include polymer-protein conjugates (Gauthier and Klok 2010), polymer-drug conjugates (Khandare and Minko 2006), and polymeric micelles (Kwon and Okano 1996). Numerous types of polymeric structures have found application in biomedicine, including multivalent polymers (van Dongen et al. 2014), branched polymers (Roovers 1985), graft polymers (Ratner 1980), dendrimers (Vögtle et al. 2000), dendronized polymers (A. Zhang et al. 2003), block copolymers (Hamley and Hamley 1998), and peptide derivatives (Sakakibara et al. 2000). The most widely and commercially used polymers in the initial therapeutic approaches were poly(ethylene glycol)-protein conjugates (Alconcel et al. 2011). The conjugation of proteins or peptides with synthetic polymers offers several benefits in the resulting structures. A protein/peptide-polymer conjugate can synergistically enhance the necessary traits of both the materials (Messina et al. 2019). These conjugates can also overcome the

*Corresponding author: ktlim@kangwon.ac.kr

shortcomings of either of the materials. Conjugation of protein to a polymer can stabilize or even enhance their biological functions, such as enzymatic activity and receptor recognition (Vandermeulen and Klok 2004). Moreover, the polymer shields the degradation of conjugated protein by proteolytic enzymes. An increase in the solubility and targeted localization of the protein, as a result of its conjugation, is also a desirable property (Veronese and Morpurgo 1999). Improving the pharmacokinetics of a therapeutic protein by its conjugation to a polymer through PEGylation is well known to significantly elongate the protein's half-life and subsequently extend its proper circulation in the body and efficient bioactivity (Harris et al. 2001). An advanced class of functionalized proteins generated by their conjugation with stimuli-responsive polymers is being synthesized intensively because of the similarity of these bioconjugates to mimic biomolecules. Proteins functionalized with stimuli-responsive polymers respond to environmental stimuli such as pH and temperature, resulting in controlled reversible changes in their physicochemical properties. Consequently, protein-polymer conjugates play a substantial role in many aspects of biotechnology; however, their importance is most clearly highlighted in therapeutic formulations. The highest growth area for pharmaceutical research and development is the development of biologics (Cobo et al. 2015).

The most widely used polymer has been poly(-ethylene glycol) (PEG), which has resulted in numerous commercially available drugs approved by the U.S. Food and Drug Administration (Vo et al. 2015). However, anti-PEG antibodies have arisen in a significant subset of the population as a result of its prevalent use (Chapman 2002, Gaberc-Porekar et al. 2008). These antibodies challenge the therapeutic effectiveness of the pharmaceutical formulations concerned. Therefore, several other polymers based on polyacrylates, polynorbornenes, and polyglycerol have emerged to overcome the anti-PEG antibody effect. The physicochemical properties of these polymers play a crucial role in the ability to prevent proteins from triggering immune responses without altering their natural function. This chapter briefly describes the strategies of protein functionalization and some of its biological as well as non-biological applications (Pokorski and Hore 2019).

2. Methods of protein-polymer conjugation

PEGylation was the extensively adopted method of protein-polymer conjugation (Abuchowski et al. 1977) until the more modern technique of living radical polymerization or LRP (also called controlled radical polymerization, CRP) was developed (Pelegri-O'Day et al. 2014). The CRP technique of polymer synthesis is further divided into three fundamental techniques: reversible addition/fragmentation chain transfer polymerization (RAFT), atom transfer radical polymerization (ATRP), and nitroxide-mediated polymerization (NMP). Advancements in these synthetic polymerization techniques promoted the development of high-performance protein-polymer conjugates because they could accurately regulate the physical properties and chemical functionalities of the polymer to achieve *in vivo* compatibility (Grajales et al. 2012). The two most widely used methods for the preparation of protein-polymer conjugates are the convergent (or grafting-to) and the divergent (including grafting-from and grafting-through) strategies (Dehn et al. 2011, Meißig et al. 2016). An example of protein-polymer conjugate formation is schematically represented in Fig. 1.

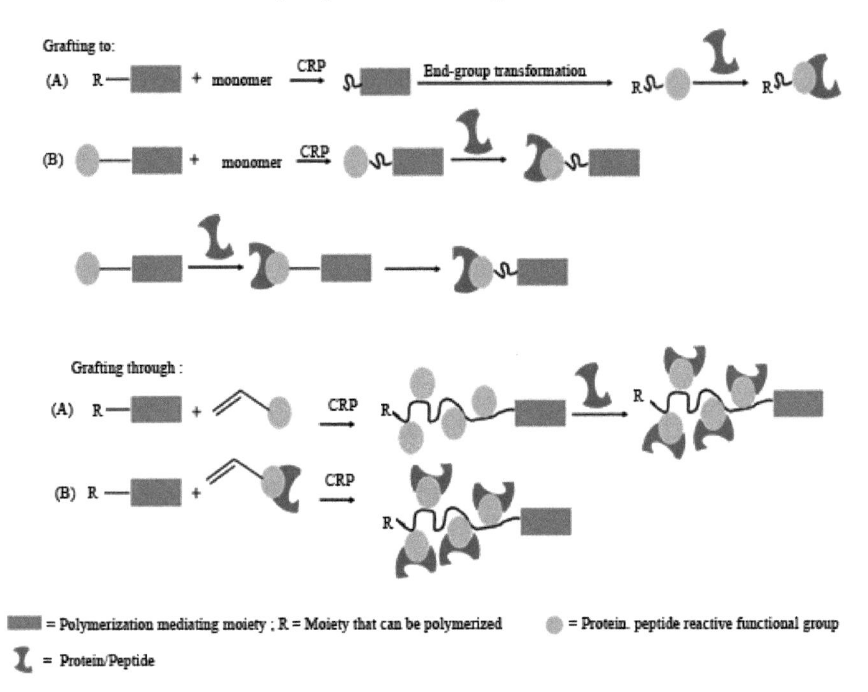

Fig. 1. Schematic representation of grafting-to, grafting-from, and grafting-through strategies of protein-polymer conjugation

2.1. Convergent strategy

Grafting-to approach

In the grafting-to approach, the desired polymer is first synthesized with high α- or ω-chain end-group. A protein can then be attached in several ways, including covalent attachment of a reactive functional group of the polymer to the target amino acid side chain or vice-versa. A ligand-apoprotein interaction strategy is also applied in several instances (Minko et al. 2002). Naturally available functional groups on protein surfaces, as well as the introduction of unnatural amino acids, play a significant role in conjugating proteins and polymers through the grafting-to strategy. The amino acid residue strongly determines the ligation strategies of the protein-polymer conjugation at the attachment site (Stephanopoulos and Francis 2011). The commonly used amino acid residues as attachment sites of proteins are lysine, cysteine, tyrosine, glutamine, tryptophan, histidine, aspartic acid, and glutamic acid, based on their intrinsic biochemical properties. The specific physiochemical traits of individual amino acids and the chemical properties of the polymers strongly determine the strategies for the synthesis of protein-polymer conjugates. For a grafting-to approach, nucleophilic amines present at the N-terminus and in the side chains of lysine residues are much exploited for their reactivity towards functional groups such as carboxylic acids through carbodiimide coupling, activated esters or thiazoline-2 thione groups through nucleophilic substitution, and isocyanates by nucleophilic addition. Besides, the thiol group of cysteine residue is preferred for site-specific modifications. Thiol groups usually undergo disulfide bond metathesis with pyridyl disulfide functional polymers to obtain reductively labile protein/peptide conjugates (Bontempo et al. 2004). Unnatural amino acids have also been utilized to achieve the site-selective ligation of proteins and polymers (Connor and Tirrell 2007) (Fig. 2).

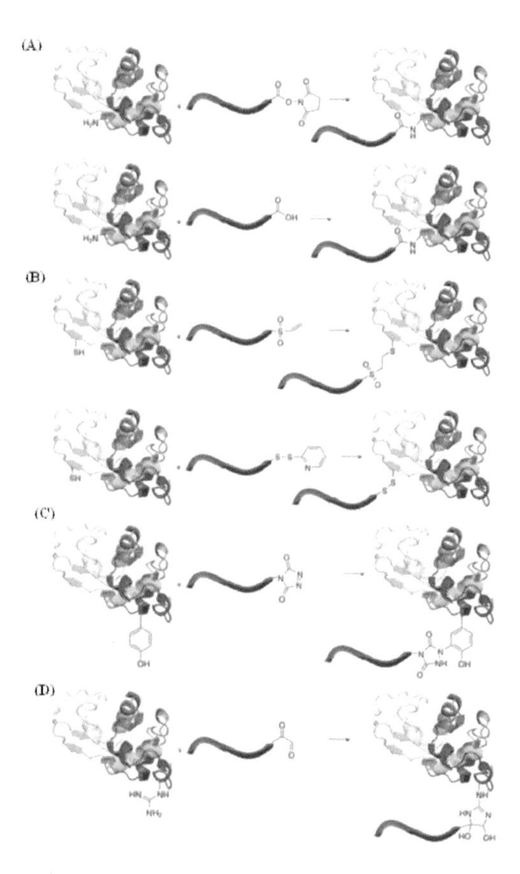

Fig. 2. Chemistry of protein-polymer conjugation at indicated lysine residues (A), cysteine residues (B), tyrosine residues (C), and arginine residues (D) with the amino-terminus.

2.2. Divergent strategy

Grafting-from approach

The grafting-from approach is an indirect method of protein-polymer conjugation. In this case, an amino acid side chain is introduced with a moiety that can initiate the polymerization process. This results in the formation of a macro-initiator, which acts as a nucleation site for the polymer chain to grow (B.S. Sumerlin 2012). RAFT polymerization has proved an ideal synthetic process for the divergent strategy of protein-polymer conjugation (Bulmus 2011). This is mainly because of the ability of the monomers to polymerize without the need for a catalyst, which eliminates the need for further purification after the polymerization reaction. However, the RAFT approach does need the use of an external radical source and a stable chain transfer agent (CTA) (Gody et al. 2015, Lowe and McCormick 2007). The immobilization of CTA on the target protein or peptide is mandatory to implement the grafting-from approach using RAFT polymerization. The immobilization of CTA on protein or peptide is done via R-group or Z-group (B. Sumerlin 2008). When R-group is used, the protein or peptide acts as the initiator, while the polymeric chain remains attached to the protein/peptide throughout the entire process. The conjugates synthesized by this method are more stable; however, termination of the propagating radical produces dead chains attached to the biomolecules in which further polymerization can not be reinitiated. If CTA is attached via the Z-group, the growth of the polymer is favored in the solution and attaches to the protein/peptide carrying the CTA only during the chain transfer event (Gody et al. 2014).

Grafting-through approach

In a grafting-through approach, several protein reactive groups are incorporated within a growing polymer chain by the use of monomeric units that react with the peptides or proteins before or after their polymerization. In this approach, a polymer chain can bind to more than one peptide/protein (Wright 2019). For a grafting-through approach, monomers are modified with desirable functionality, including alkyne, epoxide, aldehyde/ketone, and oxime modifications (Carter et al. 2016).

Some non-covalent protein-polymer conjugation modifications use the interactions between proteins and natural or synthetic ligands incorporated into the polymer (Murata et al. 2020). The most commonly used natural ligands include sugars that can bind to lectins or biotin and have an affinity for streptavidin. The synthetic ligands mostly exploit electrostatic interactions, metal-ligand affinity, and host-guest interactions to bind proteins to polymers.

3. Applications of functionalized proteins

Polymer-protein conjugates find several applications as polymeric therapeutics, polymer conjugates, amphiphiles, and other products. A few of the applications of protein-polymer conjugates are mentioned in this section.

3.1. Biological applications

Among the numerous biological applications of protein-polymer conjugates, polymer-templated protein nanoballs (PTPNBs) synthesized from nickel-complexed nitrilotriacetic acid (NTA)-end-functionalized polycaprolactone with His-GFP have been significant in the drug delivery of the pharmaceutically important drug doxorubicin (DOX) (Jose et al. 2020). Moreover, the peptide-polymer hybrid of mannosylated cell-penetrating peptides (CPPs) and low-molecular-weight polyethyleneimine (LPEI) serves as an efficient hybrid vector for the delivery of pTRAIL (plasmid tumor necrosis factor-related apoptosis-induced ligand) for the treatment of colon cancer (Pan et al. 2017). Additionally, protein functionalized polymer brushes have been used in the development of protein microarrays. Polymer brushes of poly[oligo (ethylene glycol) methacrylate] POEGMA functionalized with O^6-alkylguanine-DNA-alkyltransferase (AGT) fusion proteins have been prepared via surface-initiated ATRP for the development of protein microarrays. The AGT-mediated immobilization serves as the principle mechanism of selective functionalization of the polymer brushes with proteins at specific orientations, and the surface density of the proteins could be controlled by varying the amount of AGT substrate O^6-benzylguanine (BG) present on the surface of the brushes (Tugulu et al. 2005). Receptor-like synthetic materials such as molecularly imprinted polymers (MIP) are a great alternative to antibody-mediated immunoassays. The copolymerization of a monomer prepares MIPs with a crosslinker in the presence of a removable template molecule, and leaves cavities in the polymer matrix complementary in size, shape, and chemical functionality to the template molecule. The pioneering work on molecular imprinting was the preparation of polymer scaffolds with the cavities decorated with functional groups that acted as synthetic receptors (Wulff et al. 1973). Carbon nanotubes are useful supports for the surface imprinting of proteins. A protein molecularly imprinted membrane, multi-walled carbon nanotubes (PMIM/MWCNTs), was synthesized with bovine serum albumin (BSA) as the template molecules. The conjugate was reported to have a 2.6-fold higher affinity for BSA compared to the non-imprinted control set (M. Zhang et al. 2010). An artificial antibody for troponin T (TnT) was synthesized on the surface of MWCNT. The sensitivity of this biomaterial was assessed by using it as an electroactive compound in a PVC/plasticizer mixture, coating a wire of gold, silver and titanium. Among these, the gold wire coated with the MI-based membrane produced an anionic slope of 50 mV $decade^{-1}$ when dipped in HEPES buffer of pH 7. The limit of sensitivity was found to be 0.16 μg/ml (Moreira et al. 2011). A few widely used protein-polymer conjugates in clinical development are presented in Table 1.

Table 1. Polymer-protein conjugates in clinical development (Ekladious et al. 2019)

Name (company)	*Polymer carrier*	*Drug*	*Indication(s)*	*Stage (clinical trials, gov. identifier)*	*References*
Turoctocog alfa pegol (Novo Nordisk)	PEG	Factor VIII	Hemophilia A	Pre-registration (NCT01480180)	Murata et al. 2020
Calaspargase pegol (Shire)	PEG	Asparaginase	Acute lymphoblastic leukemia and lymphoblastic lymphoma	Pre-registration (NCT01574274)	Angiolillo et al. 2014
Elapegademase (Leadiant Biosciences)	PEG	ADA	ADA-SCID	Pre-registration (NCT01420627)	Bowser et al. 2017
Pegvorhyaluronidase alfa (Halozyme Therapeutics)	PEG	Hyaluronidase	Pancreatic cancer	Phase III (NCT02715804)	Hingorani et al. 2018
TransCon Growth Hormone (Ascendis Pharma)	PEG	Human growth hormone	Growth hormone deficiency	Phase III (NCT03344458)	Gilfoyle et al. 2018
Pegilodecakin (ARMO BioSciences)	PEG	IL-10	Pancreatic cancer	Phase III (NCT02923921)	Naing et al. 2016
Pegargiminase (Polaris Pharmaceuticals)	PEG	Arginine deiminase	Mesothelioma	Phase II/III (NCT02709512)	Szlosarek et al. 2017
BCT-100 (Bio-Cancer Treatment International)	PEG	Arginase 1	Acute myeloid leukemia	Phase II (NCT02899286)	Mussai et al. 2015
Pegsiticase (Selecta Biosciences)	PEG	Uricase	Chronic gout	Phase II (NCT02959918)	Azeem et al. 2018
Sanguinate (Prolong Pharmaceuticals)	PEG	Carboxyhemoglobin	Sickle cell disease	Phase II (NCT02411708)	Misra et al. 2014

Pegzilarginase (Aeglea BioTherapeutics)	PEG	Arginase 1	Arginase 1 deficiency	Phase II (NCT03378531)	Burrage et al. 2015
BMS-986036 (Bristol-Myers Squibb)	PEG	FGF21	Nonalcoholic steatohepatitis	Phase II (NCT03486899)	Sanyal et al. 2018
Dapirolizumab pegol (UCB Pharma)	PEG	Anti-CD40L Fab′	Systemic lupus erythematosus	Phase II (NCT02804763)	Chamberlain et al. 2017
Zimura (Ophthotech Corporation)	PEG	Aptamer complement C5 inhibitor	Neovascular age-related macular degeneration	Phase II (NCT03362190)	Drolet et al. 2016
NKTR-214 (Nektar Therapeutics)	PEG	IL-2	Solid tumors	Phase I/II (NCT02869295)	Charych et al. 2016
BIVV001 (Bioverativ Therapeutics)	XTEN	Recombinant factor VIII Fc–von Willebrand factor	Hemophilia A	Phase I/II (NCT03205163)	Seth Chhabra et al. 2015
Olaptesed pegol (NOXXON Pharma)	PEG	Anti-CXCL12 aptamer	Colorectal cancer and pancreatic cancer	Phase I/II (NCT03168139)	Ludwig et al. 2017
Fovista (Ophthotech Corporation)	PEG	Anti-PDGFB aptamer	Ocular von Hippel–Lindau syndrome	Phase I/II (NCT02859441)	Jaffe et al. 2016
BMS-986171 (Bristol-Myers Squibb)	PEG	FGF21	Nonalcoholic steatohepatitis	Phase I (NCT02538874)	Wu et al. 2016
NKTR-358 (Nektar Therapeutics)	PEG	IL-2	Systemic lupus erythematosus	Phase I (NCT03556007)	Langowski et al. 2017

Smart polymer conjugates are yet another class of conjugates that responds dramatically under various stimuli such as changes in pH, temperature, and light, thereby causing changes in the physical properties of the polymer and the associated biomolecule. This ability of smart polymers to respond to environmental changes is highly desirable in pharmaceutical applications, protein sensors, and catalysis. Polyethylene glycol monomethyl ether (mPEG) is one of the first polymers ever used for therapeutic proteins to change their drug-like characteristics. This polymer was selected for its hydrophilicity and it was one among the few defined semitelechelic polymers available. "PEGylation" of proteins is given considerable attention, and many PEGylated proteins are available clinically (Grigoletto et al. 2016, Jevševar et al. 2010). Various processes have incorporated peptides in hydrogels in, for example, tissue engineering (González-Henríquez et al. 2017, Malmsten 2011) or surgical wound dressings (Babavalian et al. 2015, Xie et al. 2015). Depending on the requirement, these approaches have a promising future for drug delivery systems. An interdependent relationship exists among the conjugates that can be optimized by tuning their physical properties (Sun et al. 2018, Wilson 2017). In one study, nisin was loaded into the dental adhesive 'single Bond 2' by nonspecific adsorption (de Arauz et al. 2009, Su et al. 2018). This showed an inhibitory effect against *Streptococcus mutans*, which was found to grow with increasing concentration of nisin. However, in the control formulation, no significant differences were noted. An example of drug delivery by protein-polymer conjugates is shown in Fig. 3.

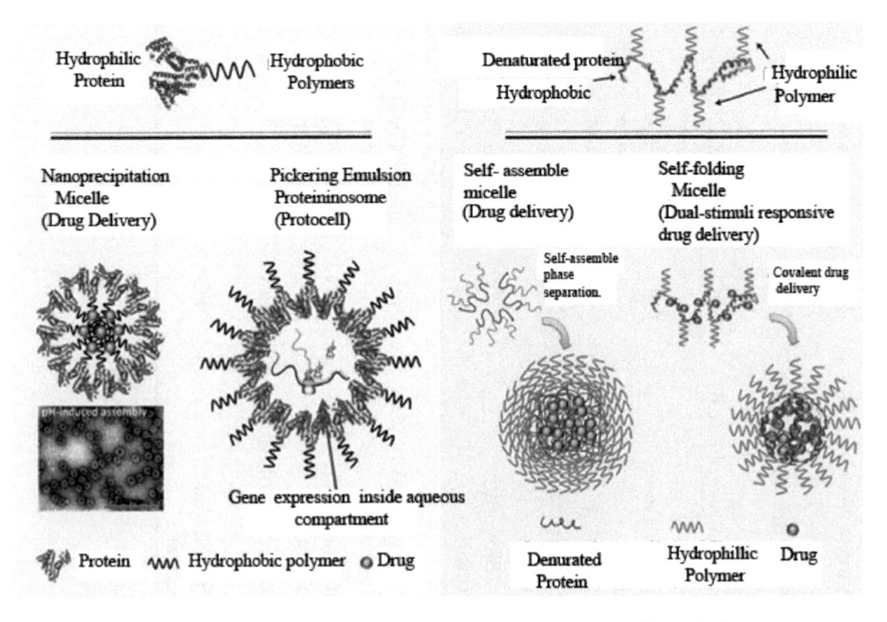

Fig. 3. Application of protein-polymer conjugates in drug delivery

3.2. Non-biological applications

Protein-polymer conjugates have numerous applications as catalyst in gas separation and optoelectronics. Cyclic peptide to polymer conjugates have been developed to generate porous thin films containing subnanometer channels that can act as selective membranes for separation and protective coatings (Xu et al. 2011). Highly stable nanotapes of polyethylene oxide (PEO)-peptide (Val-Thr-Val-Thr-dimethylGly)-PEG conjugate have been used for the formulation of well-ordered template-directed nucleation and growth of silver nanoparticles. This AgNP array proved to be a potential light trapping candidate owing to its unique optical and electronic properties. (Díez et al. 2010). Encapsulation of glucose oxidase (GOx) in horseradish peroxidase-polystyrene (HRP-PS) and horseradish peroxidase–poly(methyl methacrylate) (HRP-PMMA) amphiphile vesicles has been used in high-performance nanobioreactors for the cascade oxidation of glucose and 2,2′-azido-bis(3-ethylbenzothiazoline-6-sulfonic acid (ABTS) (Delaittre et al. 2009). An artificial cell made of liposomes embedded within a poly(dopamine) (PDA) carrier shell has been created using a multi-compartment encapsulation strategy. These capsosomes were the first ever compartmentalized systems used in encapsulated catalysis, performing numerous simultaneous reactions in parallel and proving to be suitable as biosensors and therapeutic carriers (Hosta-Rigau et al. 2014).

Table 2. summarizes a few other notable applications of functionalized proteins.

4. Perspectives and conclusion

Protein-polymer conjugates are an important biomaterial with biological and non-biological applications. Numerous approaches have been adopted to synthesize protein-polymer conjugates. A fundamental understanding of both the protein biochemistry and the chemistry of the polymers is crucial for the development of conjugates. Alteration in the chemical nature of the polymers, the active sites of the proteins, and various solvent systems used in the synthesis of the conjugates are useful in the advancement of protein to polymer conjugation. Several challenges still remain to be addressed.

Importantly, the bulk of the polymer-protein conjugates so far reported, including approved products, are synthesized with stable linkages. The primary concern, however, is the preservation of protein activity, which in general is significantly reduced after conjugation, because of steric hindrance provided by polymer strands. Site-directed conjugation most likely can be the best approach to avoid such limitation by targeting a protein surface area far from the activity sites. There were several attempts to use the prodrugs concept to make polymer-protein conjugates in which trigger-based polymer chain *in vivo* release is intended. In this approach, native protein can regain its activity even if the conjugate loses starting protein activity. Thus, the number of sites of polymer chain coupling is not crucial. The most critical

Table 2. Some notable applications of functionalized proteins

Protein-polymer conjugate	*Technique*	*Conjugate property*	*Application*	*References*
Bovine serum albumin-poly (*N*-isopropylacrylamide)	Grafting-from approach, R-group/ RAFT	Thermal responsiveness	Modulation of bioactivity	De et al. 2008
Protein A-poly(N-isopropylacrylamide)	Co-polymerization	Affinity precipitation separation	Separation of human immunogammaglobulin	Chen and Huffman 1990
Biodegradable protein-poly (ε-caprolactone)	Grafting-from; enzymatic ring opening polymerization (eROP)	Amphiphilic graft copolymers	Time-dependent therapeutic application	Bao et al. 2020b
Lipase-polyacrylamide conjugates	Atom transfer radical polymerization (ATRP)	Formation of spherical nanoparticles based on polymerization-induced self-assembly in aqueous medium	Protein delivery and enzyme immobilization	Bao et al. 2020a
Cytochrome c- polyethylene glycol	PEGylation	Acid-responsive nanoparticles	Drug delivery	Steiert et al. 2020
Amyloidogenic Alzheimer peptide, $A\beta_{1\text{-}40}$- poly(oligo(ethylene glycol)$_m$ acrylates)	Resin-based synthesis, solution-based coupling chemistries, RAFT	Enhanced aggregation of protein	Development of effective strategies for purification of the conjugate	Evgrafova et al. 2020
Functionalized bovine serum albumin (BSA)- phenylboronic acid functionalized poly(N-isopropylacrylamide) (PNIPAAm)	RAFT	pH-sensitive protein-based nanoconjugate	Drug delivery for cancer therapy	Zhou et al. 2019
Lipase–palladium (Pd) nanohybrids	Covalent linkage via Schiff base reaction	Catalytic activity	Construction of enzyme–metal hybrid catalysts	Li et al. 2019

part is fine-tuning of protein kinetics and conjugate clearance for developing a polymer-protein prodrug. We envision that polymer-protein conjugates of site-specific conjugation will be accompanied by further refinement of controlling stability and lability of critical linkages. The future progress of these conjugates will rely significantly on the advancement in synthetic polymers and the understanding of the biochemical aspects of proteins and diseases.

Acknowledgements

This work was supported by the Basic Science Research Program through the National Research Foundation of Korea (NRF) funded by the Ministry of Education (NRF-2018R1A6A1A03025582) and the National Research Foundation of Korea (NRF-2019R1D1A3A03103828).

References

Abuchowski, A., T. Van Es, N. Palczuk and F. Davis. 1977. Alteration of immunological properties of bovine serum albumin by covalent attachment of polyethylene glycol. *Journal of Biological Chemistry* 252(11): 3578-3581.

Alconcel, S.N., A.S. Baas and H.D. Maynard. 2011. FDA-approved poly (ethylene glycol)-protein conjugate drugs. *Polymer Chemistry* 2(7): 1442-1448.

Angiolillo, A.L., R.J. Schore, M. Devidas, M.J. Borowitz, A.J. Carroll, J.M. Gastier-Foster, N.A. Heerema, T. Keilani, A.R. Lane, M.L. Loh and G.H. Reaman. 2014. Pharmacokinetic and pharmacodynamic properties of calaspargase pegol Escherichia coli L-asparaginase in the treatment of patients with acute lymphoblastic leukemia: Results from children's oncology group study AALL07P4. *Journal of Clinical Oncology* 32(34): 3874.

Azeem, R., E. Sands, L. Johnston, W. DeHaan, A.J. Kivitz, T.K. Kishimoto, J. Park and S. Nicolaou. 2018. Initial Phase 2 Clinical Data of SEL-212 in Symptomatic Gout Patients: Measurement of Dissolution of Urate Deposits Associated with Monthly Dosing of a Pegylated Uricase (Pegadricase) with SVP-Rapamycin by Dual Energy Computed Tomography (abstract). Paper presented at the Arthritis & Rheumatology. 70(suppl 10). https://acrabstracts.org/abstract/initial-phase-2-clinical-data-of-sel-212-in-symptomatic-gout-patients-measurement-of-dissolution-of-urate-deposits-associated-with-monthly-dosing-of-a-pegylated-uricase-pegadricase-with-svp-rapamyc/

Babavalian, H., A.M. Latifi, M.A. Shokrgozar, S. Bonakdar, S. Mohammadi and M.M. Moghaddam. 2015. Analysis of healing effect of alginate sulfate hydrogel dressing containing antimicrobial peptide on wound infection caused by methicillin-resistant Staphylococcus aureus. *Jundishapur Journal of Microbiology* 8(9).

Bao, C., J. Chen, D. Li, A. Zhang and Q. Zhang. 2020a. Synthesis of lipase-polymer conjugates by Cu (0)-mediated reversible deactivation radical polymerization: Polymerization vs degradation. *Polymer Chemistry* 11(7): 1386-1392.

Bao, C., X.-L. Xu, J. Chen and Q. Zhang. 2020b. Synthesis of biodegradable protein-poly (ε-caprolactone) conjugates via enzymatic ring opening polymerization. *Polymer Chemistry* 11(3): 682-686.

Bontempo, D., K.L. Heredia, B.A. Fish and H.D. Maynard. 2004. Cysteine-reactive polymers synthesized by atom transfer radical polymerization for conjugation to proteins. *Journal of the American Chemical Society* 126(47): 15372-15373.

Bowser, J.L., J.W. Lee, X. Yuan and H.K. Eltzschig. 2017. The hypoxia-adenosine link during inflammation. *Journal of Applied Physiology* 123(5): 1303-1320.

Bulmus, V. 2011. RAFT polymerization mediated bioconjugation strategies. *Polymer Chemistry* 2(7): 1463-1472.

Burrage, L.C., Q. Sun, S.H. Elsea, M.-M. Jiang, S.C. Nagamani, A.E. Frankel, E. Stone, S.E. Alters, D.E. Johnson, S.W. Rowlinson and G. Georgiou. 2015. Human recombinant arginase enzyme reduces plasma arginine in mouse models of arginase deficiency. *Human Molecular Genetics* 24(22): 6417-6427.

Carter, N.A., X. Geng and T.Z. Grove. 2016. Design of self-assembling protein-polymer conjugates. *Protein-based Engineered Nanostructures* 179-214. Springer.

Chamberlain, C., P.J. Colman, A.M. Ranger, L.C. Burkly, G.I. Johnston, C. Otoul, C. Stach, M. Zamacona, T. Dörner, M. Urowitz and F. Hiepe. 2017. Repeated administration of dapirolizumab pegol in a randomised phase I study is well tolerated and accompanied by improvements in several composite measures of systemic lupus erythematosus disease activity and changes in whole blood transcriptomic profiles. *Annals of the Rheumatic Diseases* 76(11): 1837-1844.

Chapman, A.P. 2002. PEGylated antibodies and antibody fragments for improved therapy: A review. *Advanced Drug Delivery Reviews* 54(4): 531-545.

Charych, D.H., U. Hoch, J.L. Langowski, S.R. Lee, M.K. Addepalli, P.B. Kirk, D. Sheng, X. Liu, P.W. Sims, L.A. VanderVeen and C.F. Ali. 2016. NKTR-214, an engineered cytokine with biased IL2 receptor binding, increased tumor exposure, and marked efficacy in mouse tumor models. *Clinical Cancer Research* 22(3): 680-690.

Chen, J.P. and A.S. Huffman. 1990. Polymer-protein conjugates: II. Affinity precipitation separation of human immunogammaglobulin by a poly (N-isopropylacrylamide)-protein: A conjugate. *Biomaterials* 11(9): 631-634.

Cobo, I., M. Li, B.S. Sumerlin and S. Perrier. 2015. Smart hybrid materials by conjugation of responsive polymers to biomacromolecules. *Nature Materials* 14(2): 143-159.

Connor, R.E. and D.A. Tirrell. 2007. Non-canonical amino acids in protein polymer design. *Journal of Macromolecular Science, Part C: Polymer Reviews* 47(1): 9-28.

Courtney, J., L. Irvine, C. Jones, S. Mosa, L. Robertson and S. Srivastava. 1993. Biomaterials in medicine – A bioengineering perspective. *The International Journal of Artificial Organs* 16(3): 164-171.

de Arauz, L.J., A.F. Jozala, P.G. Mazzola and T.C.V. Penna. 2009. Nisin biotechnological production and application: A review. *Trends in Food Science & Technology* 20(3-4): 146-154.

De, P., M. Li, S.R. Gondi and B.S. Sumerlin. 2008. Temperature-regulated activity of responsive polymer-protein conjugates prepared by grafting - from via RAFT polymerization. *Journal of the American Chemical Society* 130(34): 11288-11289.

Dehn, S., R. Chapman, K.A. Jolliffe and S. Perrier. 2011. Synthetic strategies for the design of peptide/polymer conjugates. *Polymer Reviews* 51(2): 214-234.

Delaittre, G., I.C. Reynhout, J.J. Cornelissen and R.J. Nolte. 2009. Cascade reactions in an all-enzyme nanoreactor. *Chemistry – A European Journal* 15(46): 12600-12603.

Díez, I., H. Hahn, O. Ikkala, H.G. Börner and R.H. Ras. 2010. Controlled growth of silver nanoparticle arrays guided by a self-assembled polymer-peptide conjugate. *Soft Matter* 6(14): 3160-3162.

Drolet, D.W., L.S. Green, L. Gold and N. Janjic. 2016. Fit for the eye: Aptamers in ocular disorders. *Nucleic Acid Therapeutics* 26(3): 127-146.
Ekladious, I., Y.L. Colson and M.W. Grinstaff. (2019). Polymer–drug conjugate therapeutics: Advances, insights and prospects. *Nature Reviews Drug Discovery* 18(4): 273-294.
Evgrafova, Z., S. Rothemund, B. Voigt, G. Hause, J. Balbach and W.H. Binder. 2020. Synthesis and aggregation of polymer-amyloid β conjugates. *Macromolecular Rapid Communications* 41(1): 1900378.
Gaberc-Porekar, V., I. Zore, B. Podobnik and V. Menart. 2008. Obstacles and pitfalls in the PEGylation of therapeutic proteins. *Current Opinion in Drug Discovery and Development* 11(2): 242.
Gauthier, M.A. and H.-A. Klok. 2010. Polymer-protein conjugates: An enzymatic activity perspective. *Polymer Chemistry* 1(9): 1352-1373.
Gilfoyle, D., E. Mortensen, E.D. Christoffersen, J.A. Leff and M. Beckert. 2018. A first-in-man phase 1 trial for long-acting TransCon Growth Hormone. *Growth Hormone & IGF Research* 39, 34-39.
Gody, G., R. Barbey, M. Danial and S. Perrier. 2015. Ultrafast RAFT polymerization: Multiblock copolymers within minutes. *Polymer Chemistry* 6(9): 1502-1511.
Gody, G., T. Maschmeyer, P.B. Zetterlund and S.b. Perrier. 2014. Exploitation of the degenerative transfer mechanism in RAFT polymerization for synthesis of polymer of high livingness at full monomer conversion. *Macromolecules* 47(2): 639-649.
González-Henríquez, C.M., M.A. Sarabia-Vallejos and J. Rodriguez-Hernandez. 2017. Advances in the fabrication of antimicrobial hydrogels for biomedical applications. *Materials* 10(3): 232.
Grajales, S., G. Moad and E. Rizzardo. 2012. Controlled radical polymerization guide. Sigma-Aldrich Co. LLC. Materials Science. https://www.sigmaaldrich.com/content/dam/sigma-aldrich/docs/SAJ/Brochure/1/controlled-radical-polymerization-guide.pdf
Grigoletto, A., K. Maso, A. Mero, A. Rosato, O. Schiavon and G. Pasut. 2016. Drug and protein delivery by polymer conjugation. *Journal of Drug Delivery Science and Technology* 32: 132-141.
Hamley, I.W. and I.W. Hamley. 1998. The Physics of Block Copolymers (Vol. 19). Oxford University Press, Oxford.
Harris, J.M., N.E. Martin and M. Modi. 2001. Pegylation. *Clinical Pharmacokinetics* 40(7): 539-551.
Hingorani, S.R., L. Zheng, A.J. Bullock, T.E. Seery, W.P. Harris, D.S. Sigal, F. Braiteh, P.S. Ritch, M.M. Zalupski, N. Bahary and P.E. Oberstein. 2018. HALO 202: Randomized phase II study of PEGPH20 plus nab-paclitaxel/gemcitabine versus nab-paclitaxel/gemcitabine in patients with untreated, metastatic pancreatic ductal adenocarcinoma. *Journal of Clinical Oncology* 36(4): 359-366.
Hosta-Rigau, L., M.J. York-Duran, Y. Zhang, K.N. Goldie and Stadler, B. 2014. Confined multiple enzymatic (cascade) reactions within poly (dopamine)-based capsosomes. *ACS Applied Materials & Interfaces* 6(15): 12771-12779.
Ivanova, E.P., K. Bazaka and R.J. Crawford. 2014. New Functional Biomaterials for Medicine and Healthcare (Vol. 67). Woodhead Publishing, New Delhi, India.
Jaffe, G.J., D. Eliott, J.A. Wells, J.L. Prenner, A. Papp and S. Patel. 2016. A phase 1 study of intravitreous E10030 in combination with ranibizumab in neovascular age-related macular degeneration. *Ophthalmology* 123(1): 78-85.
Jevševar, S., M. Kunstelj and V.G. Porekar. 2010. PEGylation of therapeutic proteins. *Biotechnology Journal: Healthcare Nutrition Technology* 5(1): 113-128.

Jose, L., A. Hwang, C. Lee, K. Shim, J.K. Song, S.S.A. An and H.-j. Paik. 2020. Nitrilotriacetic acid-end-functionalized polycaprolactone as a template for polymer-protein nanocarrier. *Polymer Chemistry*. Vol n pp

Khandare, J. and T. Minko. 2006. Polymer-drug conjugates: Progress in polymeric prodrugs. *Progress in Polymer Science* 31(4): 359-397.

Kwon, G.S. and T. Okano. 1996. Polymeric micelles as new drug carriers. *Advanced Drug Delivery Reviews* 21(2): 107-116.

Langowski, K.P.J., M. Addepalli, T. Chang, V. Dixit, G. Kim and Y. Kirksey. 2017. NKTR-358: A selective, first-in-class IL-2 pathway agonist which increases number and suppressive function of regulatory T cells for the treatment of immune inflammatory disorders. *Arthritis Rheumatol* 69: 2.

Li, X., Y. Cao, K. Luo, Y. Sun, J. Xiong, L. Wang, Z. Liu, J. Li, J. Ma, J. Ge and H. Xiao. 2019. Highly active enzyme-metal nanohybrids synthesized in protein-polymer conjugates. *Nature Catalysis* 2(8): 718-725.

Lowe, A.B. and C.L. McCormick. 2007. Reversible addition-fragmentation chain transfer (RAFT) radical polymerization and the synthesis of water-soluble (co) polymers under homogeneous conditions in organic and aqueous media. *Progress in Polymer Science* 32(3): 283-351.

Ludwig, H., K. Weisel, M.T. Petrucci, X. Leleu, A.M. Cafro, L. Garderet, C. Leitgeb, R. Foa, R. Greil, I. Yakoub-Agha and D. Zboralski. 2017. Olaptesed pegol, an anti-CXCL12/SDF-1 Spiegelmer, alone and with bortezomib–dexamethasone in relapsed/refractory multiple myeloma: A Phase IIa Study. *Leukemia* 31(4): 997-1000.

Malmsten, M. 2011. Antimicrobial and antiviral hydrogels. *Soft Matter* 7(19): 8725-8736.

Meißig, M., S. Wieczorek and H.G. Börner. 2016. Synthetic Aspects of Peptide- and Protein-Polymer Conjugates in the Post-click Era. incomple

Messina, M.S., K.M. Messina, A. Bhattacharya, H.R. Montgomery and H.D. Maynard. 2019. Preparation of biomolecule-polymer conjugates by grafting-from using ATRP, RAFT, or ROMP. *Progress in Polymer Science* 101186.

Minko, S., S. Patil, V. Datsyuk, F. Simon, K.-J. Eichhorn, M. Motornov, D. Usov, I. Tokarev and M. Stamm. 2002. Synthesis of adaptive polymer brushes via "grafting to" approach from melt. *Langmuir* 18(1): 289-296.

Misra, H., J. Lickliter, F. Kazo and A. Abuchowski. 2014. PEGylated carboxyhemoglobin bovine (SANGUINATE): Results of a phase I clinical trial. *Artificial Organs* 38(8): 702-707.

Moreira, F.T., R.A. Dutra, J.P. Noronha, A.L. Cunha and M.G.F. Sales. 2011. Artificial antibodies for troponin T by its imprinting on the surface of multiwalled carbon nanotubes: Its use as sensory surfaces. *Biosensors and Bioelectronics* 28(1): 243-250.

Murata, H., S.L. Baker, B. Kaupbayeva, D.J. Lewis, L. Zhang, S. Boye, A. Lederer and A.J. Russell. 2020. Ligands and characterization for effective bio-atom-transfer radical polymerization. *Journal of Polymer Science* 58(1): 42-47.

Mussai, F., S. Egan, J. Higginbotham-Jones, T. Perry, A. Beggs, E. Odintsova, J. Loke, G. Pratt, K.P. U, A. Lo and M. Ng. 2015. Arginine dependence of acute myeloid leukemia blast proliferation: A novel therapeutic target. *Blood, The Journal of the American Society of Hematology* 125(15): 2386-2396.

Naing, A., K.P. Papadopoulos, K.A. Autio, P.A. Ott, M.R. Patel, D.J. Wong, G.S. Falchook, S. Pant, M. Whiteside, D.R. Rasco and J.B. Mumm. 2016. Safety, antitumor activity, and immune activation of pegylated recombinant human interleukin-10 (AM0010) in patients with advanced solid tumors. *Journal of Clinical Oncology* 34(29): 3562.

Pan, Z., X. Kang, Y. Zeng, W. Zhang, H. Peng, J. Wang, W. Huang, H. Wang, Y. Shen and Y. Huang. 2017. A mannosylated PEI-CPP hybrid for TRAIL gene targeting delivery for colorectal cancer therapy. *Polymer Chemistry* 8(35): 5275-5285.

Pelegri-O'Day, E.M., E.-W. Lin and H.D. Maynard. 2014. Therapeutic protein-polymer conjugates: Advancing beyond PEGylation. *Journal of the American Chemical Society* 136(41): 14323-14332.

Pokorski, J.K. and M.J. Hore. 2019. Structural characterization of protein-polymer conjugates for biomedical applications with small-angle scattering. *Current Opinion in Colloid & Interface Science*. 42: 157-168.

Ratner, B.D. 1980. Characterization of graft polymers for biomedical applications. *Journal of Biomedical Materials Research* 14(5): 665-687.

Roovers, J. 1985. Branched polymers. *Wiley-Interscience, Encyclopedia of Polymer Science and Enginering* 2, 478-499.

Sakakibara, K., M. Gondo, K. Miyazaki, T. Ito, A. Sugimura and M. Kobayashi. 2000. Peptide derivatives: Google Patents.

Sanyal, A., E.D. Charles, B.A. Neuschwander-Tetri, R. Loomba, S.A. Harrison, M.F. Abdelmalek, E.J. Lawitz, D. Halegoua-DeMarzio, S. Kundu, S. Noviello and Y. Luo. 2018. Pegbelfermin (BMS-986036), a PEGylated fibroblast growth factor 21 analogue, in patients with non-alcoholic steatohepatitis: A randomised, double-blind, placebo-controlled, phase 2a trial. *The Lancet* 392(10165): 2705-2717.

Seo, M.-D., H.-S. Won, J.-H. Kim, T. Mishig-Ochir and B.-J. Lee. 2012. Antimicrobial peptides for therapeutic applications: A review. *Molecules* 17(10): 12276-12286.

Seth Chhabra, E., N. Moore, C. Furcht, A.M. Holthaus, J. Liu, T. Liu, V. Schellenberger, J. Kulman, J. Salas and R. Peters. 2015. Evaluation of enhanced in vitro plasma stability of a novel long acting recombinant FVIIIFc-VWF-XTEN fusion protein. *Blood* 126(23): 2279.

Steiert, E., J. Ewald, A. Wagner, U.A. Hellmich, H. Frey and P.R. Wich. 2020. pH-responsive protein nanoparticles via conjugation of degradable PEG to the surface of cytochrome c. *Polymer Chemistry* 11(2): 551-559.

Stephanopoulos, N. and M.B. Francis. 2011. Choosing an effective protein bioconjugation strategy. *Nature Chemical Biology* 7(12): 876.

Su, M., S. Yao, L. Gu, Z. Huang and S. Mai. 2018. Antibacterial effect and bond strength of a modified dental adhesive containing the peptide nisin. *Peptides* 99, 189-194.

Sumerlin, B. 2008. ACS Macro Lett., 2012, 1, 141-145;(d) S. S. Sheiko, BS Sumerlin and K. Matyjaszewski. *Prog. Polym. Sci, 33*, 759-785.

Sumerlin, B.S. 2012. Proteins as initiators of controlled radical polymerization: Grafting-from via ATRP and RAFT. ACS Publications.

Sun, H., Y. Hong, Y. Xi, Y. Zou, J. Gao and J. Du. 2018. Synthesis, self-assembly, and biomedical applications of antimicrobial peptide-polymer conjugates. *Biomacromolecules* 19(6): 1701-1720.

Szlosarek, P.W., J.P. Steele, L. Nolan, D. Gilligan, P. Taylor, J. Spicer, M. Lind, S. Mitra, J. Shamash, M.M. Phillips and P. Luong. 2017. Arginine deprivation with pegylated arginine deiminase in patients with argininosuccinate synthetase 1–deficient malignant pleural mesothelioma: A randomized clinical trial. *JAMA Oncology* 3(1): 58-66.

Tugulu, S., A. Arnold, I. Sielaff, K. Johnsson and H.-A. Klok. 2005. Protein-functionalized polymer brushes. *Biomacromolecules* 6(3): 1602-1607.

van Dongen, M.A., C.A. Dougherty and M.M. Banaszak Holl. 2014. Multivalent polymers for drug delivery and imaging: The challenges of conjugation. *Biomacromolecules* 15(9): 3215-3234.

Vandermeulen, G.W. and H.A. Klok. 2004. Peptide/protein hybrid materials: Enhanced control of structure and improved performance through conjugation of biological and synthetic polymers. *Macromolecular Bioscience* 4(4): 383-398.

Veronese, F. and M. Morpurgo. 1999. Bioconjugation in pharmaceutical chemistry. *Il Farmaco* 54(8): 497-516.

Vo, Q.-A.T., J.K. Lin and L.M. Tong. 2015. Efficacy and safety of argatroban and bivalirudin in patients with suspected heparin-induced thrombocytopenia. *Annals of Pharmacotherapy* 49(2): 178-184.

Vögtle, F., S. Gestermann, R. Hesse, H. Schwierz and B. Windisch. 2000. Functional dendrimers. *Progress in Polymer Science* 25(7): 987-1041.

Wilson, P. 2017. Synthesis and applications of protein/peptide-polymer conjugates. *Macromolecular Chemistry and Physics* 218(9): 1600595.

Wright, T.A., R.C. Page and D. Konkolewicz. 2019. Polymer conjugation of proteins as a synthetic post-translational modification to impact their stability and activity. *Polymer Chemistry* 10(4): 434-454.

Wu, C.K., E.D. Charles, A. Bui, R. Christian and M. Abu Tarif. 2016. Phase 1 study of BMS-986171 (PEGylated FGF21) in healthy obese subjects. Paper presented at the Hepatology.

Wulff, G., A. Sarhan and K. Zabrocki. 1973. Enzyme-analogue built polymers and their use for the resolution of racemates. *Tetrahedron Letters* 14(44): 4329-4332.

Xie, Z., N.V. Aphale, T.D. Kadapure, A.S. Wadajkar, S. Orr, D. Gyawali, . . . J. Yang. 2015. Design of antimicrobial peptides conjugated biodegradable citric acid derived hydrogels for wound healing. *Journal of Biomedical Materials Research Part A* 103(12): 3907-3918.

Xu, T., N. Zhao, F. Ren, R. Hourani, M.T. Lee, J.Y. Shu, S. Mao and B.A. Helms. 2011. Subnanometer porous thin films by the co-assembly of nanotube subunits and block copolymers. *ACS Nano* 5(2): 1376-1384.

Zhang, A., L. Shu, Z. Bo and A.D. Schlüter. 2003. Dendronized polymers: Recent progress in synthesis. *Macromolecular Chemistry and Physics* 204(2): 328-339.

Zhang, M., J. Huang, P. Yu and X. Chen. 2010. Preparation and characteristics of protein molecularly imprinted membranes on the surface of multiwalled carbon nanotubes. *Talanta* 81(1-2): 162-166.

Zhou, P., S. Wu, M. Hegazy, H. Li, X. Xu, H. Lu and X. Huang. 2019. Engineered borate ester conjugated protein-polymer nanoconjugates for pH-responsive drug delivery. *Materials Science and Engineering: C* 104: 109914.

Zinzalla, G. and D.E. Thurston. 2009. Targeting protein-protein interactions for therapeutic intervention: A challenge for the future. *Future Med. Chem.* 1(1): 65-93. doi: 10.4155/fmc.09.12.

CHAPTER

7

Functionalization of Cellulose—Chemical Approach

Merin Sara Thomas[1,2], Prasanth K.S. Pillai[2,3], Sabu Thomas[2,3] and Laly A. Pothen[2*]

[1] Department of Chemistry, Mar Thoma College, Kuttapuzha P.O., Tiruvalla, Kerala - 689103, India

[2] International and Interuniversity Centre for Nanoscience and Nanotechnology, Mahatma Gandhi University, Priyadarsini Hills P.O., Kottayam, Kerala - 686 560, India

[3] School of Chemical Sciences, Mahatma Gandhi University, Priyadarsini Hills P.O., Kottayam, Kerala - 686 560, India

1. Introduction

Cellulose is one of the most abundant natural renewable biopolymer fibers on the surface of the earth. It can easily be obtained by the processing of agricultural waste, and hence it is more economical than any other source of fibers currently in use (Sharma et al. 2019). There is a growing trend to use cellulose, cellulose nanofibers and nanocrystalline cellulose as fillers in plastics and composites, and cellulose-reinforced plastics are increasingly gaining acceptance in structural applications. It has several advantages over traditional reinforcement materials such as glass fibers, carbon and talc: low density, low cost, good specific mechanical properties, reduced tool wear and biodegradability (Ilyas et al. 2018). Chemically it is a homopolysaccharide consisting of β-D glucopyranose units linked together by β-1, 4-linkage (Fig. 1). The excellent mechanical properties, low density, high crystallinity, safer handling protocols and working conditions of cellulose made it a green alternative to synthetic fibers.

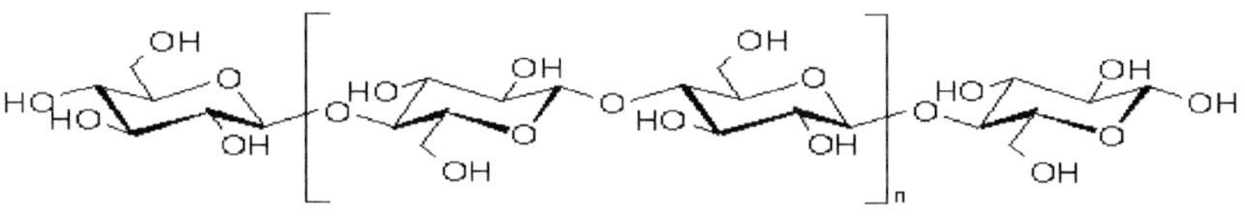

Fig. 1. The basic chemical structure of cellulose (Habibi et al. 2010).

*Corresponding author: lapothan@gmail.com

Nanosized cellulose fillers, with dimension of the cellulose microfibrils varying with respect to their source, are found to be very promising reinforcing elements for making functional composites (Siqueira et al. 2010). Depending on the nature of the source and the processing techniques used, the nanosized cellulose can be further classified as cellulose nanocrystals (CNCs) and cellulose nanofibrils (CNFs), in which both the crystalline domains and amorphous (disordered) regions are intermixed (Habibi et al. 2010). Acid hydrolysis treatment is one method to remove the amorphous phase from the cellulose microfibrils to yield highly crystalline cellulose whiskers or nanocellulose (E. Abraham et al. 2013, Mahmud et al. 2019, Wulandari et al. 2016). Sources used for the cellulose extraction and hydrolysis parameters such as time, temperature, ultrasound treatment, and purity of materials play an important role in determining the total dimensions of the nanocrystals. It was reported before that the usual dimensions of whiskers range from 5 to 10 nm in diameter and from 100 to 500 nm in length (George and Sabapathi 2015). Cellulose whiskers are also known as cellulose nanocrystals, nanorods or nano-whiskers and can be obtained from different sources, for example, microcrystalline cellulose (Bondeson et al. 2006), bacterial cellulose (Hirai et al. 2009), algal cellulose (valonia) (Sucaldito and Camacho 2017), hemp (R.E. Abraham et al. 2016), tunicin (Mathew and Dufresne 2002), cotton (Kale et al. 2018), ramie (Marinho et al. 2020), sisal (Mondragon et al. 2018), sugar beet (Salari et al. 2019), and wood (Agarwal et al. 2018). Certain reports mentioned this as nano-fibrillated cellulose (NFC), which may have been obtained by further treatment of micro-fibrillated cellulose (MFC), whose properties are closely linked to the degree of fibrillation. At lower degree of fibrillation there is a possibility that MFC will precipitate out from aqueous media; however, with desired processing conditions it can be modified (by breaking the structure) to make films that are suitable as a coating or packaging material for food products (Chinga-Carrasco and Syverud 2010). Figure 2 shows transmission electron micrographs of the colloidal suspensions after acidic treatment of cellulose and drying to produce aggregates of needle-shaped particles.

The surface functionality of cellulosic materials triggers their interactions with different chemicals, giving them potential for structural, environmental, biomedical, food and pharmaceutical applications. The skeletal structure (Fig. 1) of monomeric unit of cellulose contains three alcoholic hydroxyl groups. These hydroxyl groups, the primary hydroxyl group at C-6 and the two secondary ones at C-2 and C-3, can take part in all the classical reactions of the alcoholic hydroxyl group, such as esterification, etherification, and oxidation reactions. The presence of hydroxyl groups allows the formation of hydrogen bonds, which influences the crystalline structure as well as physical properties of cellulose. Because of the presence of these hydroxyl groups, the cellulose fibers are highly hydrophilic in nature, which is associated with a low interfacial compatibility with hydrophobic polymeric matrices, as well as with a loss of mechanical properties after moisture uptake.

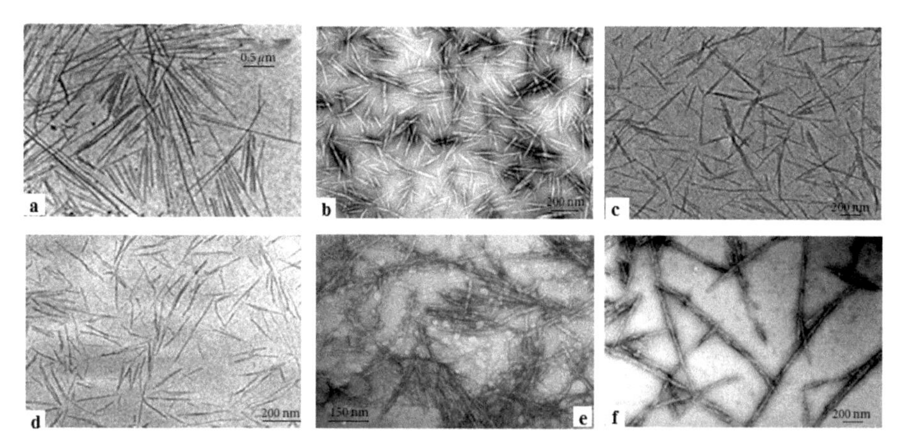

Fig. 2. Transmission electron micrographs of dilute suspensions of hydrolysed (a) tunicin (Elazzouzi-Hafraoui et al. 2008), (b) ramie (Habibi et al. 2008), (c) cotton (Dong et al. 1996), (d) sugar beet (Saïd Azizi Samir et al. 2004), (e) MCC (Kvien et al. 2005), and (f) bacterial cellulose (Grunert and Winter 2002).

These difficulties can be overcome by modification of the cellulosic material by oxidation, esterification, etherification, urethanization, amidation, and non-covalent modifications. This chapter covers some of the important chemical modifications of cellulosic materials and their nanostructures.

2. Surface modification by oxidation

Surface modification of cellulosic materials by oxidation is a conventional technique, and the oxidized cellulose is widely used in agriculture, cosmetics and medicine (Coseri et al. 2013). It is highly suitable for making functional hemostatic scaffolding material that is bioresorbable and easily degradable under different physiological conditions. The hydroxyls present in the cellulosic structures are oxidized into the carboxylic acid or aldehyde functional groups, though for the carboxylic moiety, nitroxyl radicals-based oxidation is required, and for aldehyde, periodate oxidation is required (Isogai et al. 2018). Periodate oxidation selectively cleaves the vicinal diols, resulting in two aldehyde functionalities. This type of oxidation reaction cellulose suspension is performed using sodium metaperiodate (NaIO4) solution at elevated temperatures, which proceeds through the formation of a cyclic intermediate of diol and periodate anion to form aldehydes (Kim and Choi 2014, Syamani et al. 2018). 2,2,6,6-Tetramethylpiperidinyloxyl radical (TEMPO) oxidation of cellulose is an alternative promising route to convert surface hydroxyl of cellulose into charged carboxyl units. Several studies were reported based on TEMPO oxidation functionalization. This method selectively oxidizes the primary alcoholic group only; the secondary alcoholic group remains intact. Usually, sodium hypochlorite or sodium chlorite or sodium bromide is used as a secondary oxidant to recycle the TEMPO.

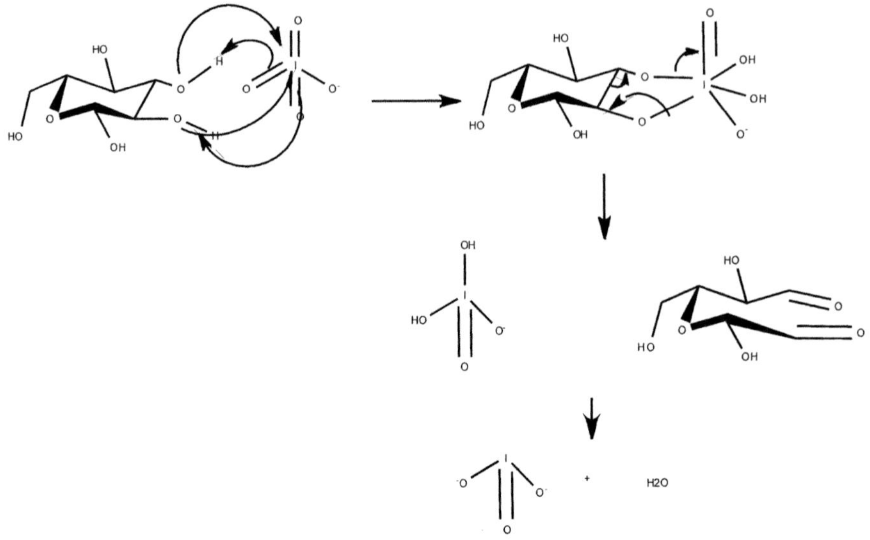

Fig. 3. Mechanism of periodate oxidation of cyclic 2,3-diols (Bunton and Shiner 1960).

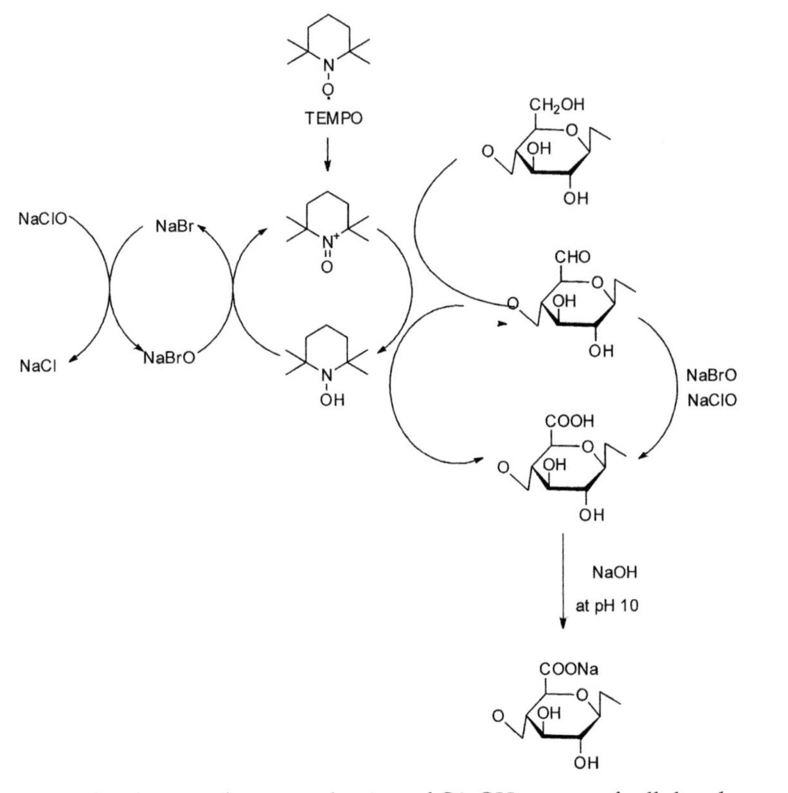

Fig. 4. Catalytic oxidation mechanism of C6–OH groups of cellulose by TEMPO/NaBr/NaClO in water at pH 10 (Isogai et al. 2011).

The TEMPO-mediated oxidation of cellulose was first studied by De Nooy et al. (1994), and they found that the primary hydroxyl groups of polysaccharides were oxidized, whereas the secondary hydroxyls remained unaffected. Tang et al. (2017) used a two-step method of oxidation protocol for cellulose, in which they treated cellulose with NaOH/urea solution to obtain cellulose powder with decreased crystallinity, and the cellulose powder further used for the TEMPO oxidation process. The product obtained by this method was completely soluble in water, indicating its suitability as polyelectrolytes, and has the oxidative conversion yield of 91%. Recently, a new rapid and low-cost method for cellulosic oxidation using Oxone ($KHSO_5$) has received much attention for the conversion of microcelluloses (1-20 microns) (Dachavaram et al. 2020). In this method, Oxone was combined with TEMPO and NaOCl in aqueous $NaHCO_3$ solution, at pH 7-8.5 and treated with microcellulose with the aid of microwave irradiation (Fig. 5), which yields two different forms of nano TEMPO-cellulose, i.e., a water-insoluble form (Form-I, 14 nm) and a water-soluble form (Form-II, 8 nm).

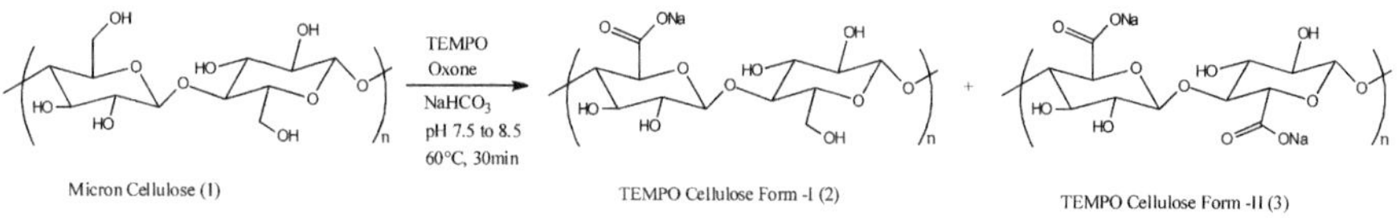

Fig. 5. Microcellulose oxidation with TEMPO/NaOCl/Oxone® under microwave irradiation conditions (Dachavaram et al. 2020).

3. Surface modification by esterification

Esterification, homogeneous and heterogeneous modes, is one of the most reliable chemical modifications for making cellulose derivatives, such that the reaction proceeds in such a way that either the whole –OH group can be esterified, or the surface hydroxyls of cellulose fibers get modified, leaving the cellulose crystalline structure in the interior intact. However, the heterogeneous mode is generally used for the surface modification of native cellulose. The traditional approach of esterification was on the preparation of cellulose nitrate (Klemm et al. 1998). Mainly there are two types of cellulose esters, organic and inorganic (e.g., nitrates, sulfates, phosphates, xanthates). The most important functionalization of cellulose esters is by acylation using carboxylic acid or more effectively using acid derivatives such as acid chlorides or acid anhydrides in the presence of acid catalysts. Figure 6 shows schematic representations of the molecular structure of cellulose, the molecular structure of cellulose ester, and the surface esterification of nanocellulose.

Keyrilainen et al. (Willberg-Keyriläinen and Ropponen 2019) synthesized long chain cellulosic esters of altered chain lengths using different esterification methods and found that the acyl chloride method was the most effective esterification method, and cellulose esters prepared

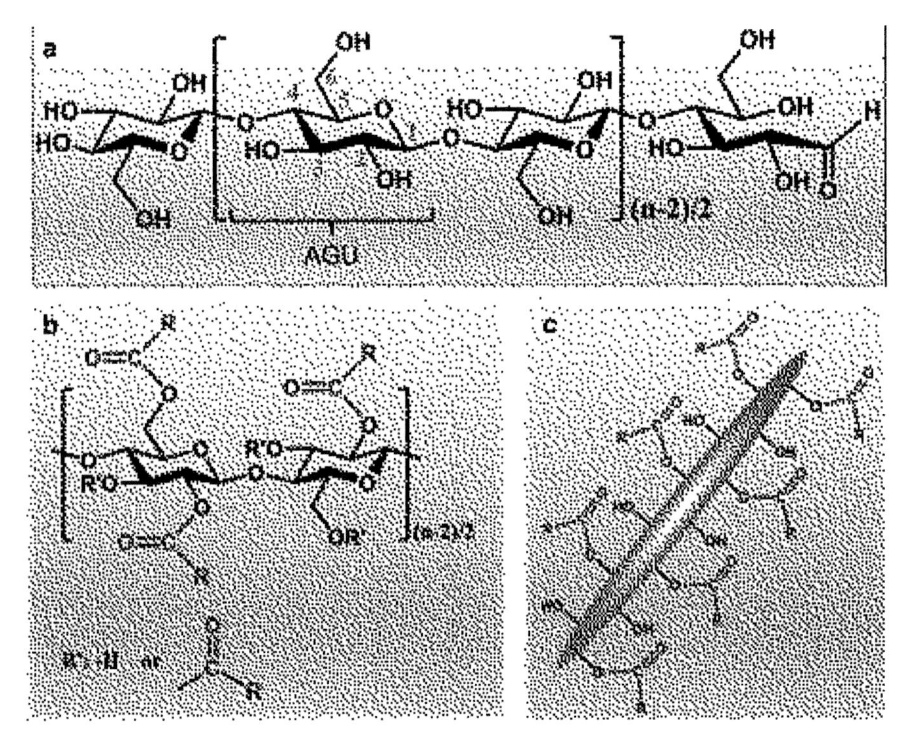

Fig. 6. Schematic representations of (a) the molecular structure of cellulose, (b) the molecular structure of cellulose ester, and (c) the surface esterification of nanocellulose. The numbers in (a) show the numbering system for carbon atoms within one anhydroglucose unit (AGU) of cellulose (Wang et al. 2018).

using that method have the highest degree of substitution values and better conversion ratio. Esterification, by trans esterification approach, is a common method used to modify cellulose nanocrystals with poly(3-caprolactone) or poly(lactide) by ring-opening polymerization (Stepanova et al. 2019). All the above-mentioned esterification methods can be adopted for nanocellulose modification; however, the main challenge is to modify the surface hydroxyls without limiting its crystallinity, which can be controlled by the adoption of mild reaction conditions (Habibi et al. 2010).

4. Surface modification by amidation

The amidation of cellulose was commonly carried out on oxidized cellulose with carboxyl functional groups. Figure 7 shows the mechanism of thermal amidation of carboxylate group in cellulosic materials (Pettignano et al. 2019). Gomez et al. (2017) reported a one-pot synthetic approach for TEMPO oxidized cellulose nanofibrils (TOCN) surface amidation by coupling up to 70% of superficial carboxylic units in TOCN with long alkyl chain primary amines (dodecylamine and octadecylamine) using TBTU [O-(1H-benzotriazol-1-yl)-*N,N,N′,N′*-tetramethyluronium tetrafluoroborate]

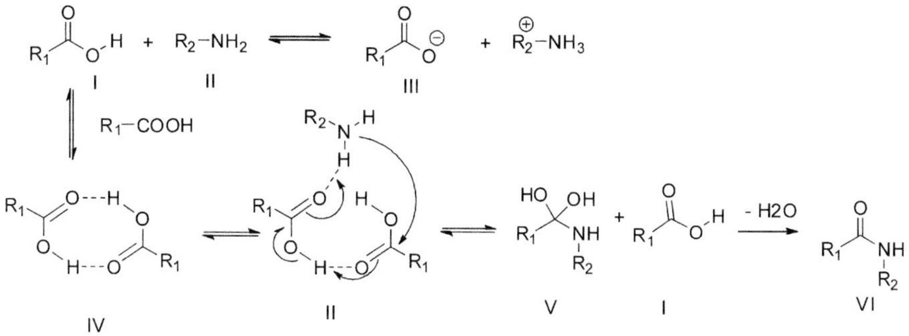

Fig. 7. Mechanism of thermal amidation of carboxylic acid (Pettignano et al. 2019).

uronium salt as coupling agent. The advantage of this procedure is that it can be performed in aqueous media.

Pettignano et al. (2019) synthesized amidated carboxymethyl cellulose derivatives at high temperature, in the absence of solvent, catalysts and coupling agents, resulting in a more sustainable alternative to conventional amidation procedures. The efficacy of the amidation method is highly dependent on the structure of the substituent, with amines presenting a long aliphatic chain showing a lower reactivity towards the desired amide.

5. Surface modification by etherification

Etherification is the second most favorable method for the functionalization of cellulose nanocrystals. Glycidyl trimethylammonium chloride (GTMAC) and its derivatives are commonly used for the etherification of cellulose nanocrystals. Epichlorohydrin etherification agent is also widely used for the etherification of cellulosic materials. Gu et al. (2019) prepared cellulose nanoparticles from waste maize-stalk pith via a sono-assisted GTMAC-etherification followed by an ultrasonication treatment. The etherification modification improved the ultrasound disintegration of cellulose from sheet-like parenchyma cells into nanoparticles. They found that the stronger electrostatic repulsion of the highly cationic charged nanoparticles led to better dispersion among CNPs, good stability and acceptable thermostability. Homogeneous etherification of cellulose was performed using a binary ionic liquid mixture as the solvent and catalyst under mild conditions by Kakibe et al. (2017). They found that the cellulose samples were successfully epoxidized in the presence of glycidol and the degree of substitution was controlled by varying the reaction time and $[BImC_4SO_3H][HSO_4]$ composition.

6. Surface modification by silylation

Chemical modification by silylation procedure is relatively cheap since silylation agents are effective and not costly. Silane coupling agents are usually adopted for creating a perfect bonding by improving the degree of crosslinking in the interface region. The number of cellulose hydroxyl groups at the fiber–matrix interface may be reduced by the action of silane coupling agents. In the presence of moisture, hydrolysable alkoxy groups lead to the formation of silanols, which further react with the hydroxyl group of the fiber, forming a stable covalent bond with the cell wall. Yu et al. (2019) fabricated hydrophobic cellulosic materials via gas–solid silylation reaction for oil/water separation. Laitinen et al. (2017) prepared low-cost, ultralight, highly porous, hydrophobic, and reusable super absorbing cellulose nanofibril aerogel via silylation of cellulose nanofibrils. Two important silylating agents that were used for the silylation process are methyltrimethoxysilane and hexadecyltrimethoxysilane (Laitinen et al. 2017). More recently, ene- or thiol-functionalized silanes were also covalently grafted on NFC-based films and consequently clicked with appropriate moieties through thiol-ene click chemistry (Huang et al. 2014).

Goussé et al. (2004) applied a mild silylation procedure to surface silylate dispersed microfibrils from parenchymal cell cellulose. These modified fibers had the same morphological features as those of the un-derivatized samples, which were dispersible into non-polar solvent to yield stable suspensions. Note that these suspensions will not flocculate and hence the microfibrils obtained acquire an inherent flexibility. Boyer et al. (2020) developed an injectable, self-hardening, mechanically reinforced hydrogel (Si-HPCH) composed of silanized hydroxypropymethyl cellulose (Si-HPMC) mixed with silanized chitosan. They found that Si-HPCH is a self-setting and cytocompatible hydrogel able to support the *in vitro* and *in vivo* viability and activity of human adipose stromal cells. It also supported the regeneration of osteochondral defects in dogs when implanted alone or with adipose stromal cells. Chantereau et al. (2019) fabricated bioactive amine functions onto the surface of bacterial cellulose nanofibrils, via a simple silylation treatment in water. The results showed that silylated material remained highly porous and hygroscopic and displayed sufficient thermal stability to support the sterilization treatments generally required in medical applications. The nanofibrillar structure and macro-porosity of the cellulose membranes was preserved after the silylation treatment, although a decrease in specific surface area was attributed to a thickening of the fibrils after the silane condensation.

7. Surface modification by urethanization

Urethanization is a useful strategy for functionalization in which the isocyanate (–NCO) group reacts with groups such as amino ($-NH_2$), hydroxyl (–OH), and carboxylic acid (–COOH). Even though urethanization is obviously a useful approach for surface modification, only a few published reports are available. Urethanization in cellulose is the process of creating a urethane linkage on the hydroxyl groups of cellulose by reaction with isocyanates. Palencia et al. (2019) introduced urethane group into the structure of cellulose by modifying it with 4,4′-Methylene-bis-phenylisocyanate to develop reinforcing agents based on cellulose fibers modified by insertion of end-alkyl groups from pyrolytic bio-oil. Gwon et al. (2016) carried out the urethanization strategy for CNC surface modification using toluene diisocyanate (TDI) for the preparation of mCNC-filled polylactic acid (PLA) nanocomposites. They found that there are two types of interactions between the mCNC surface and the PLA skeleton covalent interactions between the PLA end groups (–OH and –COOH) and free isocyanate groups (–NCO) on the mCNC surface, which can form urethane (–NHCOO–) and amide (–CONH–) crosslinks and non-covalent interactions between the PLA backbone and the mCNC surface (hydrogen bonding and van der Waals interactions). The mechanical studies revealed that the chemical tuning of CNCs with TDI can improve the tensile strength of PLA nanocomposites due to various intermolecular interactions and the enhanced dispersibility of the mCNCs.

8. Non-covalent surface modifications

The non-covalent surface modifications of cellulose are typically made via adsorption of surfactants. Heux et al. (2000) introduced the approach of non-covalent functionalization to nanocellulose, using surfactants consisting of the mono- and diesters of phosphoric acid having alkylphenol tails for the modification. They used CNC dispersion in non-polar organic solvents such as toluene and cyclohexane with phosphoric ester surfactant. After cationic surface modification with hexa-decyl-trimethylammonium bromide (HDTMA), enhanced dispersion of CNC was found in tetrahydrofuran. Another example of non-covalent approach is the electrostatic attraction between the negatively charged (OSO_3^-) groups and a cationic surfactant. Zakeri et al. (2018) reported a non-covalent surface modification of CNCs using polyethyleneimine (PEI), through a low-cost process and without the use of organic solvents. The surface modification was gained by a simple physical adsorption approach. The change in apparent size of the CNC and its surface potential indicate the degree of modification as a function of the amount of PEI. The addition of PEI to the CNC suspension results in almost an instantaneous formation of aggregates and a resulting phase separation of the CNC particles. Aïssa et al. (2019) used carbohydrate binding modules (CBMs) to introduce new non-covalent interaction on cellulose surfaces in a one-pot reaction in aqueous media. CBMs have a strong affinity for crystalline cellulose and functionalized with an alkyne at the terminal amine position. Click reaction was found to be the reason for this non-covalent interaction. The CBM-PEG modification of cellulose surfaces increased CNC redispersion after drying and improved suspension stability based on steric interactions. Basic adsorption of hydrophobic polymers in aprotic solvents was introduced as a platform technology to modify exclusively the surfaces of cellulose nanopapers by Kontturi et al. (2017). The non-covalently modified nanopapers are useful in applications such as moisture buffers where water repellence and moisture sorption are desired qualities.

Surface modification on cellulose nanofibers was achieved by Zhang et al. (2019) by TiO_2 coating. The TiO_2 modification changed the surface texture from smooth to rough and reversed the surface charge (from negative to positive). The TiO_2-coated cellulose nanofibers as adsorbent gave a significantly high capture efficiency (~100%) for the Au nanoparticles (5 nm), by effectively using the electrostatic interaction between the negatively charged Au nanoparticles and positively charged cellulose nanofibers after the TiO_2 surface modification. Figure 8 shows the FE-SEM micrograph of a single TiO_2-coated cellulose nanofiber after filtrating Au colloidal suspension, with a locally enlarged FE-SEM image and EDX element mapping, and a schematic illustration explaining the mechanism of electrostatic interaction to achieve a high capture efficiency for the Au nanoparticles after the TiO_2 surface coating (Zhang et al. 2019).

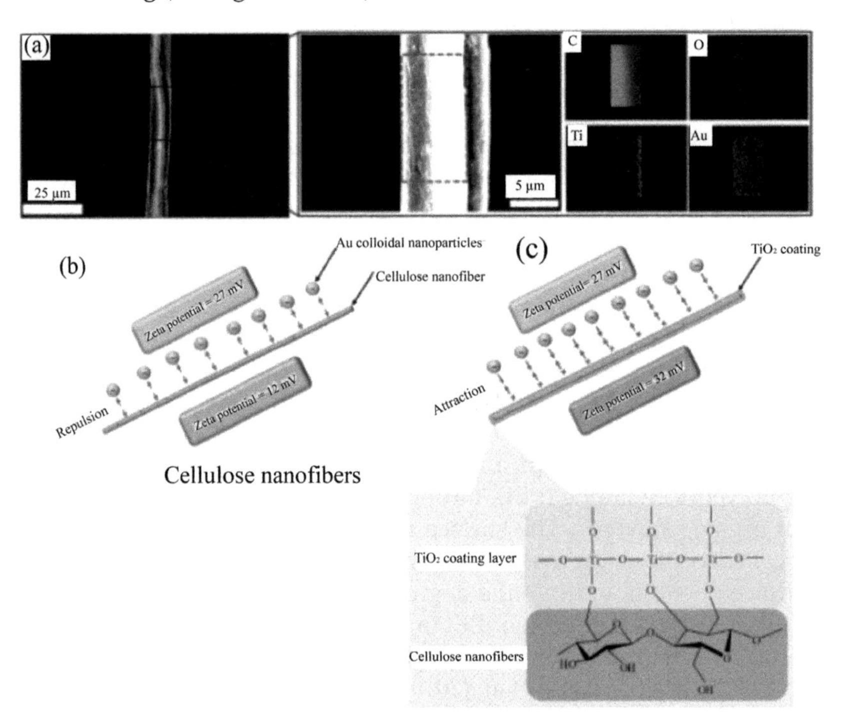

Fig. 8. (a) FE-SEM micrograph of a single TiO_2-coated cellulose nanofiber after filtrating Au colloidal suspension, with a locally enlarged FE-SEM image and EDX element mapping. (b) and (c) A schematic illustration explaining the mechanism of electrostatic interaction to achieve a high capture efficiency for the Au nanoparticles after the TiO_2 surface coating (Zhang et al. 2019).

9. Conclusion

Cellulose and nanocellulose are used as materials for a wide range of applications. The hydrophilic nature of cellulose limits its applications. Chemical modification of cellulose opens a new era in the history of cellulose because it can maintain a balance between the hydrophilicity and hydrophobicity of cellulose. This will improve the performance of cellulose or make it suitable for some specific uses. In addition, surface modification improved the compatibility of cellulose with polymer matrices.

References

Abraham, Eldho, Merin S. Thomas, Cijo John, L.A. Pothen, O. Shoseyov and S. Thomas. 2013. Green nanocomposites of natural rubber/nanocellulose: Membrane transport, rheological and thermal degradation characterisations. *Industrial Crops and Products*. 51: 415-424.

Abraham, Reinu E., Cynthia S. Wong and Munish Puri. 2016. Enrichment of cellulosic waste hemp (Cannabis sativa) hurd into non-toxic microfibres. *Materials* 9(7): 562.

Agarwal, Umesh P., Sally A. Ralph, Richard S. Reiner, Christopher G. Hunt, Carlos Baez, Rebecca Ibach and Kolby C. Hirth. 2018. Production of high lignin-containing and lignin-free cellulose nanocrystals from wood. *Cellulose* 25(10): 5791-5805.

Aïssa, Kevin, Muzaffer A. Karaaslan, Scott Renneckar and Jack N. Saddler. 2019. Functionalizing cellulose nanocrystals with click modifiable carbohydrate-binding modules. *Biomacromolecules* 20(8): 3087-3093.

Bondeson, Daniel, Aji Mathew and Kristiina Oksman. 2006. Optimization of the isolation of nanocrystals from microcrystalline cellulose by acid hydrolysis. *Cellulose* 13(2): 171.

Boyer, Cécile, G. Réthoré, P. Weiss, C. d'Arros, J. Lesoeur, C. Vinatier, B. Halgand, O. Geffroy, M. Fusellier, G. Vaillant and P. Roy. 2020. A self-setting hydrogel of silylated chitosan and cellulose for the repair of osteochondral defects: From in vitro characterization to preclinical evaluation in dogs. *Frontiers in Bioengineering and Biotechnology* 8: 23.

Bunton, C.A. and V.J. Shiner. 1960. 321 periodate oxidation of 1, 2-diols, diketones, and hydroxy-ketones: The use of oxygen-18 as a Tracer. *Journal of the Chemical Society (Resumed)*: 1593-1598.

Chantereau, Guillaume, Nettie Brown, Marie-Anne Dourges, Carmen S.R. Freire, Armando J.D. Silvestre, Gilles Sebe and Véronique Coma. 2019. Silylation of bacterial cellulose to design membranes with intrinsic anti-bacterial properties. *Carbohydrate Polymers* 220: 71-78.

Chinga-Carrasco, Gary and Kristin Syverud. 2010. Computer-assisted quantification of the multi-scale structure of films made of nanofibrillated cellulose. *Journal of Nanoparticle Research* 12(3): 841-851.

Coseri, Sergiu, Gabriela Biliuta, Bogdan C. Simionescu, Karin Stana-Kleinschek, Volker Ribitsch and Valeria Harabagiu. 2013. Oxidized cellulose—Survey of the most recent achievements. *Carbohydrate Polymers* 93(1): 207-215.

Dachavaram, S.S., J.P. Moore, S. Bommagani, N.R. Penthala, J.L. Calahan, S.P. Delaney, E.J. Munson, J. Batta-Mpouma, J.W. Kim, J.A. Hestekin and P.A. Crooks. 2020. A facile microwave assisted TEMPO/NaOCl/Oxone ($KHSO_5$) mediated micron cellulose oxidation procedure: Preparation of two nano TEMPO-cellulose forms. *Starch-Stärke* 72(1-2): 1900213.

De Nooy, A.E.J., A.C. Besemer and H. Van Bekkum. 1994. Highly selective TEMPO mediated oxidation of primary alcohol groups in polysaccharides. *Recueil des Travaux Chimiques des Pays-Bas* 113(3): 165-166.

Dong, Xue Min, Tsunehisa Kimura, Jean-François Revol and Derek G. Gray. 1996. Effects of ionic strength on the isotropic-chiral nematic phase transition of suspensions of cellulose crystallites. *Langmuir* 12(8): 2076-2082.

Elazzouzi-Hafraoui, Samira, Yoshiharu Nishiyama, Jean-Luc Putaux, Laurent Heux, Frédéric Dubreuil and Cyrille Rochas. 2008. The shape and size distribution of crystalline nanoparticles prepared by acid hydrolysis of native cellulose. *Biomacromolecules* 9(1): 57-65.

George, Johnsy and S.N. Sabapathi. 2015. Cellulose nanocrystals: Synthesis, functional properties, and applications. *Nanotechnology, Science and Applications* 8: 45.

Gómez, F.N., M.Y. Combariza and C. Blanco-Tirado. 2017. Facile cellulose nanofibrils amidation using a 'One-Pot' approach. *Cellulose* 24(2): 717-730.

Goussé, C., H. Chanzy, M.L. Cerrada and E. Fleury. 2004. Surface silylation of cellulose microfibrils: Preparation and rheological properties. *Polymer* 45(5): 1569-1575.

Grunert, Maren and William T. Winter. 2002. Nanocomposites of cellulose acetate butyrate reinforced with cellulose nanocrystals. *Journal of Polymers and the Environment* 10(1-2): 27-30.

Gu, Huiming, Xin Gao, Heng Zhang, Keli Chen and Lincai Peng. 2019. Fabrication and characterization of cellulose nanoparticles from maize stalk pith via ultrasonic-mediated cationic etherification. *Ultrasonics Sonochemistry*: 104932. https://doi.org/10.1016/j.ultsonch.2019.104932.

Gwon, Jae-Gyoung, Hye-Jung Cho, Sang-Jin Chun, Soo Lee, Qinglin Wu and Sun-Young Lee. 2016. Physiochemical, optical and mechanical properties of poly (lactic acid) nanocomposites filled with toluene diisocyanate grafted cellulose nanocrystals. *RSC Advances* 6(12): 9438-9445.

Habibi, Youssef, Anne-Lise Goffin, Nancy Schiltz, Emmanuel Duquesne, Philippe Dubois and Alain Dufresne. 2008. Bionanocomposites based on poly (ε-caprolactone)-grafted cellulose nanocrystals by ring-opening polymerization. *Journal of Materials Chemistry* 18(41): 5002-5010.

Habibi, Youssef, Lucian A. Lucia and Orlando J. Rojas. 2010. Cellulose nanocrystals: Chemistry, self-assembly, and applications. *Chemical Reviews* 110(6): 3479-3500.

Heux, L., G. Chauve and C. Bonini. 2000. Nonflocculating and chiral-nematic self-ordering of cellulose microcrystals suspensions in nonpolar solvents. *Langmuir* 16(21): 8210-8212.

Hirai, Asako, Osamu Inui, Fumitaka Horii and Masaki Tsuji. 2009. Phase separation behavior in aqueous suspensions of bacterial cellulose nanocrystals prepared by sulfuric acid treatment. *Langmuir* 25(1): 497-502.

Huang, Jian-Lin, Chao-Jun Li and Derek G. Gray. 2014. Functionalization of cellulose nanocrystal films via 'Thiol-Ene' click reaction. *RSC Advances* 4(14): 6965-6969.

Ilyas, R.A., S.M. Sapuan, M. Lamin Sanyang, M. Ridzwan Ishak and E.S. Zainudin. 2018. Nanocrystalline cellulose as reinforcement for polymeric matrix nanocomposites and its potential applications: A review. *Current Analytical Chemistry* 14(3): 203-225.

Isogai, Akira, Tuomas Hänninen, Shuji Fujisawa and Tsuguyuki Saito. 2018. Catalytic oxidation of cellulose with nitroxyl radicals under aqueous conditions. *Progress in Polymer Science* 86: 122-148.

Isogai, Akira, Tsuguyuki Saito and Hayaka Fukuzumi. 2011. TEMPO-oxidized cellulose nanofibers. *Nanoscale* 3(1): 71-85.

Kakibe, Takeshi, Satoshi Nakamura, Waki Mizuta and Hajime Kishi. 2017. Etherification of cellulose in binary ionic liquid as solvent and catalyst. *Chemistry Letters* 46(5): 737-739.

Kale, Ravindra D., Prabhat Shobha Bansal and Vikrant G. Gorade. 2018. Extraction of microcrystalline cellulose from cotton sliver and its comparison with commercial microcrystalline cellulose. *Journal of Polymers and the Environment* 26(1): 355-364.

Kim, Joo Yong and Hyung-Min Choi. 2014. Cationization of periodate-oxidized cotton cellulose with choline chloride. *Cellul Chem Technol* 48(1-2): 25-32.

Klemm, Dieter, B. Philpp, Thomas Heinze, U. Hewinze and Wolfgang Wagenknecht. 1998. Comprehensive Cellulose Chemistry. Volume 2: Functionalization of Cellulose. Wiley-VCH Verlag GmbH.

Kontturi, Katri S., Karolina Biegaj, Andreas Mautner, Robert T. Woodward, Benjamin P. Wilson, Leena-Sisko Johansson, Koon-Yang Lee, Jerry Y.Y. Heng, Alexander Bismarck and Eero Kontturi. 2017. Noncovalent surface modification of cellulose nanopapers by adsorption of polymers from aprotic solvents. *Langmuir* 33(23): 5707-5712.

Kvien, Ingvild, Bjørn S. Tanem and Kristiina Oksman. 2005. Characterization of cellulose whiskers and their nanocomposites by atomic force and electron microscopy. *Biomacromolecules* 6(6): 3160-3165.

Laitinen, Ossi, Terhi Suopajärvi, Monika Österberg and Henrikki Liimatainen. 2017. Hydrophobic, superabsorbing aerogels from choline chloride-based deep eutectic solvent pretreated and silylated cellulose nanofibrils for selective oil removal. *ACS Applied Materials and Interfaces* 9(29): 25029-25037.

Mahmud, Md Musavvir, Asma Perveen, Rumana A. Jahan, Md Abdul Matin, Siew Yee Wong, Xu Li and M. Tarik Arafat. 2019. Preparation of different polymorphs of cellulose from different acid hydrolysis medium. *International Journal of Biological Macromolecules* 130: 969-976.

Marinho, Nelson Potenciano, Pedro Henrique Gonzalez de Cademartori, Silvana Nisgoski, Valcineide Oliveira de Andrade Tanobe, Umberto Klock and Graciela Inés Bolzon de Muñiz. 2020. Feasibility of ramie fibers as raw material for the isolation of nanofibrillated cellulose. *Carbohydrate Polymers* 230: 115579.

Mathew, Aji P. and Alain Dufresne. 2002. Morphological investigation of nanocomposites from sorbitol plasticized starch and tunicin whiskers. *Biomacromolecules* 3(3): 609-617.

Mondragon, G., A. Santamaria-Echart, Maria Eugenia Victoria Hormaiztegui, A. Arbelaiz, C. Peña-Rodriguez, V. Mucci, M. Corcuera, Mirta Ines Aranguren and A. Eceiza. 2018. Nanocomposites of waterborne polyurethane reinforced with cellulose nanocrystals from sisal fibres. *Journal of Polymers and the Environment* 26(5): 1869-1880.

Palencia, M., M.A. Mora, T.A. Lerma, N. Afanasjeva and J.H. Isaza. 2019. Reinforcing agents based on cellulose fibers modified by insertion of end-alkyl groups obtained from pyrolytic bio-oil of sugarcane bagasse. *Polymer Bulletin* 1-14.

Pettignano, Asja, Aurélia Charlot and Etienne Fleury. 2019. Solvent-free synthesis of amidated carboxymethyl cellulose derivatives: Effect on the thermal properties. *Polymers* 11(7): 1227.

Saïd Azizi Samir, My Ahmed, Fannie Alloin, Michel Paillet and Alain Dufresne. 2004. Tangling effect in fibrillated cellulose reinforced nanocomposites. *Macromolecules* 37(11): 4313-4316.

Salari, Mahdieh, Mahmood Sowti Khiabani, Reza Rezaei Mokarram, Babak Ghanbarzadeh and Hossein Samadi Kafil. 2019. Preparation and characterization of cellulose nanocrystals from bacterial cellulose produced in sugar beet molasses and cheese whey media. *International Journal of Biological Macromolecules* 122: 280-288.

Sharma, Amita, Manisha Thakur, Munna Bhattacharya, Tamal Mandal and Saswata Goswami. 2019. Commercial application of cellulose nano-composites – A review. *Biotechnology Reports* e00316.

Siqueira, Gilberto, Julien Bras and Alain Dufresne. 2010. Cellulosic bionanocomposites: A Review of preparation, properties and applications. *Polymers* 2(4): 728-765.

Stepanova, M., I. Averianov, M. Serdobintsev, I. Gofman, N. Blum, N. Semenova, Y. Nashchekina, T. Vinogradova, V. Korzhikov-Vlakh, M. Karttunen and E. Korzhikova-Vlakh. 2019. PGlu-modified nanocrystalline cellulose improves mechanical properties, biocompatibility, and mineralization of polyester-based composites. *Materials* 12(20): 3435.

Sucaldito, Melvir R. and Drexel H. Camacho. 2017. Characteristics of unique HBr-hydrolyzed cellulose nanocrystals from freshwater green algae (Cladophora rupestris) and its reinforcement in starch-based film. *Carbohydrate Polymers* 169: 315-323.

Syamani, F.A., Y.D. Kurniawan and L. Suryanegara. 2018. Oxidized cellulose fibers for reinforment in poly (lactic acid) based composite. *Asian Journal of Chemistry* 30(7): 1435-1440.

Tang, Zuwu, Wenyan Li, Xinxing Lin, He Xiao, Qingxian Miao, Liulian Huang, Lihui Chen and Hui Wu. 2017. TEMPO-oxidized cellulose with high degree of oxidation. *Polymers* 9(9): 421.

Wang, Yonggui, Xiaojie Wang, Yanjun Xie and Kai Zhang. 2018. Functional nanomaterials through esterification of cellulose: A review of chemistry and application. *Cellulose* 25(7): 3703-3731.

Willberg-Keyriläinen, Pia and Jarmo Ropponen. 2019. Evaluation of esterification routes for long chain cellulose esters. *Heliyon* 5(11): e02898.

Wulandari, W.T., A. Rochliadi and I.M. Arcana. 2016. Nanocellulose prepared by acid hydrolysis of isolated cellulose from sugarcane bagasse. *In*: IOP Conference Series: Materials Science and Engineering, IOP Publishing, 12045.

Yu, Lisha, Zeming Zhang, Hongding Tang and Jinping Zhou. 2019. Fabrication of hydrophobic cellulosic materials via gas-solid silylation reaction for oil/water separation. *Cellulose* 26(6): 4021-4037.

Zakeri, Bahareh, Dhriti Khandal, Quentin Beuguel, Bernard Riedl, Jason Robert Tavares, Pierre J. Carreau and Marie-Claude Heuzey. 2018. Non-covalent surface modification of cellulose nanocrystals by polyethyleneimine. *The Journal of Science and Technology for Forest Products and Processes* 7(4): 6-12.

Zhang, Chenning, Tetsuo Uchikoshi, Izumi Ichinose and Lihong Liu. 2019. Surface modification on cellulose nanofibers by TiO_2 coating for achieving high capture efficiency of nanoparticles. *Coatings* 9(2): 139.

CHAPTER
8

Functionalized Polymers Processed by 3D Printing

Narendra Pal Singh Chauhan[1*], Mahrou Sadri[2], Behnaz Sadat Eftekhari[3,4], Farzin Sahebjam[5] and Mazaher Gholipourmalekabadi[6,7]

[1] Department of Chemistry, Bhupal Nobles' University, Udaipur - 313002, Rajasthan, India

[2] Department of Nutrition and Health Sciences, University of Nebraska-Lincoln, Lincoln, NE, USA

[3] Department of Physiology and Institute for Medicine and Engineering, University of Pennsylvania, Philadelphia, PA, USA

[4] Department of Biomedical Engineering, Amirkabir University of Technology, Tehran, Iran

[5] School of Veterinary Science, Massey University, Palmerston North, New Zealand

[6] Cellular and Molecular Research Centre, Iran University of Medical Sciences, Tehran, Iran

[7] Department of Tissue Engineering & Regenerative Medicine, Faculty of Advanced Technologies in Medicine, Iran University of Medical Sciences, Tehran, Iran

1. Introduction

Three-dimensional printing can be used in various fields, including fabrications of polymer and metal-based materials, tissue engineering, and chemical reactions (Yue et al. 2015, Patra and Young 2016, Zarek et al. 2016). From the technological aspect, 3D printing can be categorized as vat photopolymerization (Vaezi et al. 2013), powdered bed fusion methodology (Tolochko et al. 2000, Kumar 2003), additives, binder jetting (Park et al. 2007), sheet lamination manufacturing process (Weisensel et al. 2004), as well as fused deposition modeling (FDM) (Nikzad et al. 2010). FDM is mostly employed in 3D printing technology because of its low cost, superb compatibility, and simpler process for quantities of printing materials (Chen et al. 2017).

One polymer that has a high global demand is acrylonitrile butadiene styrene (ABS), because of its FDM property. ABS is made up of polybutadiene,

*Corresponding author: narendrapalsingh14@gmail.com

butadiene copolymer grafted with acrylonitrile and styrene copolymer (SAN), and thermoplastic SAN matrix (Meincke et al. 2004).

3D printing technology is broadly applied in bioengineering (Derby 2012, Gou et al. 2014) and microfluidics (Anderson et al. 2013, Shallan et al. 2014). Similarly, additive manufacturing is applied in the development of 3D printing, which makes possible the layer-by-layer development in 3D material for computer-aided design (CAD) (Gross et al. 2014). Stereolithography (SL) technology uses photosensitive liquid polymer material for crosslinked coating by a light source (Vaezi et al. 2013, Lee et al. 2015). In a digital lighting processing (DLP), light is passed from the base level of the tank, consisting of resin, and basic level is dipped directly into the resin from the irradiation system above, which is ideally designed with LED lamp source and a digital mirror gadget (Vaezi et al. 2013, Gross et al. 2014).

Poly(ethylene glycol) (PEG) is a synthetic hydrogel used in DLP and SL for its hydrophilicity and biocompatibility properties (Harris 2013). PEG-based photocurable (meth) acrylic oligomers are employed for biomedical and biotechnological applications (Nguyen and West 2002, Tibbitt and Anseth 2009).

Nowadays, hexagonal boron nitride (hBN) is used because of its electrical insulation and thermal conductivity (Harris 2013). The honeycomb-type framework of hBN is prepared in the form of planar 2D layers, consisting of conjugation between boron and nitrogen atoms. Boron and nitrogen atoms are stacked alongside one another by van der Waals forces as well as electrostatic interactions (Nguyen and West 2002, Tibbitt and Anseth 2009). 3D printing devices are manufactured to produce 3D scaffolds, which are composed of synthetic and natural polymers (Hofmann 2014, Chia and Wu 2015). A blend ratio of the gelatin and poly(vinyl alcohol) (PVA) is used to create a scaffold, which affects the bone growth (Taniguchi et al. 2016). Stimulus-responsive hydrogel material shows hydrophilic polymeric network type morphology due to reaction conditions such as pH and temperature (Cohen Stuart et al. 2010, Ionov 2013). Poly(*N*-isopropylacrylamide) hydrogel is a temperature-sensitive hydrogel used in the preparation of microfluidic devices (Richter et al. 2009, Benito-Lopez et al. 2014) and drug delivery vehicles (Fundueanu et al. 2008, Hoare et al. 2009).

Projection micro-stereolithography (PμSL) is a lithography-based additive synthesis approach currently used because of its speed, low cost, and adaptability in material selection (Sun et al. 2005, Zheng et al. 2012). Adhesive bonding is an excellent source of joining materials because of its low cost and light weight (Paroissien et al. 2007, Dillard 2010, Kumar and Mittal 2013). 3D printing is applied in the production of robotic arms (Bartlett et al. 2015), multilayer tooth (Imbeni et al. 2005), and orthopedic implants (Murr et al. 2010, Mäkitie et al. 2013). Direct-write 3D printing is a technique in which a syringe using a nozzle is moved across a surface area as it dispenses ink (Guvendiren et al. 2012, Peterson et al. 2015).

2. Synthetic methods

2.1. 3D Materials based on acrylonitrile butadiene styrene (ABS)

Acrylonitrile butadiene styrene (ABS), rubberized particle size, poly(methyl methacrylate) (PMMA), methacrylate-butadiene-styrene (MBS), lubricants (magnesium stearate and dimethicone) and antioxidant were employed to fabricate 3D materials. MBS powder and PMMA and ABS pellets were dried in the oven at 100°C for 5 h. ABS, PMMA, other polymer aids, and MBS were mixed using a twin-screw extruder (Chen et al. 2018). The monofilament of various blends can be employed to print the multiple forms using the FDM 3D printers. Figure 1 shows the manufacture of 3D structure used as commercial FDM 3D printer explained by a flowchart. A model is constructed using digital model-based software SketchUp, then it is converted to standard triangle language (STL) file. These STL files are modified with the printer software (Maxform) and exported as g-code file designs. The specimens were printed layer by layer. MBS and PMMA were in a position to enhance the transmittance as well as surface area gloss of ABS. Pure ABS was prepared by emulsion polymerization and resulted in a milk gray or slightly yellow compound. The graph shows that PMMA enhanced the transmittance quality of ABS throughout the duration for which light was visible (Fig. 2). The addition of haziness played a vital role in developing the exact transparency for multiphase blends polymer, and its refractive index depended on the rubber particle size (Ren et al. 2014). The refractive index of rubber material was not a match for SAN and PMMA containing ABS mainly because rubbers had a poor refractive index. MBS showed no effect on the transmittance of ABS/PMMA. When the mass of MBS, ABS, and PMMA was 70, 30, and 6, the surface patina extended to around 95.5, which typically resulted in the product of MBS containing ABS/MBS/PMMA blends.

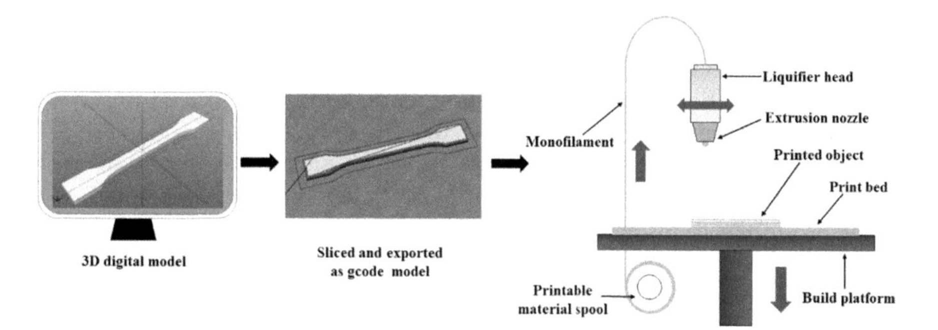

Fig. 1. The manufacturing process of 3D system, depending on an industrial FDM 3D printer. With permission from ACS (Chen et al. 2018).

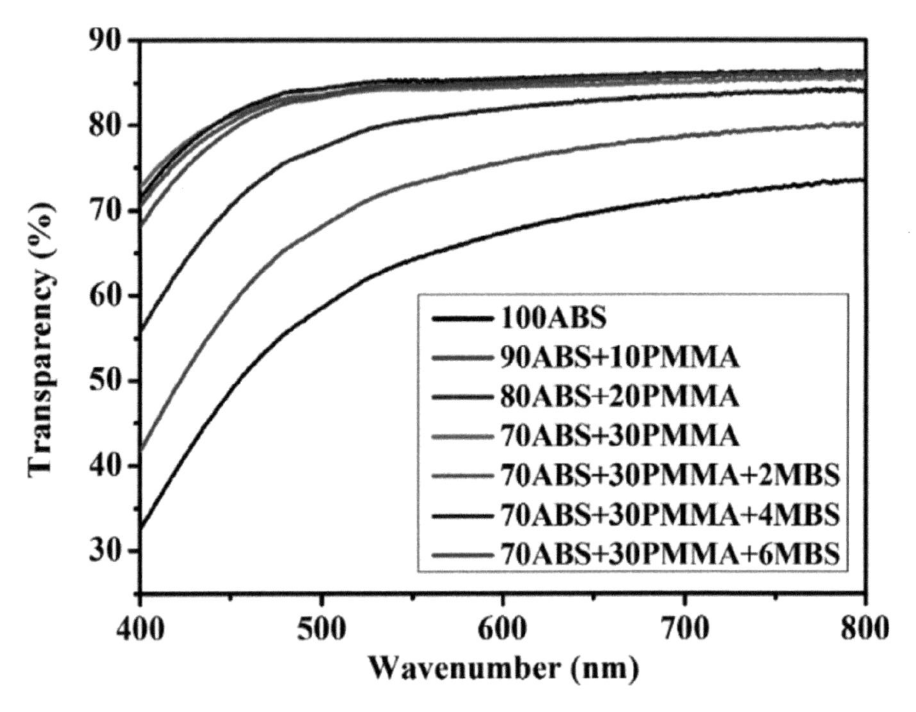

Fig. 2. The transmittance of pure ABS, ABS/PMMA binary blends, and ABS/PMMA/MBS ternary blends in the wavelength ranging from 400 nm to 800 nm. With permission from ACS (Chen et al. 2018).

2.2. Nanocomposite with PEG hybrid

Poly(ethyleneglycol) diacrylate (PEGDA), tetraethyl orthosilicate (TEOS), 3(trimethoxysilyl) propyl methacrylate, MEMO, bis-(2,4,6-trimethylbenzoyl) phenylphosphineoxide (Irgacure 819, BASF), and 2-hydroxy-2-methyl-1-phenyl-propan-1-(Irgacure 1173, BASF) were used. Mixtures containing PEGDA, MEMO 5 phr, and TEOS (0, 20, 30, 40 phr) with two photoinitiators were prepared. A 3DLPrinter-HD 2.0 equipped with a projector was employed for the printing. The recommended printing procedures were followed, and a medium pressure mercury lamp was used.

Hydrolysis and condensation reactions of silica precursors, as well as PEGDA samples, were performed at 80°C using the acidic vapors as a catalyst known as sol-gel post-treatment method. Computer-aided design and style data were used to print the various 3D items ranging from tensile test specimens to honeycomb constructions (Fig. 3). The color chart suggested green as nominal, red as an additional content in the beneficial direction, and blue as less content in the negative direction (Fig. 3C) (Chiappone et al. 2016). The average distance between positive and negative values was 0.045 mm. DSC measurements were taken to address 3D samples and in contrast to the pristine PEGDA constructions. When MEMO was included in the PEGDA

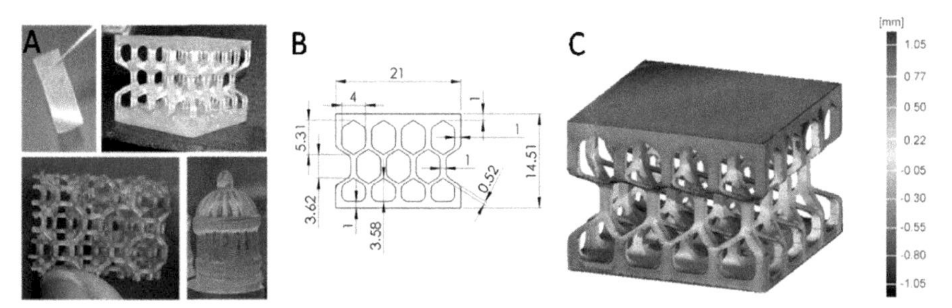

Fig. 3. (A) 3D printed structures. (B) Section of the CAD picture of the alveolar framework with dimensions. (C) Map of final resolution and fidelity to the CAD data evaluated by 3D checking. With permission from ACS (Chiappone et al. 2016).

resin, it was observed without deviation in the tag value. When the quantity of silica precursor within the photocurable formulation increased, a rise of glass transition temperature (T_g) values was observed. The increase in T_g values was due to the existence of the silica phase produced after the sol-gel method, which in turn hindered the mobility of this polymer chain. When MEMO was used as a coupling agent, immediate chemical bonds between inorganic and organic domains developed.

2.3. 3D Nanocomposites containing hexagonal boron nitride (HBN)

3D printable hBN polymer ink was used in preparation of graphene, carbon nanotubes (Jakus et al. 2015a), and alloys (Jakus et al. 2015b). The 3D printable hBN inks included only two components, poly(lactic-co-glycolic acid), and hBN (PLGA). PLGA is an elastomer that works as a polymer binder for hBN particles. hBN particles and the PLGA were dissolved and dispersed, respectively, within a trisolvent system; dibutylphthalate was used as a plasticizer, and ethylene glycol monobutyl ether was used as a surfactant to disperse the hBN particles. Additionally, dichloromethane (DCM) makes a higher volatility solvent that dissolves the PLGA and triggers the mechanism for solidification throughout the printing activity. The thermal conductivities are illustrated in Fig. 4a (Guiney et al. 2018). The thermal conductivity of the 3D printed materials developed with increasing hBN. The thermal conductivity decreased as the particles of hBN increased, owing to changes that occurred in the microstructure of the 3D printed content. As the content of hBN improved, DCM evaporated and the porosity of the printed framework increased quickly (Fig. 4b). Excessive porosity presented in a huge amount of hBN particles, which can be prevented by optimizing the drying rate of the inks. For instance, by increasing the quantity of surfactant ratios and plasticizer in the ink formulation or even by thoroughly controlling the ambient environment (e.g., temperature and humidity), the evaporation rate was slowed, reducing the resulting porosity in addition to the increase of thermal conductivity of hBN.

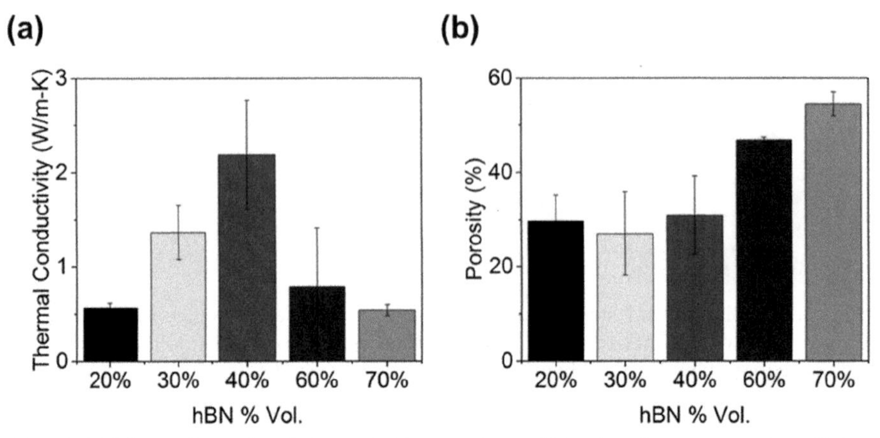

Fig. 4. (a) Thermal conductivity and (b) porosity of 3D printed constructs ($n = 3$) for different hBN loading. With permission from ACS (Guiney et al. 2018).

Although no difference was found between the volumes of 20%, 60% and 70% hBN samples, the thermal conductivity values for the 20%, 30% and 40% volume of hBN sample were significantly different from each other ($p < 0.05$). The porosity was larger ($p < 0.05$) for the 60% and 70% volume hBN constructs, but it showed no significant difference between the 20%, 30% and 40% volume hBN constructs.

2.4. Fabrication of gelatin/PVA scaffolds

Gelatin and poly(vinyl alcohol) were employed to fabricate gelatin/scaffolds. Gelatin/PVA scaffolds were fabricated using a low-temperature 3D printing process (Fig. 5). Different weight fractions of gelatin/PVA solution were used—G10, G7P3, G5P5—in which G represented the gelatin, P indicated PVA, and the numbers indicated the weight fractions. Thus, P10 and G10 scaffolds were composed of pure PVA and pure gelatin, respectively. The application of low-temperature 3D printing was crucial in this experiment because of the reduced viscosity of the gelatin/PVA solution (Kim et al. 2009).

The printability of the different gelatin/PVA solutions was determined by the temperature of applied pressure and the stage (Fig. 6). The enhanced processing conditions were noticeable in the red array. Besides, the temperature of the stage was varied from –15°C to –5°C, and the pneumatic pressure was varied from 10 to 405 kPa. To identify the optimization of this process, the printed structures were evaluated (Fig. 6). The structures were categorized into three groups: non-structure (x), non-porous framework (Δ), and porous framework (O). PVA solution had higher viscosity than the gelatin solution (Kim et al. 2018). Therefore, as the content of PVA increased, it required higher pressure. Ultimately, the optimized pressures were 15, 20, 40, 220, and 400 kPa for G10, G7P3, G5P5, G3P7, and P10, respectively.

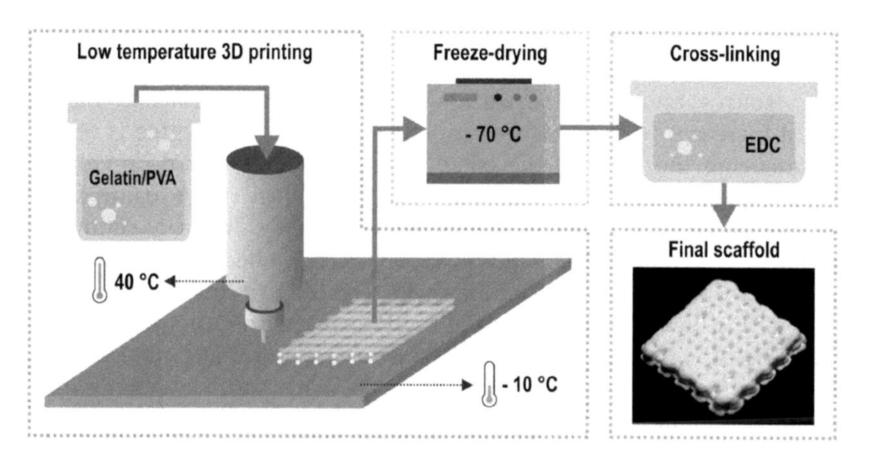

Fig. 5. Sequential representation of the fabrication activity. With permission from Elsevier (Kim et al. 2018).

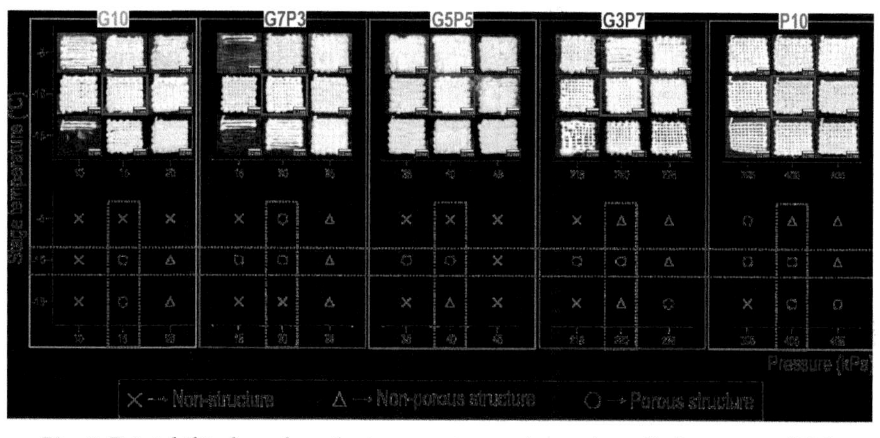

Fig. 6. Printability based on the temperature point and applied pressure. With permission from Elsevier (Kim et al. 2018).

2.5. 3D Printing of PNIPAAm hydrogel

For 3D printing, *N,N'*-Methylene-bis(acrylamide), N isopropylacrylamide (NIPAAm) resin as a crosslinker, Phenylbis(2,4,6-tri-methylbenzoyl) phosphine oxide, Methacryl amidopropyl trimethyl ammonium chloride (MAPTAC), ethanol, and a fluorescent dye Rhodamine B were employed. Photocurable PNIPAAm and resins were developed by dissolving NIPAAm (monomer) and *N,N'*-Methylene-bis(acrylamide) as a crosslinker, along with Phenylbis(2,4,6-tri-methylbenzoyl) phosphine oxide as a photo-initiator, due to light absorption with the help of photo-polymerization. Ethanol was used as a solvent, and the photo-absorber (PA) material was used to regulate the penetration level of the light. The dye material, particularly Rhodamine B, was used for visualization of 3D printed PNIPAAm. The degree of photopolymerization and curing degree could be varied by changing the light as energy source. Since polymerization begins from the surface of the glass slide, it increases with exposure time, where the curing degree is obtained by measuring the level of the ensuing rectangular pattern. Previously, a similar experiment was conducted with PNIPAAm resins using various concentrations of PA (Han et al. 2018). As shown in Fig. 7b, it was claimed that the curing degree of PNIPAAm hydrogel was controlled by the light energy dosage and focus of PA within the resin. The 3D printing of PNIPAAm is shown in Fig. 7c. After printing, the PNIPAAm micro-structure was stored at 20°C in DI water for post-printing rinsing. During that time, the ethanol within the structure was replaced with DI water, and an uncrosslinked polymer was eliminated in the resin from the printed structure. The 3D PNIPAAm hydrogel structure was swollen at 10°C as the height increased (Fig. 7c). Its structure shrank at 50°C when the height decreased. Reversible swelling distortion was observed when the temperature declined to 10°C. It demonstrated on the swelling deformation of 3D PNIPAAm hydrogel printed by PμSL, which was reversible by changing the ambient temperature. As the concentration of PA increased, the growth of a level decreased with the rise of energy dosage. The temperature was responsible for swelling of 3D printed PNIPAAm hydrogen structure. PNIPAAm hydrogel possesses the ability to change its degree of swelling with respect to the temperature. When the swelling ratio was greater than 1, the PNIPAAm hydrogel swelled and

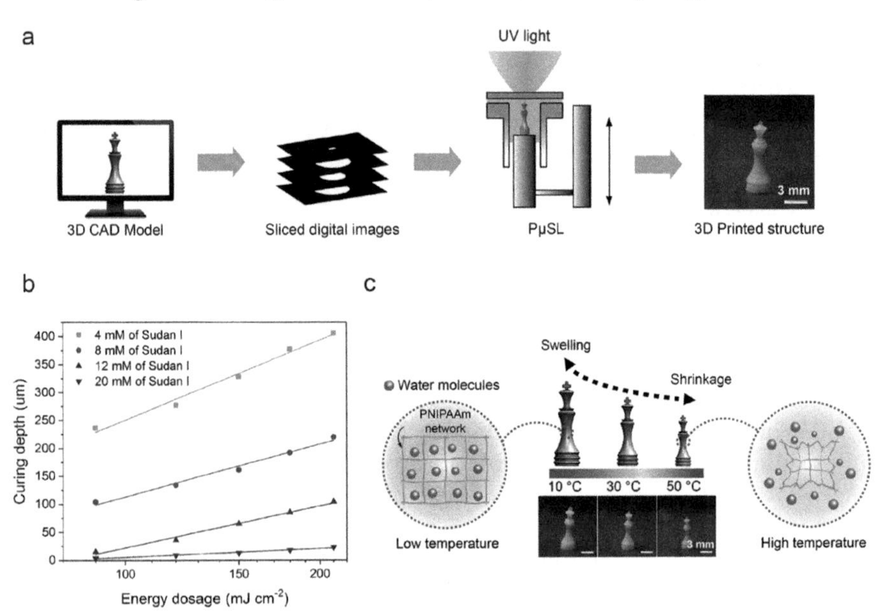

Fig. 7. 3D printing of the temperature-responsive hydrogel using PμSL: (a) schematic illustration of the PμSL process; (b) result of the curing degree study. The curing depth tended to decrease with increasing PA concentration. With permission from Springer-Nature (Han et al. 2018).

became bigger than the printed dimension at the lower temperature. When the swelling ratio was less than 1, the PNIPAAm hydrogel could shrink and become smaller than the printed dimension at a higher temperature.

The swelling ratio at 10°C was 1.50, as shown in Fig. 8a. It could increase in length compared to the original dimensions belonging to the printed disk sample. The swelling ratio at 50°C is 0.80. It showed a reduction in the length of the printed dimension of the sample. The temperature was changed every 3 h by 5°C to establish the swelling equilibrium state. It showed that, in cooling cycles and heating, the swelling ratios overlapped. After multiple periods of cooling and heating, the maximum and minimum swelling ratios remained constant, indicating that the swelling behavior of PNIPAAm was repeatable and reversible (Chester and Anand 2011).

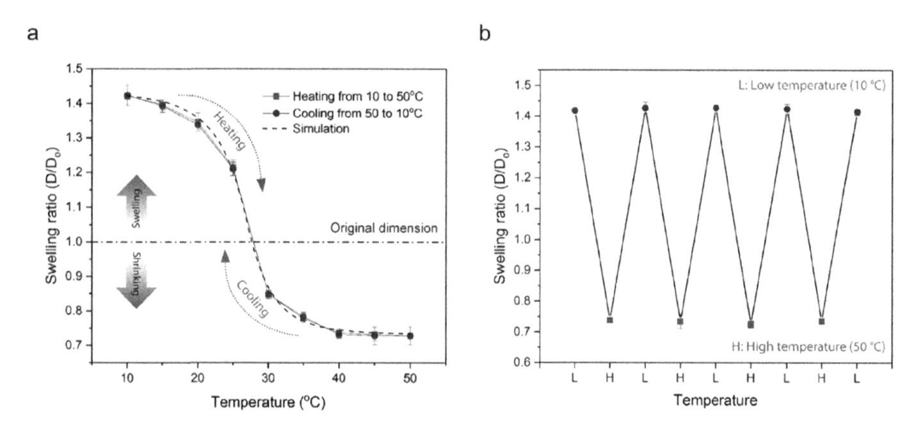

Fig. 8. The thermal effect of PNIPAAm hydrogel depends on swelling/shrinkage reversibility. (a) The swelling effect of PNIPAAm was dependent on heat. The magnitude of swelling of PNIPAAm throughout a heating cycle complements perfectly with values during the cooling period. (b) Swelling ratio of PNIPAAm remained constant at high and low temperature of 50°C and 10°C respectively. With permission from Springer-Nature (Han et al. 2018).

2.6. Diblock and triblock polymer using 3D printing

Diblock (AB) and triblock (ABA) copolymers that include at least one poly(alkyl glycidyl ether) block form the thermally responsive hydrogels (Labbé et al. 2007, Lee et al. 2011). Dual stimuli-responsive hydrogels consist of poly(isopropyl glycidyl ether)-*block* PEG-*block*-poly(isopropyl glycidyl ether) (PiPrGE-*b*-PEG-b-PiPrGE) for direct-write 3D printing. The polymers were synthesized using organocatalytic ring-opening polymerization of glycidyl ether. The resulting polymers were similar to *Pluronic F127* block copolymer (PF127) in their thermal and shear effects, but in comparison to PF127 they had thermo-responsive A blocks rather than the thermo-responsive B block contained in PF127. These polymers formed flower shapes as micelles and bridged micelles inside the solvents picky for the B obstruct

(Yang et al. 2011), thereby promoting an extended and well-percolated physical network. Isopropyl glycidyl ether (iPrGE), PEG, isopropylglycidyl ether, phosphazene foundation P4-t-Bu, and anhydrous toluene were used.

Triblock copolymers, i.e., PiPrGE-*b*-PEG-*b*-PiPrGE, were polymerized inside a glove package. Then, PEG and toluene were poured directly into a round-bottomed flask. Tert-butyl-P4 catalyst and isopropyl glycidyl ether were added to the response combination, and the reaction was stirred for 10 h at 60°C until it became yellow-orange. The polymer was collected via centrifugation, and the supernatant was discarded. The polymer was dissolved directly into dichloromethane, filtered and finally dried at 70°C in the response of combination. The micellization conduct of these block copolymers was dependent on the temperature. ^{1}H NMR spectroscopic analysis was used to estimate the critical micellar concentration (CMC). Lacelle and co-workers calculated the free power, enthalpy, and entropy of essential temperatures of micelle and micellization, as well as concentrations for a PEO-PPO-PEO triblock copolymer (Cau and Lacelle 1996). Spin-lattice relaxation period (T_1) was obtained by ^{1}H NMR spectroscopy that established that the approximate CMC for polymer 1 and polymer 2 at 27°C was 0.8 and 0.22 wt%, respectively (Fig. 9) (Zhang et al. 2015).

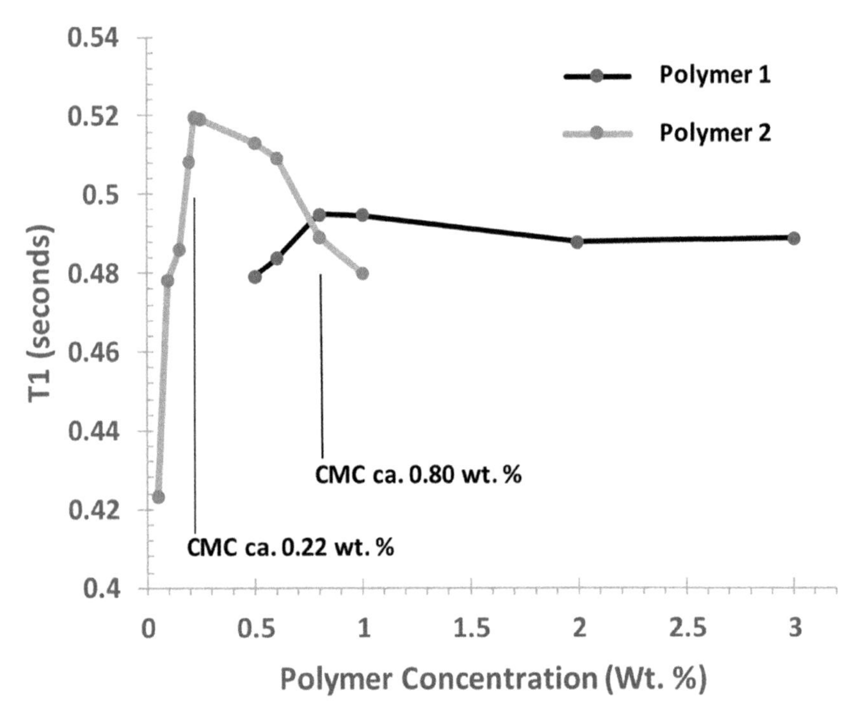

Fig. 9. ^{1}H NMR T_1 study for determining the value of critical micelle concentration (CMC) of polymers 1 and 2 in D_2O at 27°C. With permission from ACS (Zhang et al. 2015).

3. Enhancing the properties of bonded joints (SLJ) using 3D printing

A single-lap joint (SLJ) with far-field tensile stress σ_∞ is made up of an eccentric load-path. Both the shear and peel stresses/strains are generated in the adhesive, which includes high gradients and comparatively high absolute values at the ends of the adhesive (Fig. 10a). The distribution of the adhesive stresses over the bond length in an SLJ with a constant modulus adhesive is shown in Fig. 10a. The increasing compliance of the adhesive requires a larger length of the bond line to participate in the transfer of shear stress, which has the positive effect of reducing the concentrations at the ends. The compliance tailoring strategy uses 3D printing abilities to regulate the relative volume fraction of polymer so as to modify the compliance of the adhesive (Dimas et al. 2013). The ends were 3D printed using a rigid polymer Vero White Plus TM RGD835 (VW) (Liljenhjerte et al. 2016, Lin et al. 2014). The various configurations tested are illustrated in Fig. 10b and include stiffness (within the center), compliance (at the ends), and the base adhesive by printing widthwise-constant circular features into the adhesive in a staggered pattern. All were aligned to the study, in which the 3D printed SLJs are analyzed using Zwick-Roell tensile testing machine (Kumar et al. 2017).

Digital image correlation was employed to calculate strain in the joint, which depended on the load. Some random speckle patterns were formed on the specimen's surface area before testing by applying a thin coating of white acrylic paint, then spraying random dots of black acrylic paint with an airbrush. A monochrome 5.0 MP camera was used to capture the speckle

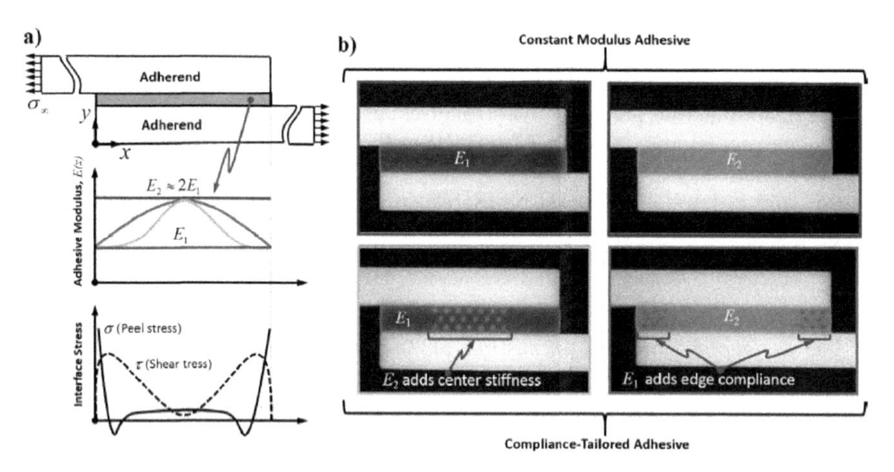

Fig. 10. Spatially tailored adhesive compliance in single-lap joint (SLJ). (a) Tensile stress σ_∞ along with various adhesives shows modulus adhesive (either E_1 or E_2). (b) Optical images of 3D printed joints with edge compliance-tailored cases. With permission from ACS (Kumar et al. 2016).

images during the test. Vic-2D software was used to compute the evolution of the strain field as a characteristic of load.

4. Conclusion

3D printing will revolutionize the future and will give rise to new applications and demands. So far, various polymers have been produced by 3D printing that demonstrate unique properties, e.g., water-proof property, tensile toughness, bending strength and elasticity modulus. When hBN is released directly into 3D printable polymer nanocomposite, the outcome possesses high thermal conductivity. Also, gelatin and cell proliferation can be effectively promoted by PVA, which shows a promising future in 3D printing for biomedical purposes.

Conflicts of interest

The author declares no conflict of interest.

References

Anderson, K.B., S.Y. Lockwood, R.S. Martin and D.M. Spence. 2013. A 3D printed fluidic device that enables integrated features. *Analytical Chemistry* 85(12): 5622-5626. https://doi.org/10.1021/ac4009594

Bartlett, N.W., M.T. Tolley, J.T.B. Overvelde, J.C. Weaver, B. Mosadegh, K. Bertoldi, G.M. Whitesides and R.J. Wood. 2015. A 3D-printed, functionally graded soft robot powered by combustion. *Science* 349(6244): 161-165. https://doi.org/10.1126/science.aab0129

Benito-Lopez, F., M. Antoñana-Díez, V.F. Curto, D. Diamond and V. Castro-López. 2014. Modular microfluidic valve structures based on reversible thermoresponsive ionogel actuators. *Lab on a Chip* 14(18): 3530-3538. https://doi.org/10.1039/c4lc00568f

Cau, F. and S. Lacelle, 1996. 1H NMR relaxation studies of the micellization of a poly(ethylene oxide)–poly(propylene oxide)–poly(ethylene oxide) triblock copolymer in aqueous solution, Macromolecules 29 (1): 170–178. doi.org/10.1021/ma950976+

Chen, S., Q. Zhang and J. Feng. 2017. 3D printing of tunable shape memory polymer blends. *Journal of Material Chemistry C* 5(33): 8361-8365. https://doi.org/10.1039/c7tc02534c

Chen, S., Jiabao, L. and J. Feng. 2018. 3D-printable ABS blends with improved scratch resistance and balanced mechanical performance. *Industrial & Engineering Chemistry Research* 57(11): 3923-3931. http://doi 10.1021/acs.iecr.7b05074.

Chester, S.A. and L. Anand. 2011. A thermo-mechanically coupled theory for fluid permeation in elastomeric materials: Application to thermally responsive gels. *Journal of the Mechanics and Physics of Solids* 59(10): 1978-2006. https://doi.org/10.1016/j.jmps.2011.07.005

Chia, H.N. and B.M. Wu. 2015. Recent advances in 3D printing of biomaterials. *Journal of Biological Engineering* 9(1): 4. https://doi.org/10.1186/s13036-015-0001-4

Chiappone, A., E. Fantino, I. Roppolo, M. Lorusso, D. Manfredi, P. Fino, C.F. Pirri and F. Calignano. 2016. 3D Printed PEG-based hybrid nanocomposites obtained by sol-gel technique. *ACS Applied Materials & Interfaces* 8(8): 5627-5633. https://doi.org/10.1021/acsami.5b12578.

Cohen Stuart, M.A., W.T.S. Huck, J. Genzer, M. Müller, C. Ober, M. Stamm, G.B. Sukhorukov, I. Szleifer, V.V. Tsukruk, M. Urban, F. Winnik, S. Zauscher, I. Luzinov and S. Minko. 2010. Emerging applications of stimuli-responsive polymer materials. *Nature Materials*, 9(2): 101-113. https://doi.org/10.1038/nmat2614

Derby, B. 2012. Printing and prototyping of tissues and scaffolds. *Science* 338(6109): 921-926. https://doi.org/10.1126/science.1226340

Dillard, D.A. (Ed.). 2010. Advances in Structural Adhesive Bonding. Elsevier. ISBN: 9781845694357

Dimas, L.S., G.H. Bratzel, I. Eylon and M.J. Buehler. 2013. Tough composites inspired by mineralized natural materials: Computation, 3D printing, and testing. *Advanced Functional Materials* 23(36): 4629-4638. https://doi.org/10.1002/adfm.201300215

Fundueanu, G., M. Constantin and P. Ascenzi. 2008. Preparation and characterization of pH- and temperature-sensitive pullulan microspheres for controlled release of drugs. *Biomaterials* 29(18): 2767-2775. https://doi.org/10.1016/j.biomaterials.2008.03.025

Gou, M., X. Qu, W. Zhu, M. Xiang, J. Yang, K. Zhang, Y. Wei and S. Chen. 2014. Bio-inspired detoxification using 3D-printed hydrogel nanocomposites. *Nature Communications* 5: 3774. https://doi.org/10.1038/ncomms4774

Gross, B.C., J.L. Erkal, S.Y. Lockwood, C. Chen and D.M. Spence. 2014. Evaluation of 3D printing and its potential impact on biotechnology and the chemical sciences. *Analytical Chemistry* 86(7): 3240-3253. https://doi.org/10.1021/ac403397r

Guiney, L.M., N.D. Mansukhani, A.E. Jakus, S.G. Wallace, R.N. Shah and M.C. Hersam. 2018. Three-dimensional printing of cytocompatible, thermally conductive hexagonal boron nitride nanocomposites. *Nano Letters* 18(6): 3488-3493. https://doi.org/10.1021/acs.nanolett.8b00555

Guvendiren, M., H.D. Lu and J.A. Burdick. 2012. Shear-thinning hydrogels for biomedical applications. *Soft Matter* 8(2): 260-272. https://doi.org/10.1039/c1sm06513k

Han, D., Z. Lu, S.A. Chester and H. Lee. 2018. Micro 3D printing of a temperature-responsive hydrogel using projection micro-stereolithography. *Scientific Reports* 8: Article number 1963. https://doi.org/10.1038/s41598-018-20385-2

Harris, J.M. 2013. Poly (ethylene glycol) Chemistry: Biotechnical and Biomedical Applications. Springer, Boston, MA. ISBN 978-1-4899-0703-5. https://doi.org/10.1007/978-1-4899-0703-5

Hoare, T., J. Santamaria, G.F. Goya, S. Irusta, D. Lin, S. Lau, R. Padera, R. Langer and D.S. Kohane. 2009. A magnetically triggered composite membrane for on-demand drug delivery. *Nano Letters*, 9(10): 3651-3657. https://doi.org/10.1021/nl9018935

Hofmann, M. 2014. 3D printing gets a boost and opportunities with polymer materials. *ACS Macro Lett* 3(4): 382-386. https://doi.org/10.1021/mz4006556

Imbeni, V., J.J. Kruzic, G.W. Marshall, S.J. Marshall and R.O. Ritchie. 2005. The dentin-enamel junction and the fracture of human teeth. *Nature Materials* 4(3): 229-232. https://doi.org/10.1038/nmat1323

Ionov, L. 2013. Biomimetic hydrogel based actuating systems. *Adv Funct Mater* 23(36): 4555-4570. https://doi.org/10.1002/adfm.201203692

Jakus, A.E., E.B. Secor, A.L. Rutz, S.W. Jordan, M.C. Hersam and R.N. Shah. 2015. Three-dimensional printing of high-content graphene scaffolds for electronic and biomedical applications. *ACS Nano* 9(4): 4636-4648. https://doi.org/10.1021/acsnano.5b01179

Jakus, A.E., S.L. Taylor, N.R. Geisendorfer, D.C. Dunand and R.N. Shah. 2015. Metallic architectures from 3d-printed powder-based liquid inks. *Advanced Functional Materials* 25(45): 6985-6995. https://doi.org/10.1002/adfm.201503921

Kim, G., S. Ahn, H. Yoon, Y. Kim and W. Chun. 2009. A cryogenic direct-plotting system for fabrication of 3D collagen scaffolds for tissue engineering. *Journal of Material Chemistry* 19(46): 8817-8823. https://doi.org/10.1039/b914187a

Kim, H., G.H. Yang, C.H. Choi, Y.S. Cho and G.H. Kim. 2018. Gelatin/PVA scaffolds fabricated using a 3D-printing process employed with a low-temperature plate for hard tissue regeneration: Fabrication and characterizations. *International Journal of Biological Macromolecules* 120(A): 119-127. https://doi.org/10.1016/j.ijbiomac.2018.07.159

Kumar, S. 2003. Selective laser sintering: A qualitative and objective approach. *Journal of Metals* 55(10): 43-47. https://doi.org/10.1007/s11837-003-0175-y

Kumar, S. and K.L. Mittal (Eds.). 2013. Advances in Modeling and Design of Adhesively Bonded Systems. John Wiley & Sons. ISBN:9781118686379. https://doi.org/10.1002/9781118753682

Kumar, S., B.L. Wardle and M.F. Arif. 2017. Strength and performance enhancement of bonded joints by spatial tailoring of adhesive compliance via 3D printing. *ACS Applied Materials & Interfaces* 9(1): 884-891. https://doi.org/ 10.1021/acsami.6b13038

Labbé, A., S. Carlotti, A. Deffieux and A. Hirao. 2007. Controlled polymerization of glycidyl methyl ether initiated by onium salt/triisobutylaluminum and investigation of the polymer LCST. *Macromolecular Symposia* 249-250(1): 392-397. https://doi.org/10.1002/masy.200750409

Lee, B.F., M.J. Kade, J. Chute, N. Gupta, L.M. Campos, G.H. Fredrickson, J. Kramer, N. Lynd and C.J. Hawker. 2011. Poly(allyl glycidyl ether) – A versatile and functional polyether platform. *Journal of Polymer Science, Part A: Polymer Chemistry* 49(20): 4498-4504. https://doi.org/10.1002/pola.24891

Lee, M.P., G.J.T. Cooper, T. Hinkley, G.M. Gibson, M.J. and L. Cronin. 2015. Development of a 3D printer using scanning projection stereolithography. *Scientific Reports* 5: 9875. https://doi.org/10.1038/srep09875

Liljenhjerte, J., P. Upadhyaya and S. Kumar. 2016. Hyperelastic strain measurements and constitutive parameters identification of 3D printed soft polymers by image processing. *Additive Manufacturing* 11: 40-48. https://doi.org/10.1016/j.addma.2016.03.005

Lin, E., Li, Y., Ortiz, C. and M.C. Boyce. 2014. 3D printed, bio-inspired prototypes and analytical models for structured suture interfaces with geometrically-tuned deformation and failure behavior. *Journal of the Mechanics and Physics of Solids* 73: 166-182. https://doi.org/10.1016/j.jmps.2014.08.011

Luo, W., Y. Wang, E. Hitz, Y. Lin, B. Yang and L. Hu. 2017. Solution processed boron nitride nanosheets: Synthesis, assemblies and emerging applications. *Advanced Functional Materials* 27(31): 1701450. https://doi.org/10.1002/adfm.201701450

Mäkitie, A., J. Korpela, L. Elomaa, M. Reivonen, A. Kokkari, M. Malin, H. Korhonen, X. Wang, J. Salo, E. Sihvo, M. Salmi, J. Partanen, K.S. Paloheimo, J. Tuomi, T. Närhi and J. Seppälä. 2013. Novel additive manufactured scaffolds for tissue engineered trachea research. *Acta Otolaryngol* 133(4): 412-417. https://doi.org/10.3109/00016489.2012.761725

Meincke, O., D. Kaempfer, H. Weickmann, C. Friedrich, M. Vathauer and H. Warth. 2004. Mechanical properties and electrical conductivity of carbon-nanotube filled polyamide-6 and its blends with acrylonitrile/butadiene/styrene. *Polymer* 45(3): 739-748. https://doi.org/10.1016/j.polymer.2003.12.013

Murr, L.E., S.M. Gaytan, F. Medina, H. Lopez, E. Martinez, B.I. Machado, D.H. Hernandez, L. Martinez, M.I. Lopez, R.B. Wicker and J. Bracke. 2010. Next-generation biomedical implants using additive manufacturing of complex, cellular and functional mesh arrays. *Philosophical Transactions* 368(1917): 1999-2032. https://doi.org/10.1098/rsta.2010.0010

Ngo, I.L., S. Jeon and C. Byon. 2016. Thermal conductivity of transparent and flexible polymers containing fillers: A literature review. *International Journal of Heat and Mass Transfer* 98: 219-226. https://doi.org/10.1016/j.ijheatmasstransfer.2016.02.082

Nguyen, K.T. and J.L. West. 2002. Photopolymerizable hydrogels for tissue engineering applications. *Biomaterials* 23(22): 4307-4314. https://doi.org/10.1016/s0142-9612(02)00175-8

Nikzad, M., S.H. Masood, I. Sbarski and A.M. Groth. 2010. Rheological properties of a particulate-filled polymeric composite through fused deposition process. *Materials Science Forum* 654-656: 2471-2474. https://doi.org/10.4028/www.scientific.net/msf.654-656.2471

Pakdel, A., Y. Bando and D. Golberg. 2014. Nano boron nitride flatland. *Chemical Society Reviews* 43(3): 934–959. https://doi.org/10.1039/c3cs60260e

Park, J.U., M. Hardy, S.J. Kang, K. Barton, K. Adair, D.K. Mukhopadhyay, C.Y. Lee, M.S. Strano, A.G. Alleyne and J.G. Georgiadis. 2007. High-resolution electrohydrodynamic jet printing. *Nature Materials* 6(10): 782-789. https://doi.org/10.1038/nmat1974

Paroissien, E., M. Sartor, J. Huet and F. Lachaud. 2007. Analytical two-dimensional model of a hybrid (bolted/bonded) single-lap joint. *Journal of Aircraft* 44(2): 573-582. https://doi.org/10.2514/1.24452

Patra, S. and V. Young. 2016. A review of 3D printing techniques and the future in biofabrication of bioprinted tissue. *Cell Biochemistry and Biophysics* 74(2): 93-98. https://doi.org/10.1007/s12013-016-0730-0

Peterson, G.I., M.B. Larsen, M.A. Ganter, D.W. Storti and A.J. Boydston. 2015. 3D-printed mechanochromic materials. *ACS Applied Materials & Interfaces* 7(1): 577-583. https://doi.org/10.1021/am506745m

Ren, L., M. Zhang, Y. Wang, H. Na and H. Zhang. 2014. The Influence of the arrangement of styrene in methyl methacrylate/butadiene/styrene on the properties of PMMA/SAN/MBS blends. *Polymers for Advanced Technologies* 25(3): 273-278. https://doi.org/10.1002/pat.3232

Richter, A., S. Klatt, G. Paschew and C. Klenke. 2009. Micropumps operated by swelling and shrinking of temperature-sensitive hydrogels. *Lab on a Chip* 9(4): 613-618. https://doi.org/10.1039/b810256b

Shallan, A.I., P. Smejkal, M. Corban, R.M. Guijt and M.C. Breadmore. 2014. Cost-effective three-dimensional printing of visibly transparent microchips within minutes. *Analytical Chemistry* 86(6): 3124-3130. https://doi.org/10.1021/ac4041857

Sun, C., N. Fang, D. Wu and X. Zhang. 2005. Projection micro-stereolithography using digital micro-mirror dynamic mask. *Sensors and Actuators A: Physical* 121(1): 113-120. https://doi.org/10.1016/j.sna.2004.12.011

Taniguchi, N., S. Fujibayashi, M. Takemoto, K. Sasaki, B. Otsuki, T. Nakamura, T. Matsushita, T. Kokubo and S. Matsuda. 2016. Effect of pore size on bone ingrowth into porous titanium implants fabricated by additive manufacturing: An *in vivo* experiment. *Materials Science and Engineering C* 59: 690-701. https://doi.org/10.1016/j.msec.2015.10.069

Tibbitt, M.W. and K.S. Anseth. 2009. Hydrogels as extracellular matrix mimics for 3D cell culture. *Biotechnology & Bioengineering* 103(4): 655-663. https://doi.org/10.1002/bit.22361

Tolochko, N.K.K., Y.V. Khlopkov, S.E. Mozzharov, M.B. Ignatiev, T. Laoui and V.I. Titov. 2000. Absorptance of powder materials suitable for laser sintering. *Rapid Prototyping Journal* 6(3): 155-161. https://doi.org/10.1108/13552540010337029

Vaezi, M., H. Seitz and S. Yang. 2013. A review on 3D micro-additive manufacturing technologies. *The International Journal of Advanced Manufacturing Technology* 67(5-8): 1721-1754. https://doi.org/10.1007/s00170-012-4605-2

Weisensel, L., N. Travitzky, H. Sieber and P. Greil. 2004. Laminated object manufacturing (LOM) of SiSiC composites. *Advanced Engineering Materials* 6(11): 899-903. https://doi.org/10.1002/adem.200400112

Yang, D., Y. Li and J. Nie. 2007. Preparation of gelatin/PVA nanofibers and their potential application in controlled release of drugs. *Carbohydrate Polymers* 69: 538-543. https://doi.org/10.1016/j.carbpol.2007.01.008

Yang, J., X. Zheng, B. Zhang, R. Fu and X. Chen. 2011. Intrinsic fluorescence studies of conformational relaxation and its dynamics of triblock copolymer during the micellization in selective solvents. *Macromolecules* 44(4): 1026-1033. https://doi.org/10.1021/ma102243u

Yue, J., P. Zhao, J.Y. Gerasimov, M. van de Lagemaat, A. Grotenhuis, M. Rustema-Abbing, H.C. van der Mei, H.J. Busscher, A. Herrmann and Y. Ren. 2015. 3D-printable antimicrobial composite resins. *Advanced Functional Materials* 25(43): 6756-6767. https://doi.org/10.1002/adfm.201502384

Zarek, M., M. Layani, I. Cooperstein, E. Sachyani, D. Cohn and S. Magdassi. 2016. 3D printing of shape memory polymers for flexible electronic devices. *Advanced Materials* 28(22): 4449–4454. https://doi.org/10.1002/adma.201503132

Zhang, M., A. Vora, W. Han, R.J. Wojtecki, H. Maune, A.B.A. Le, L.E. Thompson, G.M. McClelland, F. Ribet, A.C. Engler and A. Nelson. 2015. Dual-responsive hydrogels for direct-write 3D printing. *Macromolecules*, 48(18): 6482-6488. DOI: 10.1021/acs.macromol.5b01550.

Zheng, X., J. Deotte, M.P. Alonso, G.R. Farquar, T.H. Weisgraber, S. Gemberling, H. Lee, N. Fang and C.M. Spadaccini. 2012. Design and optimization of a light-emitting diode projection micro-stereolithography three-dimensional manufacturing system. *Review of Scientific Instruments*, 83(12): 125001. https://doi.org/10.1063/1.4769050

CHAPTER

9

Polyvinylcarbazole Composite Membranes

Gaurav Sharma[1] and Balasubramanian Kandasubramanian[2*]

[1] Nanomaterials Characterization Lab, Centre for Converging Technologies University of Rajasthan, JLN Marg, Jaipur - 302001, Rajasthan, India

[2] Nano Surface Texturing Lab, Department of Metallurgical and Materials Engineering, DIAT(DU), DRDO, Ministry of Defence, Girinagar, Pune - 411025, India

1. Introduction

Polyvinylcarbazole (PVK) is a type of thermoplastic polymer that can be synthesized by the radical polymerization of N-vinylcarbazole. N-vinylcarbazole (9-vinylcarbazole), generally produced by the chemical reaction of acetylene with carbazole in the presence of a hydroxide alkali or by the reaction of acetylene with alkali salts of carbazole, is shown in Fig. 1 (Sandler and Karo 1996).

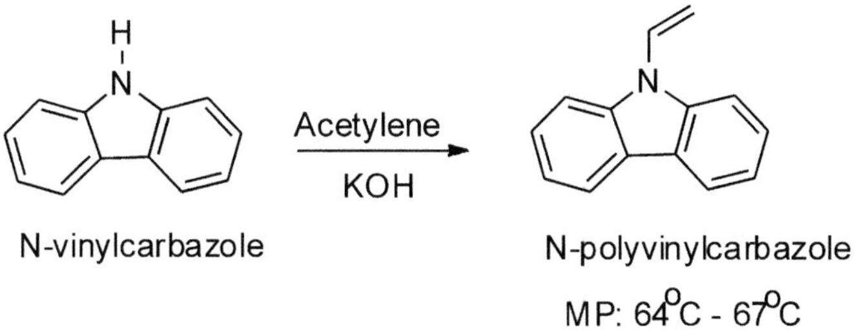

Fig. 1. Synthesis reaction of PVK using N-vinylcarbazole.

The polymer was first produced by Walter Reppe, Ernst and Eugen and was first patented by I.G. Farben in 1937. It was the first polymer whose photoconductivity was acknowledged and, because of this property, the

*Corresponding author: meetkbs@gmail.com

polymer established the fundamentals for organic LEDs and photorefractive polymers. PVK is a temperature-resistant polymer and can be used in the temperature range of 160°C-170°C. Variation in illumination changes the electrical conductivity of PVK, making it a semiconductor (photoconductor). In its chemical properties, it is soluble in aromatic hydrocarbons, HCls and ketones, where it is resistant to acids, bases, polar solvents and aliphatic hydrocarbon compounds. Combining these physical and chemical properties, PVK is an appropriate material for use in composites incorporating other materials. Since its discovery, research communities have designed novel and sturdy composites and films using PVK as the key material for various environmental and physical applications. Its optical and electrical properties enable its use in electrophotography, organic LEDs, photovoltaic devices (solar cells), holography and many other applications. Composites using PVK as base are strong, sturdy, durable, and resistant to wear and tear, temperature, weather and harsh chemicals (Wypych 2012). Some of the key advantages and general applications of PVK composites are presented in Fig. 2.

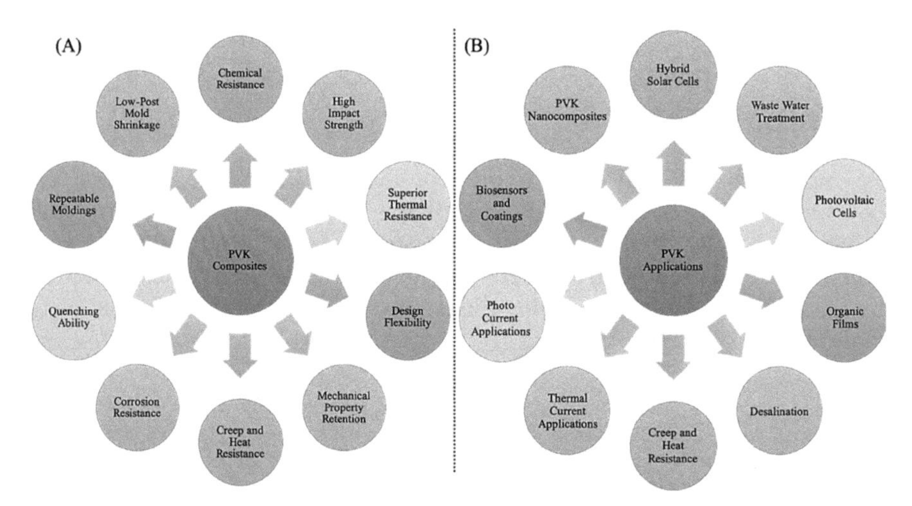

Fig. 2. (A) Key advantages of PVK composites/materials. (B) Some general applications of PVK.

2. Polyvinylcarbazole: Study of charge conduction

2.1. Charge carriers: dispersive behavior of hole transport in PVK

The continuous-time random walk model of Scher and Montroll is the most frequently used theoretical approach for defining charge carriers in PVK (Scher and Montroll 1975). This model was introduced as a generalization of physical diffusion process to efficiently describe the anomalous diffusion in polymers and thus PVK. The model assumes that charge transport in PVK is governed by a hopping rate w with a functional form:

$$w \propto e^{r/r_0} e^{-\Delta E/KT} \tag{1}$$

where w represents the hopping rate with hopping distance r, r_0 characterizes the interaction range responsible for hopping, and ΔE is the activation energy for hopping. K is Boltzmann constant with T as temperature.

In case of ordered PVK crystal, w acts as constant for carrier motion along a fixed direction with r_0 and ΔE as constants from one site to another. Ordered systems of PVK exhibit transport according to gaussian nature due to the gaussian nature of the charge carrier wave packet that moves through the material. In this case, the hopping time distribution becomes exponential:

$$\Psi(t) \propto e^{-wt} \tag{2}$$

In disordered systems of PVK, both the interaction range and activation energy can differ in statistical manner. This difference leads to a widespread distribution of hopping rate that results in continuous hopping time in the experimental scale. In this case, the distribution of hopping time can be expressed as:

$$\psi(t) \propto t^{-(1+\alpha)} \tag{3}$$

where α represents the extent of disorder, such that the lower the value of α, the higher the measure of disorder. This type of transport is known as dispersive transport of holes in PVK conduction mechanism. In experimental terms, charge carrier transport in PVK can be governed by Time-of-Flight (TOF) techniques (Kepler 1960). In an experimental setup of PVK-TOF, a panel of charge is held at one interface with powerfully absorbed radiation pulse. A schematic TOF setup for PVK sample is shown in Fig. 3 (Bos et al. 1989). Under the effect of external electric field, the charged panel is flowed through the PVK sample and leaves the sample at the end electrode. This phenomenon is used by the electrodes to determine the produced current as a function of time and tries to relate the absorbed pulse rate with the theoretical predictions.

2.2. Transfer of electronic excitation energy

The transfer of excitation energy in organic molecules such as PVK is common and well known. In organic high molecules, energy transfers from one host molecule to guest molecules such as polystyrene (Leibowitz and Weinreb 1967). Klopffer et al. (Klöpffer 1969) described the ultraviolet excited fluorescence in amorphous PVK of high purity. A similar model has been proposed by Birks for certain crystals (The relations between the fluorescence and absorption properties of organic molecules, 1963). Various quenching substances were taken into account for this study. The most

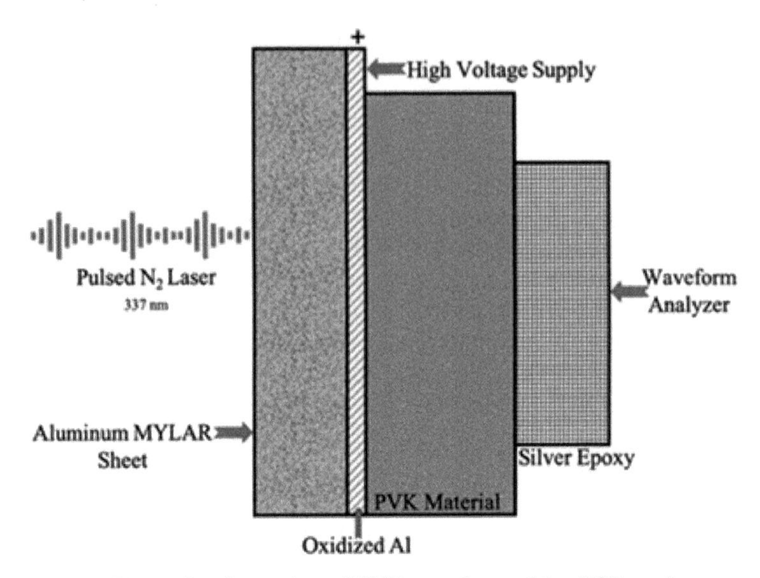

Fig. 3. Configuration of PVK sample used for TOF study.

important guest material used in the process was perylene. Perylene is a powerful quencher in terms of dipole-dipole resonance. The required PVK is prepared by following extensive purification and careful polymerization of vinyl carbazole. Beckman DK-2A spectrophotometer was used to record all the spectra. Uniform thin films of PVK were prepared by using dip coating from the solution of benzene where the thickness was given by weight and area. Finally, the benzene traces were removed from in the free environment to obtain dry and clear PVK films. The illuminated surface of the sample was used to derive fluorescent light that entered the monochromator. The sample was swapped with an uncoated material, where no comeback response was observed in the spectral region even at uppermost sensitivity. The ultraviolet absorption spectra of the PVK solution and the highest peaks found were compared with the spectrum of N-isopropyl carbazole (NIPC). The comparison studies showed that spectral positions and absorption intensities did not change while moving from monomeric NIPC to polymer. This result proves that the carbazole chromophores of the polymer work as autonomous units in the process of light absorption for both solid and solution conditions. The solid PVK presents excimer fluorescence and that transfer of energy was due to the migration of excitation and a significant contest between the guest molecules and excimer-forming spots of polymer.

3. Polyvinylcarbazole composites/nanocomposites and films

A PVK composite material is a combination of PVK with another material having different physical and chemical properties. Fiber-reinforced-polymer

(FRP) composites are mainly made from a matrix of polymer reinforced with synthetic or natural fiber or other substance. The work of the matrix is to protect the fibers from external wear and tear or environmental damage. The fibers offer strength to help the matrix resist cracks and fractures. PVK composites are being designed for multiple tasks. For example, FRP composites are being used to replace metal materials and their alloys. In this chapter, we discuss numerous novel and important nanocomposites and their applications that are widely used in various industrial scales.

3.1. Preparation of PVK nanocomposites, films and coatings

There are various chemical methods reported to prepare PVK-based nanocomposites, including chemical vapour deposition, physical vapour deposition, thermal spray coating, electrodeposition and electroless deposition, solution dispersion, and dip coatings (Nguyen-Tri et al. 2018). The most commonly used and understood methods for preparing PVK nanocomposites are sol-gel and cold spray coating. PVK nanocomposites use an organic matrix (Cui et al. 2011). The sol-gel method is appropriate to acquire films of high quality up to a thickness of a few microns. Sol-gel method has some limitations for metallic coatings but it is a reliable method for producing PVK coatings. The schematic processes of sol-gel method are shown in Fig. 4 (Sanchez et al. 2011).

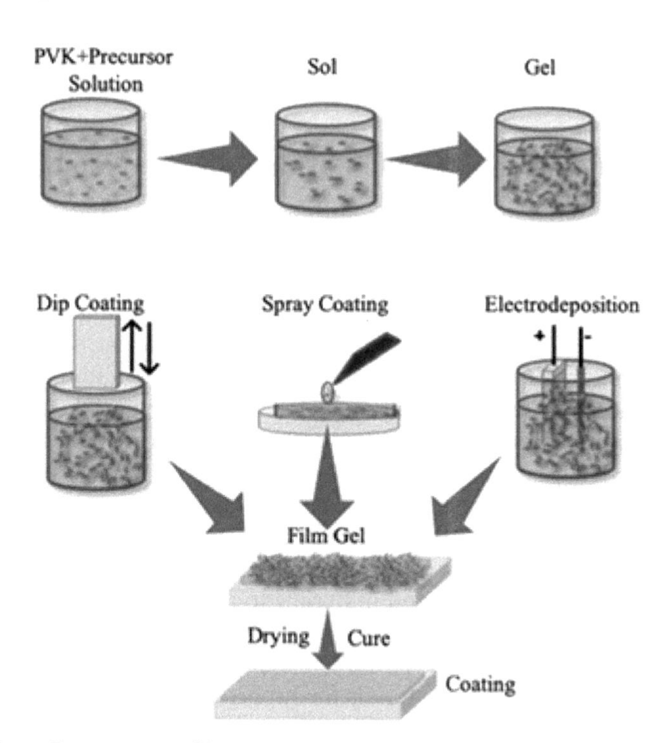

Fig. 4. Preparation of PVK-composite coatings using sol-gel methods.

Cold spraying methods of producing PVK nanocomposite films and coatings allow the designing process at lower temperatures than the actual melting point of sprayed materials (Luo and Li 2015). This method is also very reliable as it avoids the weakening of materials by oxidation and decomposition along with the phase transition. The coatings obtained by this method have low oxygen concentrations and high strength with strong adhesion. The method is suitable for producing nanocomposite films on metallic substrates such as copper, aluminum, or alloy (Kim et al. 2007, Sanpo et al. 2008, Luo and Li 2015). A schematic figure of cold spray coating is shown in Fig. 5 (Pialago and Park 2014).

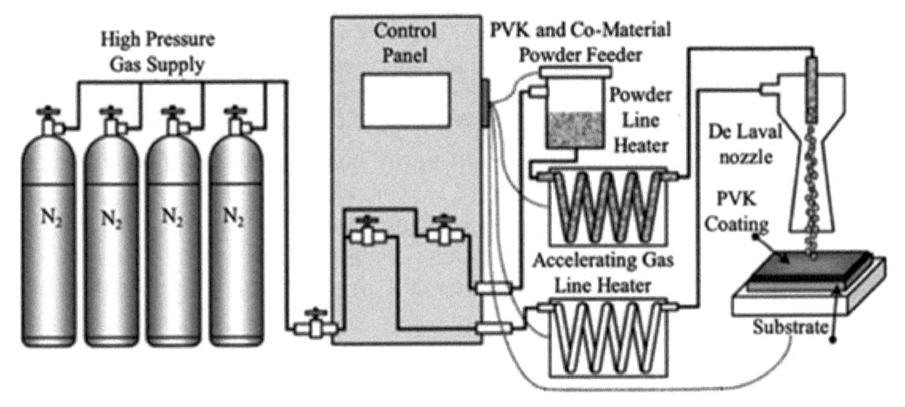

Fig. 5. Preparation of PVK-composite coating by cold spraying method.

3.2. Graphene oxide/ruthenium oxide/PVK nanocomposites

The cost of supercapacitors is an important factor in development of a practically cheaper device for use in electronic systems (Gao 2017). For reasons of doping capacity and physical morphology, electrochemical capacitors have confined specific capacitances in the typical range of 100-280 F/g (Wang et al. 2007). As an extension to this limitation, supercapacitors are a good choice as they do not possess any chemical reaction as the energy in supercapacitors being reserved on a electroactive material (Aradilla et al. 2011). Designing a ternary nanocomposite material can help improve the performance of supercapacitors. The rGO/RuO_2/PVK nanocomposites have properties that make them useful in supercapacitor applications. For this purpose, M. Ates et al. (2019) chose three principal materials: graphene oxide, ruthenium oxide and polyvinylcarbazole. RuO_2 nanoparticles can be used as a pseudo-capacitance on C-based materials to present an enhanced value of capacitance, where the protons are capable to get the inner part of RuO_2 with size declination (Yan et al. 2014). In this way, RuO_2 enhances the capacitance of nanocomposite electrode materials for supercapacitors. Furthermore, PVK is an amorphous hole-charge carrier vinyl polymer capable of having free-radicals, charge-transfer using holes and polymerization processes (Natori

2006). The nanoclusters of Ru were prepared in a controlled pH system below 4.9 in aqueous phase. Around 200 mL of distilled water was used to dissolve 41.5 mg of $RuCl_3$ with continuous stirring. Freshly prepared solution of $NaBH_4$ with DI water was added drop by drop in the stirring solution of $RuCl_3$. The chemical reaction of adding $NaBH_4$ to the $RuCl_3$ solution is shown in Equation 4 (Glavee et al. 1993), where the Ru^{+3} ions are reduced. After drop addition, the solution was stirred for more 5 min to get hydrosols of Ru where the pH was kept constant at 4.5 to increase the pore sizes of RuO_2 nanomaterials.

$$8Ru^{+3} + 3BH_4^- + 12H_2O \leftrightarrow 8Ru + 3B(OH)_4^- + 24H^+ \quad (4)$$

For incorporating Ru nanoclusters, 15 mL of distilled water was used to disperse 30 mg of graphene oxide. The mixture of Ru nanoclusters and GO was treated under ultrasonication for 10 min and stirred for 12 h; the synthesized material was rGO/RuO_2, which was then reduced under microwave methods for 10 min with 180 watt. The solution was allowed to cool and mixed under room temperature conditions. The process of rGO/RuO_2 synthesis is shown in Fig. 6.

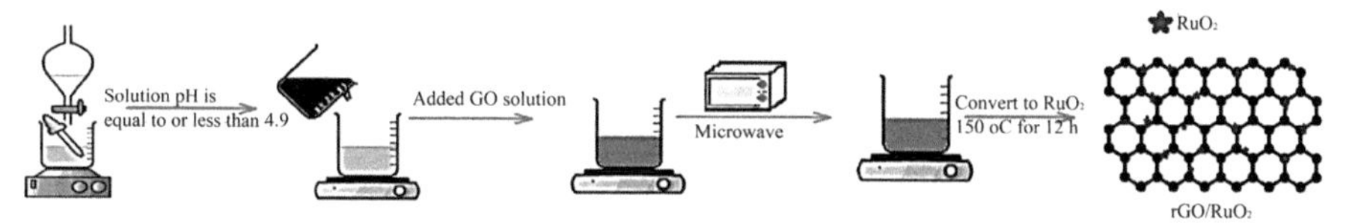

Fig. 6. Schematic representation for the synthesis of rGO/RuO_2.

After the preparation of this mixture solution, 50 mL of acetonitrile solution was used to dissolve 0.0517 mmol of 9-vinylcarbazole monomer. This solution of monomer was added to the solution of produced rGO/Ru hydrosols. In parallel, 0.01 mol of ammonium cerium nitrate was dissolved in 50 mL of acetonitrile and added to the above mixture. The polymerization was allowed to proceed for about 12 h. Finally, the precipitate of Ru/rGO was subjected to heating for 2 h at 150°C to transform Ru into RuO_2 to get rGO/RuO_2/PVK nanocomposites (Wang et al. 2016). The synthesis of this nanocomposite is shown in Fig. 7.

The major application of this nanocomposite is in the areas of electronics (supercapacitors). It offers low cost, higher electrochemical supercapacitor

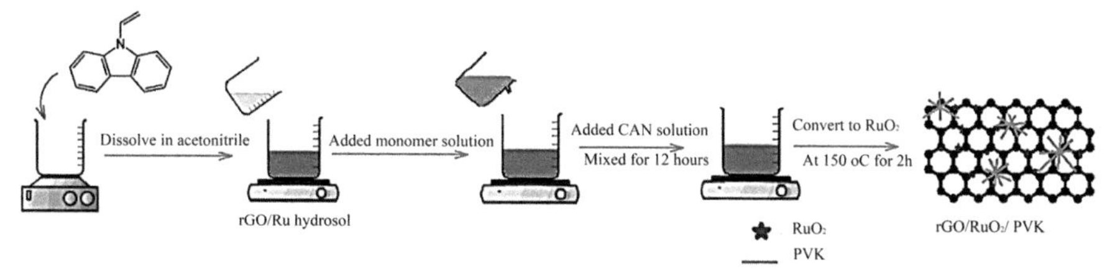

Fig. 7. Schematic representation for the synthesis of rGO/RuO_2/PVK nanocomposites.

performance and eco-friendly synthesis. Under various experimental studies, it was found that the rGO/RuO_2/PVK nanocomposites are stable during cycling and appropriate for use at lower frequency ranges.

3.3. PVK/reduced graphene oxide nanocomposite

One of the main sources of renewable energy is photovoltaic devices, which are used to convert solar energy to electricity. The high costs of silicon devices raised interest in technological advances that can focus sunlight and minimize cost (Green et al. 2014). For such photovoltaic applications, the two most important materials are graphene and ZnO. Graphene is much preferred over conventional electrodes because of its fair thermal and electrical performance (Joshi et al. 2013, Park et al. 2013, Thakur and Kandasubramanian 2019). ZnO holds good mobility for load transfer and enhanced photoactivity (Khurana et al. 2013). Mota et al. (Mota and de Fátima V. Marques 2019) described the production of a PVK nanocomposite and GO-ZnO nanostructured hybrid in which the GO was obtained by Hummers' method, ZnO and a hybrid of ZnO+GO by sol-gel methods with the resultant products ZnO SG and rGO-ZnO SG respectively. In the preparation of nanocomposite, 0.2 g of vinylcarbazole, 0.1% of nanoparticles per monomer mass and sodium (metallic) were added to 10 mL of DCM under N_2 flow in a Schlenk flask. Then the solution was subjected to sonication for 2 h. After sonication, $BF_3.OEt_2$ (0.0065 mol) was added to the solution, which changes the reaction color to dark green. The mixture was then transferred to a beaker containing 50 mL ethanol after continuous stirring for 10 min at 0°C. The polymer was precipitated, filtered, washed with methanol and then dried. The resultant compound is polyvinylcarbazole nanocomposite (PVK3-rGO-ZnOSG), and it is followed by cationic polymerization. The SEM images of produced graphene oxide, ZnO SG and rGO-ZnO SG are shown in Fig. 8.

- According to the SEM analysis of these compounds, it can be seen that GO maintains the structure of graphene lamellar (6b) but is defective because of the oxidation. The ZnO SG directed particle formation of poor morphology (6c) and agglomerates formed. The hybrid of rGO-ZnO exhibited reduced GO lamella comprising ZnO particles attached to the surface (6d).
- According to XRD analyses of GO, a diffraction peak at 26.39° was obtained, which specifies the layer spacing of the material to 0.34 nm. After analyzing the hybrid diffractogram, diffraction peak with a value of 26.5° was observed in response to the reduced graphene oxide. Furthermore, diffraction peaks were also detected at angles consistent with ZnO hexagonal wurtzite, which indicated the successful synthesis of PVK3-rGO-ZnO SG hybrid.

The above work reported by Mota et al. is an exceptional way to fuse PVK with the finest ability of photoconductors. In exhibiting a high yield of this nanocomposite, cationic polymerization was seen to be very effective in

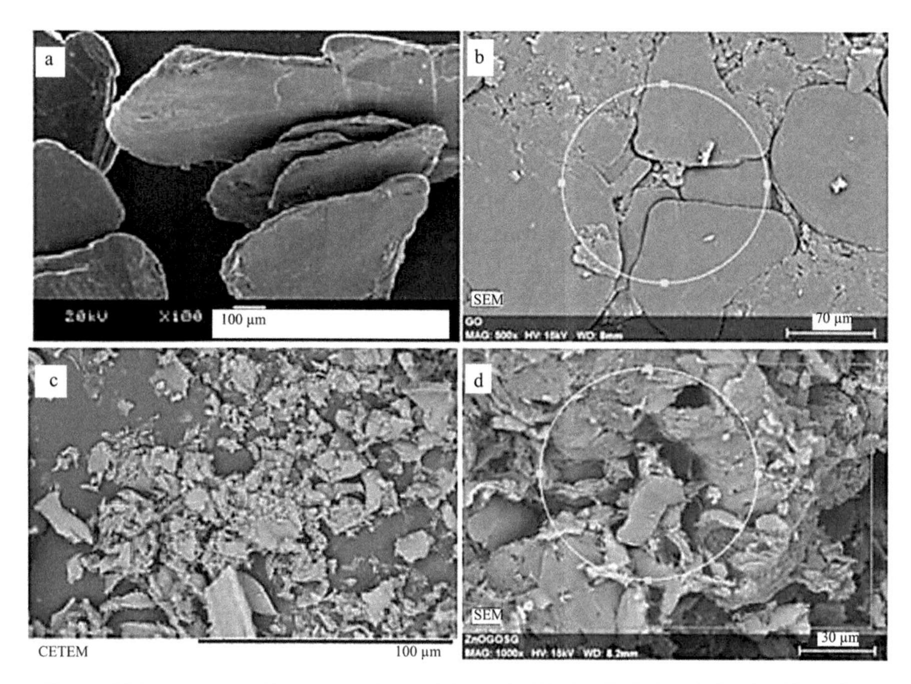

Fig. 8. SEM images of (a) original graphite; (b) GO; (c) ZnO SG; (d) rGO-ZnO SG.

this case. The fusion of nanoparticles enhanced the conducting capabilities of PVK, making it more favorable for photovoltaic applications.

3.4. CdSe/PVK nanocomposite films

Electroluminescence (EL) is the process of electron transfer by applying electric field to a particular material. EL has been reviewed in various reports and for materials based on semiconductors (Colvin et al. 1994). Thin film nanocomposites of II-VI compounds are preferable because materials supported by polymers exhibit good stability and processing (Sharma et al. 2019). PVK is a semiconductor with blue light emission and organic hole-transport polymer. It possesses excellent properties such as photoconduction, photorefraction and high refractive indices (Sanfelice et al. 2017). PVK semiconductors promise efficient charge generation and processing of organic polymer. Increasing the concentration of CdSe in PVK matrix also increases the luminescence property of the PVK nanocomposite films (Jeon et al. 2012, Sanfelice et al. 2017). The nanocomposites of CdSe/PVK show photoluminescence due to the nanoparticles of CdSe. A novel invention of CdSe/PVK nanocomposite films was reported by S. Kumari et al. (2019), where CdSe/PVK composite films were developed for significant photoluminescence. The research was based on the fact that the nanoparticles of CdSe can be finely dispersed in the organic type of solution and then can be mixed with PVK to induce the properties of charge transportation, electro- and photoluminescence for the display applications.

In the preparation of constituent materials, the source of selenium was prepared by taking 20 mL of DI water for dissolving 0.05 mol of selenium powder with the careful addition of 0.10 mol $NaBH_4$. Following magnetic stirring, the mixture was kept at room temperature for 2 h. Under these conditions, the selenium powder completely dissolved in water to give selenium solution. PVK solution was obtained by dissolving 400 mg of PVK in 10 mL of DMF on constant magnetic stirring at 60°C. On the other hand, 200.2 g of $CdCl_2$ was dissolved in 20 mL of DI water to get 0.05 mol of solution. The nanocomposite thin film of CdSe/PVK was prepared with prepared selenium, cadmium and PVK solution. The PVK solution was transferred to a beaker under magnetic stirring at 60°C. Simultaneously, 1 mL of $CdCl_2$ was added to the beaker containing PVK and stirred at the same temperature for 10 min, maintaining the pH of solution at 10 using the solutions of ammonia. After this, 1 mL of selenium solution was added drop by drop to the above prepared solution. After a careful stirring, the prepared mixture was cast on a glass substrate. Four samples of the same nanocomposite were taken for experimental studies. A SEM image of the produced CdSe/PVK nanocomposite is shown in Fig. 9. Following the FESEM, it is observed that the size of grains also increases with the increasing concentrations of CdSe in the PVK matrix.

Figure 10 presents the photoluminescence spectra and the effect of CdSe concentration on threshold voltage for CdSe/PVK nanocomposite.

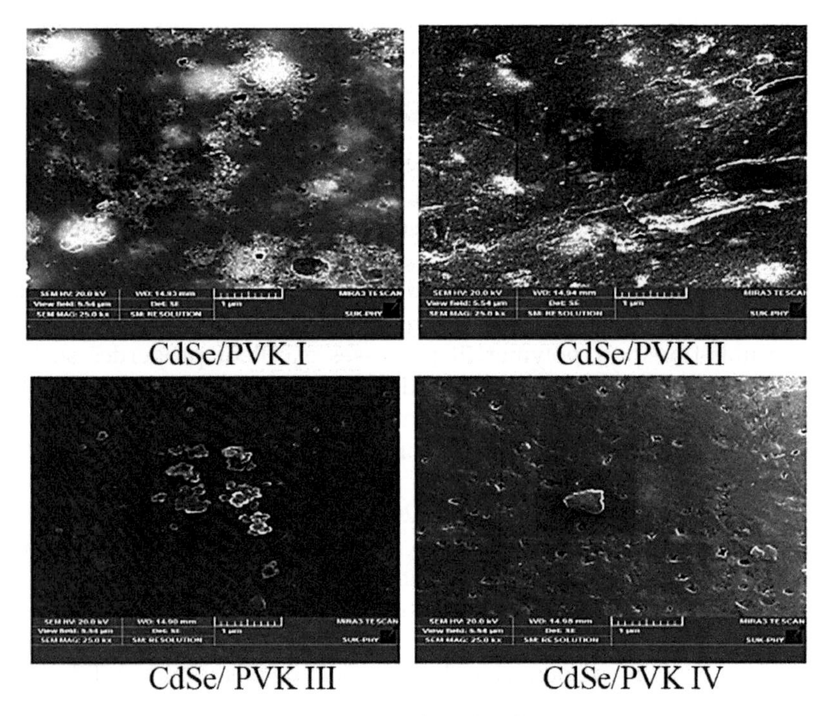

Fig. 9. FESEM images of the produced CdSe/PVK nanocomposite.

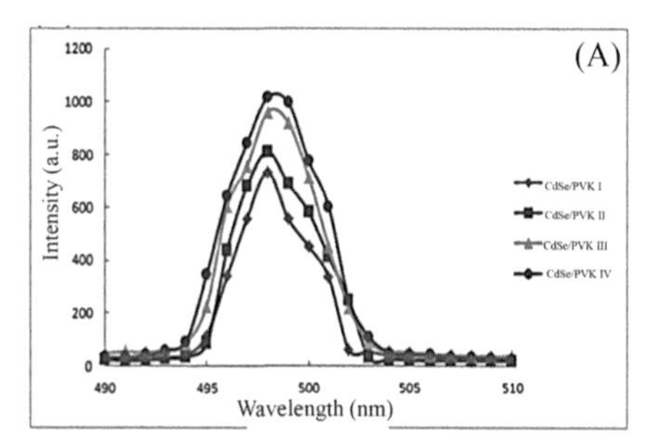

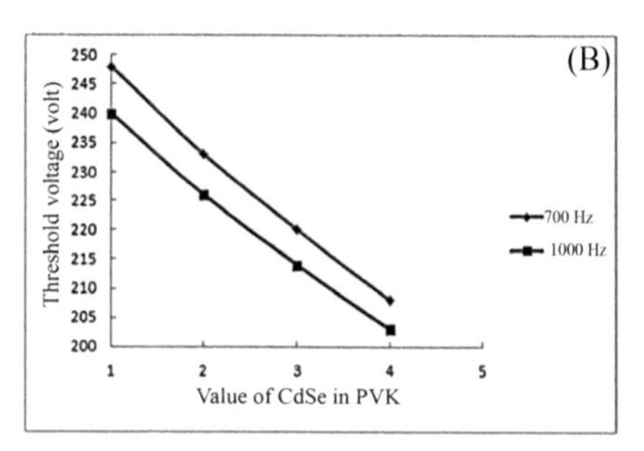

Fig. 10. (A) Photoluminescence spectra of CdSe/PVK nanocomposites. (B) Effect of increasing concentration of CdSe on the threshold voltage.

In the PL spectra, a single PL was observed at 498 nm, which shows that the luminescence is not due to the band-to-band transition, but to defect states (luminescence centers). The brightness of EL depends on the applied frequency and voltage where the emission starts only at a particular frequency and voltage, the threshold frequency and voltage. The high values of EL can be obtained by increasing the CdSe concentrations in the PVK matrix.

4. Applications of PVK-based nanocomposites

PVK is a π-conjugated saturated hydrocarbon having a helical configuration due to the presence of aromatic carbazolyl groups with raised chemical and thermal stability that favors electroluminescence and photorefractive-based devices (Cheng et al. 1997). Being a conducting polymer, PVK exhibits various properties that favor charge conduction mechanisms. The interactions of charge transfer can occur between electron-donor (carbazolyl) and the groups of electron-acceptor, both in polymer-polymer and polymer-low molecular systems (Fahrländer et al. 2003, Yu et al. 2010). We have studied and considered various examples of applications previously reported with novel finding.

4.1. Corrosion-resistant applications of PVK

Corrosion is a naturally occurring electrochemical process that can convert refined metals into chemically oxidized products and is mostly due to environmental factors. Corrosion is a dangerous and costly issue on the industrial scale. A variety of anti-corrosive coating materials are available in the market but are costly (Gore et al. 2019). In 1985, MacDiarmid suggested the use of conducting polymers for anti-corrosive applications because of their fine ambient stability and good chemical resistance to metals. Electroactive polymers such as polyaniline and polypyrrole are used in various anti-corrosive works for steel (Elangovan et al. 2018). PVK is also a promising candidate for these applications for the following reasons: (1) it can produce robust films, (2) it has good thermal, chemical and mechanical properties and charge carrier transport, and (3) it exhibits a wide use in LEDs, electrochromic devices, batteries, electrostatic discharge protections, and other applications (Frau et al. 2010). PVK coatings can be processed on various metal surfaces by the techniques of electrochemical polymerization. It is also confirmed by various studies that incorporating other favorable materials in the PVK matrix can help enhance its anti-corrosive properties. One such candidate is ZrO_2 nanoparticles, whose electrochemical and anti-corrosive influence has been studied by various scientists. Elangovan et al. (2018) recently reported the study of such nanocomposite of ZnO_2 nanoparticles and PVK for anti-corrosive coatings on steel substrates. This study showed that the nanocomposite coatings of PVK-ZrO_2 offer better protection from corrosion than PVK alone and can be considered a potential coating material for steel substrates. Frau et al. (2010) also offered a new approach for anti-corrosive coatings by electrodeposition of PVK. Potentiostatic methods were used for the electrodeposition of PVK films having crosslinked oligo- and PVK constituents. The coatings of PVK were fabricated and deposited on glass and steel substrates. There are a number of similar studies carried out by scientists for making reliable anti-corrosion coatings for valuable metals by incorporating PVK as a key element.

4.2. Water purification applications of PVK

Water purification systems often use membrane-based separation technology to remove micro-impurities from water (Sahoo and Balasubramanian 2015). The main problem with this technique is that it is sometimes compromised by bacterial adhesion on the surfaces, which leads to the formation of a bacterial film on the separation film (Khulbe et al. 2000, Hilal et al. 2004). So, there is a need to design antimicrobial separation films. Most of the available films are based on silver nanoparticles that reduce the bacterial and viral concentration from water and also improve mechanical strength and stability of the membrane along with the water flux (Peng et al. 2007, Kang et al. 2008, Aslan et al. 2010), but they are also limited by their high cost and maintenance. Similarly, single-walled nanotubes (SWNTs) are widely used in this field but are limited by their low dispersibility in aqueous solution. PVK nanocomposite coatings can be used for the modification of these membranes. A recent study by Ahmed et al. (2012) confirmed the antimicrobial effect of PVK nanocomposites on Gram (+) and Gram (–) bacteria. The multiple aromatic groups of PVK offer π-π connections with the SWNTs, making them better matched with SWNTs. PVK also exhibits outstanding thermal and biocompatible properties with low cost and easy maintenance (Ahuja et al. 2007). Cui et al. (2011) reported that the ratio of SWNT in the PVK-SWNT composite (97.3 wt% PVK:SWNT) also reduces the cost efficiency of SWNT and shows better dispersion of SWNTs in aqueous medium. A novel and first invention in designing PVK-SWNT was proposed by Farid Ahmed and his team from the University of Houston. Ahmed et al. (2013) described the antimicrobial properties of nitrocellulose membranes layered with the nanocomposites of PVK-SWNT, where PVK-SWNT nanocomposites were synthesized with highly purified and characterized SWNT of 97:3% wt ratio of PVK:SWNT. The method of dip coating was used to coat the membranes with PVK-SWNT suspension. The team also demonstrated the cytotoxicity effects of this product with mammalian fibroblast cells for the treatment of drinking water. The membrane of nitrocellulose (10 μm) was dipped in 1 mg/ml each of PVK, SWNT and PVK-SWNT suspension for coating. Each membrane was placed separately in a disk covered with 5 ml of corresponding suspension for complete covering for 30 min, then the membranes were removed and dried in a vacuum oven overnight. The resultant product is membranes that can be directly applied to the water purification process. The morphology of these membranes is well defined by their SEM images, as shown in Fig. 11.

According to the SEM analysis, Fig. 11(a) depicts a layered and porous-mat structure of the original nitrocellulose membrane, whereas Fig. 11(b) and (c) show that the coating of PVK and PVK-SWNT formed denser and homogenous layers on the nitrocellulose membrane. The antibacterial studies were performed on *E. coli* and *B. subtilis*. Observing the comparative antibacterial properties, it was found that around 91% and 81% *E. coli* cells were deactivated on PVK-SWNT and SWNT respectively and around 90% and 40% *B. subtilis* cells were deactivated on the same. These results indicated that just 3% of SWNT with the incorporation of 97% PVK can have much better antibacterial properties than 100% of SWNT-coated membranes with the increased dispersion of SWNT in the aqueous systems. A digital representation of live and dead cells of *E. coli* and *B. subtilis* is shown in Fig. 12(a) and (b) with FITC as total bacteria and TRITC as dead bacteria, where control stands for normal membrane without the use of PVK-SWNT. The graph in 12(c) shows that the inactivation of *E. coli* and *B. subtilis* is much higher in the PVK-SWNT-coated nitrocellulose membrane than in the other ones.

4.3. Optoelectronic and photovoltaic applications of PVK

There are many chemical studies of PVKs in optoelectronics but now its derivatives are also widely used in optoelectronic devices such as photovoltaics and OLEDs (Xia et al. 2002). There are a number of advantages of using PVK and its derivatives in organic electronics: low cost, enhanced performance, reliable performance, durability, and better environmental stability. The balanced process of charge carrier transportation in PVK makes it a more desirable candidate for use in optoelectronics. PVKs have high thermal stability and predictable levels of HOMO-LUMO energy levels, which can be easily enhanced by substitution of chemicals (Ates and Uludag 2016). The key advantages of using PVKs are that they can form stable radical cations very quickly and offer high mobility of charge carriers, and various other substituents can be easily attached to them (Grazulevicius et al. 2003). PVK in small emitting layer molecules is used to simulate the characters of charge transfer when attached to an acceptor. These small molecules are used to produce OLEDs with quantum efficiency above 20%. PVK also contains

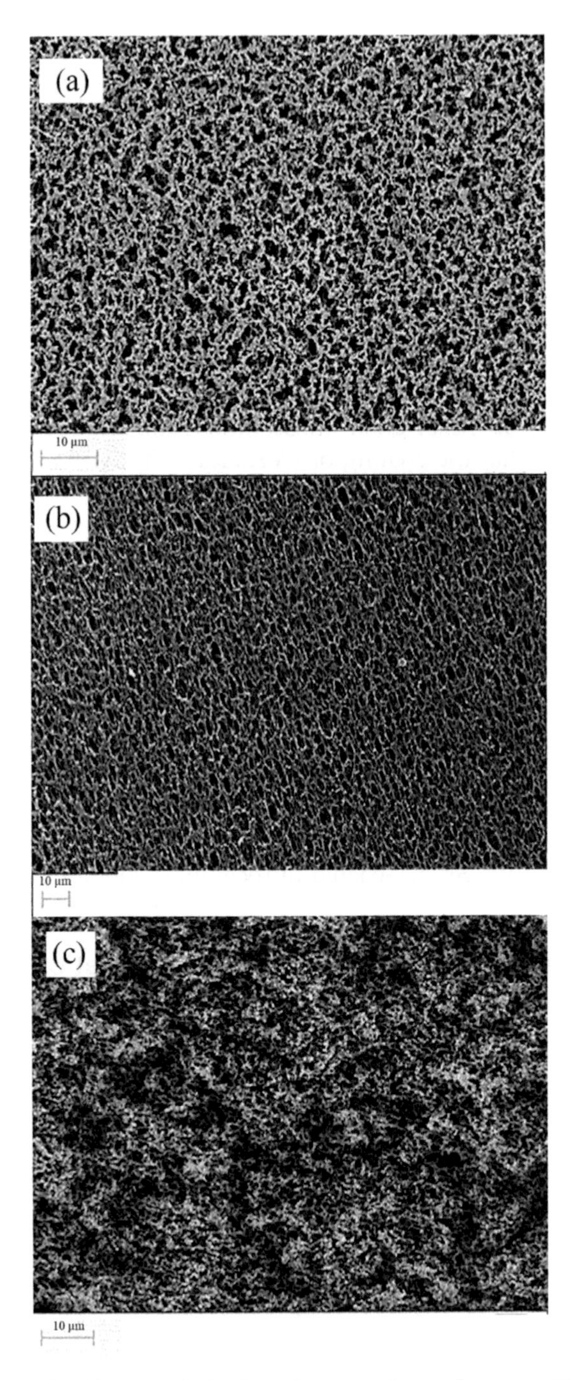

Fig. 11. SEM images for the morphologies of prepared membranes with a scale of 10 μm. (a) Simple nitrocellulose membrane; (b) NC membrane coated with SWNT; (c) NC membrane coated with PVK-SWNT.

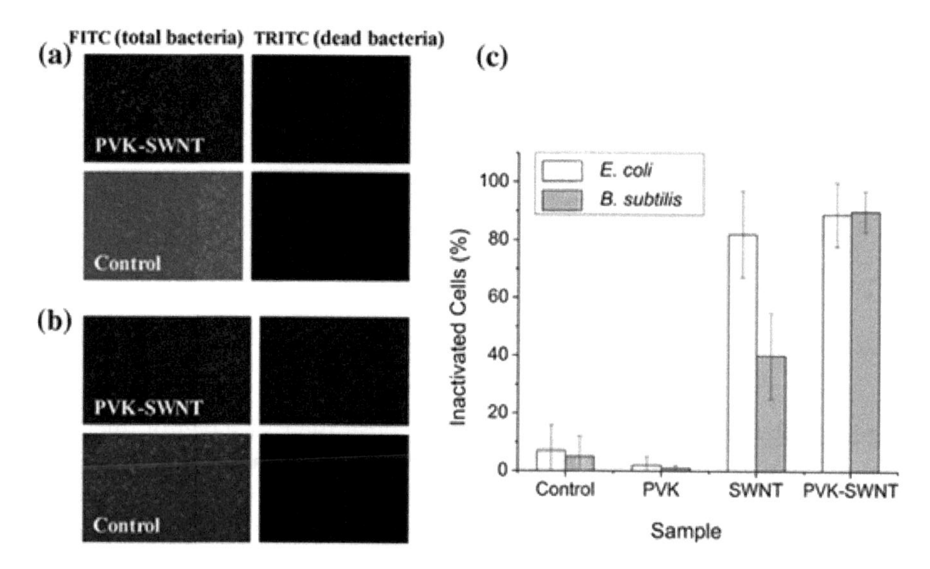

Fig. 12. (a) Digital images of *E. coli* bacteria with total and dead count in control (normal membrane), before and after PVK-SWNT–coated NC membrane. (b) Digital images of *B. subtilis* bacteria with total and dead count with control (normal membrane) before and after PVK-SWNT-coated NC membrane. (c) Percentage of bacterial inactivation with normal membrane, PVK-, SWNT- and PVK-SWNT-coated membranes.

local triplet levels that are necessary for adequate coupling in thermally activated decayed fluorescence, which will be used in OLEDs in future (Higginbotham et al. 2017).

Among photovoltaic devices, the solar cell is the most commonly used device that produces electric current in proportion to the incident radiation. PVK-based solar cells can be designed to enhance the charge conduction and efficiency of solar cells. Most of the available solar cells are inorganic and completely based on silicon substrates. Prado et al. (2008) reported a study on solar cells combining inorganic and organic materials in which PVK acted as the active layer used for converting light energy into electricity. The internal electric field mechanism and outer structure of the device are shown in Fig. 13.

The process of manufacturing a solar cell with PVK as the active layer on a glass substrate having tin oxide as conduction layer is as follows:

- **Step 1:** Chemical deposition method was used to put a a-Si:H-p (amorphous silicon hydrogenated with boron) layer upon TiO_2 conduction layer over a reactor composed of a chamber of deposition, gas flow, rotary pump and control of temperature on substrate. The process of depositing this layer was conducted at a temperature of 200°C and the thickness kept to 20 nm for collecting electrons produced on the polymer.

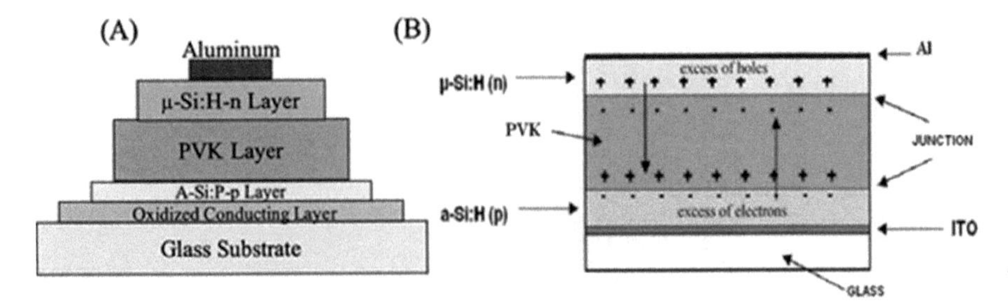

Fig. 13. (A) Structural depiction of PVK-TiO_2 dye sensitized cell. (B) Internal electric field representation of PVK-TiO_2 cell.

- **Step 2:** Casting method was used to deposit the PVK layer of thickness 1.6 μm. Then, a μ-Si:H-n (microcrystalline silicon hydrogenated) layer was placed over the PVK using chemical deposition method at 100°C providing a 50 nm thickness liable for collecting gaps produced in the polymer. Lastly, an aluminum layer was placed by thermal metallization process upon the n layer (μ-Si:H) forming the contact between glass substrate and the layers.

The cell was designed to combine inorganic and organic materials on a single substrate, ensuring easy process control and low cost. The developed material was a thermoplastic (transparent) with good chemical and thermal stability and becomes electrically conductive when exposed to sunlight. The studies carried out in the field of photovoltaic devices are meant to improve performance and implementation of these systems.

4.4. Solar energy harvesting applications of PVK

The development and organized design of PVK nanocomposites is getting good attention in solar energy harvesting research because of its novel properties in tuning the performance of photovoltaic devices (Komarneni 1992). Nanocomposites of PVK and titania are a good choice for solar energy harvesting applications. The studies of TiO_2 are generally in two phases, anatase and rutile, where anatase has a band gap of ~3.2eV, which is a photoactive phase, and rutile has a band gap of ~3eV, which is a thermodynamically stable phase (Boppella et al. 2012). PVK and titania have been extensively studied in nanodevice fabrications such as solar cells, memory cards, and photo detectors (Cho et al. 2009). Jin et al. (2006) from the University of Queensland designed a photovoltaic device using PVK-ZnS nanocomposite with open circuit of high voltage 1.65. Buschbaum et al. (Kaune et al. 2008) designed layered nanocomposites of PVK-TiO_2 for UV active photovoltaics. The use of PVK-titania nanocomposites increases the optical absorption and decreases the rate of recombination by separation of charge, which also offers the benefit of large surface area-volume ratio and enhances the performance of the device. Aashish et al. (2016) depicted a novel and self-assembled approach for PVK-titanium nanotubes (PVTs) as a photoanode for harvesting of solar energy. During polymerization, the molecules of PVK attached inside the titania nanotube walls through interaction of ions and dipoles, which enhanced the system of photoanode by higher harvesting of photon and simplifying the process of charge transfer, and ultimately the device performance. PVTs were synthesized by chemical oxidative polymerization using $FeCl_3$ as initiator. The anatase titania and rutile titania were prepared by hydrothermal reactions followed by sonication to get the white powders of nanotitania. Dye sensitized solar cells were fabricated by using the prepared nanocomposites as photoanodes. The efficiency of power conversion was observed to be 3.03%, which is the highest among the photoanode systems of PVK nanocomposites. Observing the results and significance, these approaches can be accepted to fabricate photovoltaic devices for solar energy harvesting applications. The energy transfer mechanism in PVK-TiO_2 nanocomposite is shown in Fig. 14.

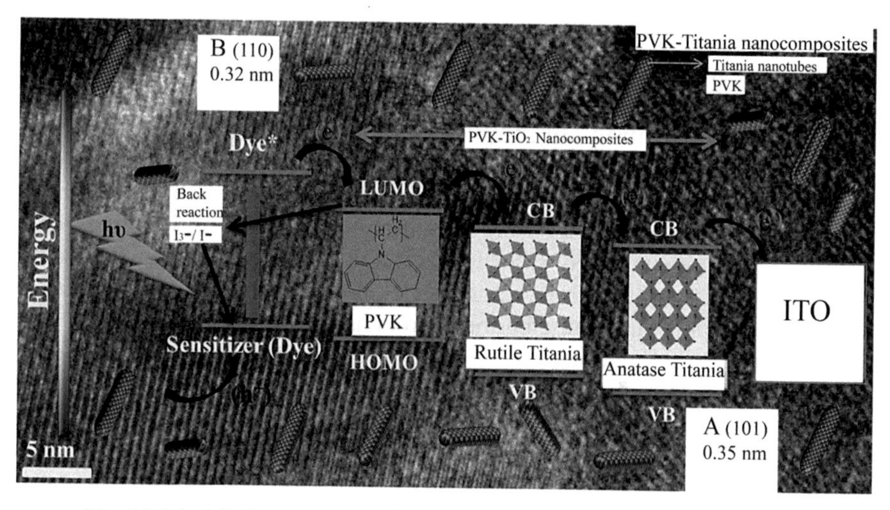

Fig. 14. Model of energy transfer mechanism in PVK-TiO_2-based DSSC.

4.5. Radioprotectors, cellular markers and antimicrobial applications of PVK

Radioprotectors: Radioprotectors (radioprotective agents) are complexes that avert injuries from radiation when used before exposure. They balance the adverse effects of radiotherapy and unintentional radiation exposure. Scientists studying this application established new tests with PVK that contained molecules of "anti-rads". González-Pérez et al. (Gonzalez-Perez et al. 2012, González-Pérez and Burillo 2013) improved the chemical structure of rubbers based on silicones to advance their resistance to radiation. Thus, PVK is added in commercial silicone films used for sealant films as they play the role of "anti-rad" and are a highly conjugated aromatic compound. The "anti-rads" of films without PVK inhibit the crosslinking upon radiation exposure, while in the case of PVK-based films, the crosslinking is supressed by the conjugated-aromatics owing to the lower sensitivity to ionizing radiation than other complexes and their ability to dissipate radiation. The efficiency of radioprotection can be assessed by $P = (D - D_0)/D$, where D is the absorbed radiation dose in the initial polymer material and D_0 is the protective agent polymer having physiochemical variations of the sealant, such as Young's modulus, elongation, density of crosslinking, and molecular weight. The finest radioprotection is considered to be achieved when the produced radiation energy swiftly moves from the initial polymer to the protective agent (Grigoras 2016).

Cellular markers: Detection technologies based on nanoparticles such as bioluminescence and fluorescence have successfully proved their capability in improving sensitivity of detection. In this technology, the optical signal resulting from the device is recorded and associated with the biomarker concentration from biological fluids. A biomarker is a quantifiable indicator of physical, functional, chemical and pathological grade of cells (Jain 2010). There are a number of scientific studies that confirm the contribution of PVK in this field. One of the most important, reported by Sun et al. (2013)· is the scheming of europium complex grafted/doped PVK dots. The group used two europium complexes: (i) Eu15F (having beta-diketone ligands attached with naphthyl groups) and (ii) EuDNM (dinaphthoylmethane complex), where the Pdots were designed using PVK and polystyrene-grafted-polyethylene glycol (PS-PEG-COOH) in THF solution. After being introduced with water, the chains of polymer distorted to form smaller nanoparticles of 20 nm diameter and entrenched the complexes of Europium inside them. PVK functioned as a host matrix that detached the complexes of Europium and reduced them. The surface reactions of bio-conjugations were then performed between streptavidin (a protein extracted from bacteria *Streptomyces avidinii*) and the Eu/PVK-Pdot complex. The *in vitro* targeting of particular cell was carried out for tumour cells and the receptors of the cell were named EpCAM (Epithelial Cell Adhesion Module). EpCAM was used as a biomarker for sensing and spotting the circulation of tumour cells. The study also reported that the incorporation of PVK-Pdots can enhance the properties of nanoparticles in cellular biomarker and imaging applications (Li et al. 2014).

Antimicrobial coatings: It is a well-known fact that bacteria have naturally high hydrophobic-surface affinity. Chemists therefore adjust different types of surfaces to produce chemistry of hydrophilicity by incorporating polymer materials for the purpose of antibacterial coating (Yadav and Balasubramanian 2013, Ramdayal and Balasubramanian 2014). Highly focused studies have been carried out on the bacteria-graphene interaction. The PVK-graphene nanocomposite films in biosensors (Prajapati and Kandasubramanian 2019), purification technologies, medical coated devices, and other products require two features: (i) low cytotoxicity for living cells and (ii) well-characterized antimicrobial properties. Santos et al.

(2011, 2012) first described the designing of PVK-graphene nanocomposite thin films for antimicrobial properties using solution mixing and sonication methods. The antibacterial studies performed on *E. coli* and *B. subtilis* showed that more than 80% of the bacteria found with unmodified films were inhibited by PVK-G-modified nanocomposite films.

5. Future trends and conclusion

The multifunctionality and improved properties of PVK nanocomposites have attracted much attention from various research communities but their use and applications have still not been applied extensively on a commercial and industrial scale. In global marketing terms, packaging applications consume around $38 billion every year and that figure continues to rise by 3.5% annually. PVK packaging films can be useful in this regard because of their low cost, recyclability and better stability than other packaging materials. PVK is suitable for use in packaging owing to its good mechanical strength, thermal, chemical and dimensional stability, and good heat and gas resistance. Coatings consisting of PVK nanocomposites not only serve as a material protection, they also play other roles by virtue of the presence of co-composite materials such as nanofillers. Antibacterial and smart coatings are the best examples of this case. We will face many challenges and perils in the coming future, mainly in the energy sector, and research on sustainable energy is expected to boom in both theoretical and experimental terms. PVK nanocomposites will be an important aspect of these developments. There are numerous expected future applications of PVK nanocomposites, for example, "self-cleaning" films that will be able to coat glass and building materials that can save energy and water used in cleaning. Smart coatings will also help in amplifying sustainable energy by their multifunctionality in photovoltaic cells, especially solar-to-fuel conversion applications (Yadav et al. 2018). PVK nanocomposites are also expected to be more numerous in systems of drug delivery, anti-corrosion barrier coatings, self-scratch repair, fire-resistant coatings, and screen effect coatings. The increasing number of patents and publications from different parts of the world and related technologies have initiated a chain reaction in the domains of PVK and other polymer-based nanocomposites. It is expected that, by the year 2025, the market of polymer-nanocomposites will rise by about 60%, and there will correspondingly be a boom in research expansion for PVK-based nanocomposites. Use of nanoparticle-reinforced PVK composites can be greatly profitable in chemical, biological, medical, electronic, and various other emerging sectors. Graphene-reinforced PVK composites are also expected to extend their applications as thermal barriers (Badhe and Balasubramanian 2014a, b, Kumar and Balasubramanian 2016) and will facilitate the understanding of thermodynamics in space science and thermal stabilities of materials in military tools and ballistics. High diversity, properties, stability, and good host-matrix affinity are the practical advantages of PVK-based nanocomposites. These multifunctional PVK composites, nanocomposites and films at reasonable cost will soon be seen in the global markets.

With advances in nanotechnology and related fields, nanocomposites are becoming smarter, cheaper and multifunctional. The incorporation of PVK with nanomaterials for designing significant composites is an exciting new field of research. We reported various studies here to correlate them with the present proceedings to make them useful for future investigations. Various physical, chemical, electronic, medical, and biological applications have been briefly described here with possible schematics and examples. Only novel preparations and applications have been reported in this chapter to highlight the foundations of the findings and experiments that can make new pathways for more innovations.

Conflicts of interest

The authors declare that there are no conflicts of interest regarding the publication of this chapter.

Acknowledgements

The authors want to express their gratitude to Dr. C.P. Ramanarayanan, Vice-Chancellor, DIAT-DRDO, Pune, for his encouragement and support. Also, the first author wants to acknowledge Shivani Rastogi and Swaroop Gharde for technical support and refinement.

References

Aashish, Ramakrishnan, J.D. Sudha, Sankaran and Krishnapriya. 2016. Self-assembled hybrid polyvinylcarbazole–titania nanotubes as an efficient photoanode for solar energy harvesting. *Solar Energy Materials and Solar Cells* 151: 169-178.

Ahmed, F., C.M. Santos, J. Mangadlao, Advincula and D.F. Rodrigues. 2013. Antimicrobial PVK:SWNT nanocomposite coated membrane for water purification: Performance and toxicity testing. *Water Research* 47(12): 3966-3975.

Ahmed, F., C.M. Santos, R.A.M.V. Vergara, M.C.R. Tria, Advincula and D.F. Rodrigues. 2012. Antimicrobial applications of electroactive PVK-SWNT nanocomposites. *Environmental Science & Technology* 46(3): 1804-1810.

Ahuja, T., I. Mir, Kumar and Rajesh. 2007. Biomolecular immobilization on conducting polymers for biosensing applications. *Biomaterials* 28(5): 791-805.

Aradilla, Estrany and C. Alemán. 2011. Symmetric Supercapacitors based on multilayers of conducting polymers. *The Journal of Physical Chemistry* 115(16): 8430-8438.

Aslan, C.Z. Loebick, S. Kang, M. Elimelech, L.D. Pfefferle and Van Tassel. 2010. Antimicrobial biomaterials based on carbon nanotubes dispersed in poly(lactic-co-glycolic acid). *Nanoscale* 2(9): 1789.

Ates, M. and N. Uludag. 2016. Carbazole derivative synthesis and their electropolymerization. *Journal of Solid State Electrochemistry* 20(10): 2599-2612.

Ates, Yildirim, Kuzgun and H. Ozkan. 2019. The synthesis of rGO, rGO/RuO_2 and rGO/RuO_2/PVK nanocomposites, and their supercapacitors. *Journal of Alloys and Compounds* 787: 851-864.

Badhe and K. Balasubramanian. 2014a. Reticulated three-dimensional network ablative composites for heat shields in thermal protection systems. *RSC Advances* 4(82): 43708-43719.

Badhe and K. Balasubramanian. 2014b. Novel hybrid ablative composites of resorcinol formaldehyde as thermal protection systems for re-entry vehicles. *RSC Advances* 4(55): 28956.

Birks, J.B. and D.J. Dyson. 1963. The relations between the fluorescence and absorption properties of organic molecules. *Proceedings of the Royal Society of London. Series A. Mathematical and Physical Sciences* 275(1360): 135-148.

Boppella, R., P. Basak and Manorama. 2012. Viable method for the synthesis of biphasic TiO_2 nanocrystals with tunable phase composition and enabled visible-light photocatalytic performance. *ACS Applied Materials & Interfaces* 4(3): 1239-1246.

Bos, F.C., T. Guion and Burland. 1989. Dispersive nature of hole transport in polyvinylcarbazole. *Physical Review B* 39(17): 12633-12641.

Cheng, N., B. Swedek and Prasad. 1997. Thermal fixing of refractive index gratings in a photorefractive polymer. *Applied Physics Letters* 71(13): 1828-1830.

Cho, B., Kim, M. Choe, Wang, Song and Lee. 2009. Unipolar nonvolatile memory devices with composites of poly(9-vinylcarbazole) and titanium dioxide nanoparticles. *Organic Electronics* 10(3): 473-477.

Colvin, Schlamp and Alivisatos. 1994. Light-emitting diodes made from cadmium selenide nanocrystals and a semiconducting polymer. *Nature* 370(6488): 354-357.

Cui, K.M., Tria, R. Pernites, Binag and Advincula. 2011. PVK/MWNT Electrodeposited conjugated polymer network nanocomposite films. *ACS Applied Materials & Interfaces* 3(7): 2300-2308.

Elangovan, S. Pugalmani, Srinivasan and N. Rajendiran. 2018. Synthesis and anticorrosive properties of novel PVK-ZrO_2 nano composite coatings on steel-substrate. *e-Journal of Surface Science and Nanotechnology* 16: 5-13.

Fahrländer, M., K. Fuchs, R. Mülhaupt and Friedrich. 2003. Linear and nonlinear rheological properties of self-assembling tectons in polypropylene matrices. *Macromolecules* 36(10): 3749-3757.

Frau, A.F., Pernites and Advincula. 2010. A conjugated polymer network approach to anticorrosion coatings: Poly(vinylcarbazole) electrodeposition. *Industrial & Engineering Chemistry Research* 49(20): 9789-9797.

Gao, Y. 2017. Graphene and polymer composites for supercapacitor applications: A review. *Nanoscale Research Letters* 12(1): 387.

Glavee, Klabunde, Sorensen and G.C. Hadjipanayis. 1993. Borohydride reduction of cobalt ions in water. Chemistry leading to nanoscale metal, boride, or borate particles. *Langmuir* 9(1): 162-169.

González-Pérez and G. Burillo. 2013. Modification of silicone sealant to improve gamma radiation resistance, by addition of protective agents. *Radiation Physics and Chemistry* 90: 98-103.

Gonzalez-Perez, G. Burillo, Ogawa and Avalos-Borja. 2012. Grafting of styrene and 2-vinylnaphthalene onto silicone rubber to improve radiation resistance. *Polymer Degradation and Stability* 97(8): 1495-1503.

Gore, P.M., S. Balakrishnan and K. Balasubramanian. 2019. Superhydrophobic corrosion inhibition polymer coatings. *In*: Superhydrophobic Polymer Coatings. pp. 223-243. Elsevier. Netherlands, UK.

Grazulevicius, J.V., Strohriegl, Pielichowski and K. Pielichowski. 2003. Carbazole-containing polymers: Synthesis, properties and applications. *Progress in Polymer Science* 28(9): 1297-1353.

Green, M.A., Ho-Baillie and Snaith. 2014. The emergence of perovskite solar cells. *Nature Photonics* 8(7): 506-514.

Grigoras, A.G. 2016. A review on medical applications of poly(N-vinylcarbazole) and its derivatives. *International Journal of Polymeric Materials and Polymeric Biomaterials* 65(17): 888-900.

Higginbotham, Karon, P. Ledwon and Data. 2017. Carbazoles in optoelectronic applications. *Display and Imaging*. 2: 207-216.

Hilal, N., V. Kochkodan, L. Al-Khatib and T. Levadna. 2004. Surface modified polymeric membranes to reduce (bio)fouling: A microbiological study using *E. coli*. *Desalination* 167: 293-300.

Jain, K.K. 2010. The Handbook of Biomarkers. Totowa, NJ: Humana Press.

Jeon, Y.P., S.J. Park and T.W. Kim. 2012. Electrical and optical properties of blue organic light-emitting devices fabricated utilizing color conversion CdSe and CdSe/ZnS quantum dots embedded in a poly(N-vibyl carbazole) hole transport layer. *Optical Materials Express* 2(5): 663.

Jin, H., Y. Hou, X. Meng and F. Teng. 2006. High open-circuit voltage in UV photovoltaic cell based on polymer/inorganic bilayer structure. *Chemical Physics* 330(3): 501-505.

Joshi, A. Bajaj, R. Singh, P.S. Alegaonkar, K. Balasubramanian and Datar. 2013. Graphene nanoribbon-PVA composite as EMI shielding material in the X band. *Nanotechnology* 24(45): 455705.

Kang, Herzberg M., D.F. Rodrigues and Elimelech. 2008. Antibacterial effects of carbon nanotubes: Size does matter. *Langmuir* 24(13): 6409-6413.

Kaune, G., W. Wang, Metwalli, M. Ruderer, R. Rober, Roth and Müller-Buschbaum. 2008. Layered TiO_2:PVK nano-composite thin films for photovoltaic applications. *The European Physical Journal E* 26(1): 73.

Kepler, R.G. 1960. Charge carrier production and mobility in anthracene crystals. *Physical Review* 119(4): 1226-1229.

Khulbe, K.C., Matsuura, Singh, Lamarche and S.H. Noh. 2000. Study on fouling of ultrafiltration membrane by electron spin resonance. *Journal of Membrane Science* 167(2): 263-273.

Khurana, S. Sahoo, S.K. Barik and Katiyar. 2013. Improved photovoltaic performance of dye sensitized solar cell using ZnO-graphene nano-composites. *Journal of Alloys and Compounds* 578: 257-260.

Kim, J.S., Y.S. Kwon, Lomovsky, Dudina, Kosarev, Klinkov, D.H. Kwon and I. Smurov. 2007. Cold spraying of in situ produced TiB_2-Cu nanocomposite powders. *Composites Science and Technology* 67(11): 2292-2296.

Klopffer, W. 1969. Transfer of electronic excitation energy in polyvinyl carbazole. *The Journal of Chemical Physics* 50(6): 2337-2343.

Komarneni. 1992. Feature article. Nanocomposites. *Journal of Materials Chemistry* 2(12): 1219.

Kumar and K. Balasubramanian. 2016. Progress update on failure mechanisms of advanced thermal barrier coatings: A review. *Progress in Organic Coatings* 90: 54-82.

Kumari, S., K.K. Kushwah, S. Dubey and M. Ramrakhiani. 2019. Studies on CdSe/PVK nanocomposites films for electroluminescent display applications. *Optical Materials* 97: 109319.

Leibowitz M. and Weinreb. 1967. Effects of fluorescence and energy transfer in polystyrene under excitation in the vacuum ultraviolet. *The Journal of Chemical Physics* 46(12): 4652-4659.

Li, Zhang, W. Sun, Yu, Wu, W. Qin and D.T. Chiu. 2014. Europium-complex-grafted polymer dots for amplified quenching and cellular imaging applications. *Langmuir* 30(28): 8607-8614.

Luo, X.-T. and C.J. Li. 2015. Large sized cubic BN reinforced nanocomposite with improved abrasive wear resistance deposited by cold spray. *Materials & Design* 83: 249-256.

Mota, I.C., V. de Fátima and M. Marques. 2019. Synthesis of polyvinylcarbazole/reduced graphite oxide-ZnO nanocomposites. *Macromolecular Symposia* 383(1): 1700081.

Natori I. 2006. Anionic polymerization of N-vinylcarbazole with alkyllithium as an initiator. *Macromolecules* 39(18): 6017-6024.

Nguyen-Tri, Nguyen, P. Carriere and Ngo Xuan. 2018. Nanocomposite coatings: Preparation, characterization, properties, and applications. *International Journal of Corrosion* 2018: 1-19.

Park, H., S. Chang, Jean, J.J. Cheng, Araujo, M. Wang, M.G. Bawendi, M.S. Dresselhaus, V. Bulovic, Kong and Gradecak. 2013. Graphene cathode-based ZnO nanowire hybrid solar cells. *Nano Letters* 13(1): 233-239.

Peng, F., C. Hu and Z. Jiang. 2007. Novel ploy(vinyl alcohol)/carbon nanotube hybrid membranes for pervaporation separation of benzene/cyclohexane mixtures. *Journal of Membrane Science* 297(1): 236-242.

Pialago and Park. 2014. Cold spray deposition characteristics of mechanically alloyed Cu-CNT composite powders. *Applied Surface Science* 308: 63-74.

Prado, Moreira, S. Possidonio and K.R. Onmori. 2008. Use of polymer poly(N-vinylcarbazole) for photovoltaic applications. Translated from University of Sao Paulo, Brazil.

Prajapati, D.G. and B. Kandasubramanian. 2019. Progress in the development of intrinsically conducting polymer composites as biosensors. *Macromolecular Chemistry and Physics* 220(10): 1800561.

Ramdayal and K. Balasubramanian. 2014. Antibacterial application of polyvinylalcohol-nanogold composite membranes. *Colloids and Surfaces A: Physicochemical and Engineering Aspects* 455: 174-178.

Sahoo and K. Balasubramanian. 2015. A nanocellular PVDF-graphite water-repellent composite coating. *RSC Advances* 5(9): 6743-6751.

Sanchez, Belleville, M. Popall and L. Nicole. 2011. Applications of advanced hybrid organic-inorganic nanomaterials: From laboratory to market. *Chemical Society Reviews* 40(2): 696.

Sandler and Karo. 1996. Polymerization reactions of N-vinylcarbazole and related monomers. *In*: Polymer Synthesis. pp. 183-201. Elsevier.

Sanfelice, L.A. Mercante, Pavinatto, N.B. Tomazio, C.R. Mendonca, Ribeiro, Mattoso and D.S. Correa. 2017. Hybrid composite material based on polythiophene derivative nanofibers modified with gold nanoparticles for optoelectronics applications. *Journal of Materials Science* 52(4): 1919-1929.

Sanpo, N., Lu Saraswati and P. Cheang. 2008. Anti-bacterial property of cold sprayed ZnO-Al coating. *In*: 2008 International Conference on BioMedical Engineering and Informatics. pp. 488-491. IEEE.

Santos, C.M., Mangadlao, F. Ahmed, A. Leon, Advincula and Rodrigues. 2012. Graphene nanocomposite for biomedical applications: Fabrication, antimicrobial and cytotoxic investigations. *Nanotechnology* 23(39): 395101.

Santos, C.M., M.C.R. Tria, Vergara, F. Ahmed, R.C. Advincula and D.F. Rodrigues. 2011. Antimicrobial graphene polymer (PVK-GO) nanocomposite films. *Chemical Communications* 47(31): 8892.

Scher, H. and E.W. Montroll. 1975. Anomalous transit-time dispersion in amorphous solids. *Physical Review B* 12(6): 2455-2477.

Sharma, P. Kumar and G. Verma. 2019. Role of shell type of core/shell nanoparticles in luminescence properties of PVK-CdS/X nanocomposite films. *Applied Physics A* 125(5): 351.

Sun, W., J. Yu, R. Deng, Y. Rong, Fujimoto, C. Wu, H. Zhang and D.T. Chiu. 2013. Semiconducting polymer dots doped with europium complexes showing ultranarrow emission and long luminescence lifetime for time-gated cellular imaging. *Angewandte Chemie International Edition* 52(43): 11294-11297.

Thakur, K. and B. Kandasubramanian. 2019. Graphene and graphene oxide-based composites for removal of organic pollutants: A review. *Journal of Chemical & Engineering Data* 64(3): 833-867.

Wang, Y. Xu, X. Chen and X. Du. 2007. Electrochemical supercapacitor electrode material based on poly(3,4-ethylenedioxythiophene)/polypyrrole composite. *Journal of Power Sources* 163(2): 1120-1125.

Wang, H. Liu, Xu, Y. Chen, J. Yang and Q. Tan. 2016. Supported ultrafine ruthenium oxides with specific capacitance up to 1099 F g^{-1} for a supercapacitor. *Electrochimica Acta* 194: 211-218.

Wypych, G. 2012. PVK poly(N-vinyl carbazole). *In*: Handbook of Polymers. pp. 617-619. Elsevier. New York, US.

Xia, X. Fan, J. Locklin and R.C. Advincula. 2002. A first synthesis of thiophene dendrimers. *Organic Letters* 4(12): 2067-2070.

Yadav and K. Balasubramanian. 2013. Egg albumin PVA hybrid membranes for antibacterial application. *Materials Letters* 110: 130-133.

Yadav, R., Subhash, Chemmenchery and K. Balasubramanian. 2018. Graphene and graphene oxide for fuel cell technology. *Industrial & Engineering Chemistry Research* 57(29): 9333-9350.

Yan, Y. Liu, Li, R. Zhuo, Wu, Ren, Li, J. Wang, Yan and Z. Geng. 2014. Synthesis and electrochemical properties of MnO_2/rGO/PEDOT:PSS ternary composite electrode material for supercapacitors. *Materials Letters* 127: 53-55.

Yu, Y., Y. Yao, L. Wang and Li. 2010. Charge-transfer interaction between Poly(9-vinylcarbazole) and 3,5-dinitrobenzamido group or 3-nitrobenzamido group. *Langmuir* 26(5): 3275-3279.

CHAPTER

10

Elastomeric and Plastomeric Materials

Mohsen Khodadadi Yazdi[1], Payam Zarrintaj[2*], Saeed Manouchehri[2], Joshua D. Ramsey[2], Mohammad Reza Ganjali[1,3] and Mohammad Reza Saeb[4]

[1] Center of Excellence in Electrochemistry, School of Chemistry, College of Science, University of Tehran, Tehran, Iran

[2] School of Chemical Engineering, Oklahoma State University, Stillwater, OK, USA

[3] Biosensor Research Center, Endocrinology and Metabolism Molecular-Cellular Sciences Institute, Tehran University of Medical Sciences, Tehran, Iran

[4] Department of Resin and Additives, Institute for Color Science and Technology, P.O. Box: 16765-654, Tehran, Iran

1. Introduction

Elastomers are highly elastic and amorphous polymers known and distinguished from other polymers for their low glass transition temperature and relatively low Young's modulus. Low intermolecular interactions in the elastomers result in high extensibility and flexibility (Bhowmick 2008). They are predominantly thermosetting elastomers (TSEs), while there is also another group, known as thermoplastic elastomers (TPEs), which can be processed properly by using conventional thermal operations such as extrusion and injection molding (Paran et al. 2019) . TSEs cover two-thirds of the global elastomer market because of their high resistance to heat and chemicals, aesthetic appearance, and good mechanical properties. Styrene-butadiene rubber (SBR) is a TSE that is widely used in tire manufacturing, automotives, medical applications, consumer goods, and industrial sectors that are the major consumers of elastomers (X. Liu et al. 2014). Common unsaturated rubbers that can be cured using vulcanization agents, especially sulfur, include synthetic polyisoprene (isoprene rubber) and its natural counterpart, polybutadiene, poly(isobutylene-co-isoprene) or butyl rubber and its halogenated derivatives (especially chlorinated and brominated), polychloroprene (chloroprene rubber), and poly(acrylonitrile-co-butadiene) or nitrile rubber. Well-known saturated rubbers are ethylene propylene

*Corresponding author: payam.zarrintaj@gmail.com

diene rubber, epichlorohydrin rubber, silicone rubber, and ethylene-vinyl acetate (Boczkowska 2012).

Plastomers that combine properties of elastomers with the processability of plastics are commonly considered equivalent to TPEs. However, sometimes the term "plastomer" is used for ethylene alpha-olefin (α-olefin) copolymers. Incorporation of low quantity of an α-olefin comonomer (e.g., 1-hexene and 1-octene) to common crystalline homopolymers (especially polyethylene or isotactic polypropylene) results in more amorphous structures in plastomers (D'orazio et al. 1982). Besides, α-olefin copolymers possess uniform composition, at both inter- and intramolecular scales (Mohammadi et al. 2014). Since the predominant monomer in plastomers is commonly ethylene or isotactic propylene, plastomers are categorized as E-plastomers and P-plastomers. The flexural modulus of plastomers is lower than that of commodity and engineering plastics. Packaging, both food and non-food, is the leading application of plastomers; besides, these copolymers are widely used to modify properties of polymers for various applications. The emerging class of plastomers consists of the newly developed bottlebrush elastomers, which are treated in this chapter. This chapter particularly deals with plastomers employed as TPEs including poly(ethylene/isotactic polypropylene-co-α-olefins) and bottlebrush elastomers (Bhowmick 2008 #288).

2. Natural and nature-inspired elastomers

Nature has gifted us various kinds of macromolecules such as polysaccharides and proteins that play crucial roles in biological systems (Mahmodi et al. 2020). Many polysaccharides (e.g., cellulose and hyaluronic acid) and proteins (e.g., collagen and elastin) act as a structural part of plants and animal bodies, while they may also serve as food storage, catalysts, and signaling molecules. Many naturally occurring elastomers are summarized in this section (Khosravi et al. 2020).

Natural rubber is usually extracted as milky and sticky fluid, known as latex, from the rubber tree, *Hevea brasiliensis*. The latex is rich in polyisoprene, which is currently used in many applications such as automotives, gloves, hoses and belts, balloons, balls, lining materials, and insulation. Natural rubber is one of the most consumed rubbers across the globe; nowadays, this elastomer is mainly manufactured through synthetic routes in large-scale petrochemical plants. However, synthetic isoprene rubber possesses a simple chemical structure compared to the more elaborate proteins that are abundant in nature (De and White 2001, Hanhi et al. 2007).

Elastin, titin, resilin, abductin, byssus, dragline silks, flagelliform silk, gluten, and fibrillin are some of the elastomeric proteins that play important roles in biological systems (Tatham and Shewry 2000). These natural proteins are physically or covalently crosslinked. Resilin, elastin, and their alloys are elastomeric proteins that have been vastly investigated in many applications such as tissue engineering (Aghaei-Ghareh-Bolagh et al. 2016). Elastin and collagen are the major proteins in animal skin that provide the skin with proper texture and shape. Resilin is a structural elastic protein found in many insects and arthropods that possesses very high resilience, low stiffness, and exceptional fatigue lifetime (L. Li and Kiick 2013). Insect flight and the outstanding jumping ability of fleas are made possible by resilin, as shown in Fig. 1.

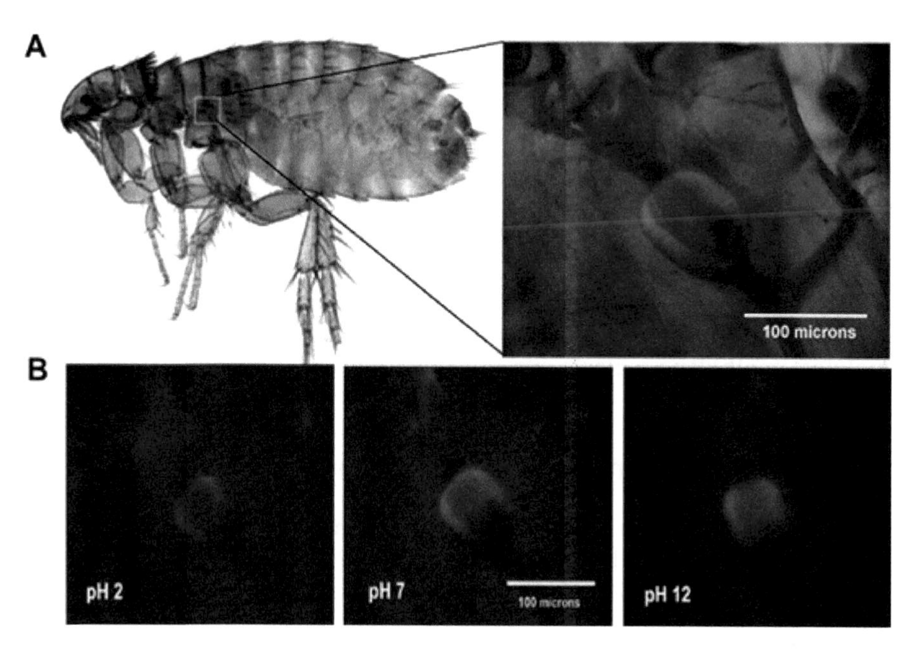

Fig. 1. (A) Resilin-rich area at the top of the hind legs of fleas under ultraviolet illumination. (B) pH-dependency of blue fluorescence originating from dityrosine crosslinks in resilin (Lyons et al. 2011).

Elastin-based materials have been used in a variety of applications such as fibrous scaffolds, hydrogels, and constructs (Yeo et al. 2015). Moreover, elastomeric polypeptides based on elastin and resilin that possess rubber-like elasticity are interesting biomaterials for tissue engineering and drug delivery (van Eldijk et al. 2011). While resilin and elastin have high resilience, there are also low-resilience elastomers in nature, such as byssus, which act as dampers that convert mechanical loads to thermal energy and are critical to survival. It is worth mentioning that hydration has a profound effect on the properties of elastomeric proteins (Bhowmick 2008).

Inspired by nature-based elastomeric proteins, scientists and engineers have been working to synthesize more effective elastomers that can mimic the dynamic mechanical behavior of biological systems such as soft tissue. In this regard, two main strategies have been used: synthetic biology and advanced polymer chemistry accompanied by engineering methods. Synthetic biology approaches have been used to make recombinant proteins based on resilin and elastin for applications in regenerative medicine (Annabi et al. 2013, Tjin et al. 2014). Sheiko Group has developed the concept of bottlebrush supersoft elastomers that can precisely mimic the complex dynamic mechanical behavior of soft tissue (Abbasi et al. 2019, Vatankhah-Varnoosfaderani et al. 2017b). Unlike recombinant proteins, bottlebrush polymers are completely synthetic polymers. These interesting elastomers are studied here in more detail.

3. Synthetic bioelastomers

There are many kinds of synthetic degradable bioelastomers that can be used in drug delivery and tissue engineering (Q. Liu et al. 2012). These biodegradable elastomers show high flexibility and elasticity, their mechanical properties mimic that of body tissue, and they have a 3D crosslinked structure. Degradable bioelastomers can be categorized as chemically and physically crosslinked bioelastomers. Some examples include poly(glycerol sebacate), poly(glycerol dodecanoate), poly(ε-caprolactone)-based bioelastomers, poly(octamethylene-maleate-citrate), polyurethane, and poly(ester amide) (Q. Liu et al. 2012).

4. Solvent-free and bottlebrush elastomer

Fabrication of biomaterials for tissue engineering applications has been the center of attention for decades (Zarrintaj et al. 2018a, Zarrintaj et al. 2018b). However, the fabrication of synthetic biomaterials that can mimic various physiochemical attributes of human tissue is a great challenge that arises from their composite nature. Conventional synthetic materials usually fail to mimic the mechanical properties of various tissues. For example, Young's modulus of most synthetic materials is higher than 100 kPa, much higher than that of soft tissues. Polymer gels such as hydrogels possess lower Young's modulus value because absorbed solvent molecules serve as plasticizers. However, solvent removal as the result of evaporation, leakage, or deformation (mechanical stresses) significantly increases their modulus.

Accordingly, design and manufacture of soft and solvent-free synthetic biomaterials can revolutionize the field of regenerative medicine (Daniel et al. 2016, Reynolds et al. 2020). Newly emerged bottlebrush polymers are plastomers that possess very low Young's modulus. These solvent-free architectures are highly elastic because their side-chains act as plasticizers that cannot migrate out, unlike solvent molecules or other common low-molecular-weight plasticizers such as dioctyl terephthalate. Bottlebrush plastomers can precisely mimic animal and human tissues such as porcine skin (Vatankhah-Varnosfaderani et al. 2018). Furthermore, they show extraordinary optical properties that enable adaptive coloration and camouflage coatings (Vatankhah-Varnosfaderani et al. 2018).

5. Smart elastomers and plastomers

Materials that can respond, controllably and repeatedly, to external stimuli (e.g., pH, temperature, light) in a dynamic mode are considered smart materials. For example, poly(N-isopropylacrylamide) is a thermo-responsive smart polymer (Aguilar and San Román 2019). Furthermore, polymers can be modified to respond to near-infrared light or/and electric/magnetic fields (Jochum and Theato 2013, Wei et al. 2017).

5.1. Dielectric elastomer

Soft and stimuli-responsive materials are important in the development of artificial muscle for soft robotics (Majidi 2019). For example, dielectric elastomer can be used for the fabrication of dielectric elastomer actuators (DEAs) (Duduta et al. 2019, Youn et al. 2020). Artificial muscles are mainly based on dielectric elastomer because of high energy density, cost, and fast response time (Qiu et al. 2019). For example, blends of waterborne polyurethane and poly(3,4-ethylenedioxythiophene) : poly(styrenesulfonate) (PEDOT:PSS) were used as transparent artificial muscles (Wang et al. 2019).

5.2. Shape memory elastomers

Shape memory polymers (SMPs) are smart polymers in which shape alters in response to external stimuli (Lendlein and Gould 2019, C. Liu et al. 2007). Thermo-shrinkable is an important subclass of SMPs whose shape changes as a result of temperature alternation. The soft segments of SMPs can bear reversible phase transformation leading to shape change. Phase transformation can occur at the glass transition temperature or melting temperature (Ratna and Karger-Kocsis 2008).

Various shape memory elastomers (SME) have been developed based on various elastomers (Maciejewska and Krzywania-Kaliszewska 2012). TPEs are among the interesting subclasses of SMEs (Mineart et al. 2016, Palza et al. 2018). For example, Mineart et al. manufactured an SME blend based on TPE triblock copolymer and a phase change material as additive (Mineart et al. 2016).

5.3. Chromogenic elastomers

Chromogenesis is a color change phenomenon related to changes in optical properties of materials (e.g., color and refractive index) in response to various stimuli. For example, the color of thermochromic and photochromic materials changes in response to temperature variation and light intensity alternation (Jenekhe and Kiserow 2004). Chromogenic polymers have gained attention as smart materials for various applications such as food packaging (Sadeghi et al. 2019).

Vatankhah-Varnosfaderani et al. managed to design and fabricate chromogenic elastomers based on a linear–bottlebrush–linear (LBBL) copolymer (Vatankhah-Varnosfaderani et al. 2018). This solvent-free elastomer showed tissue-like mechanical properties similar to porcine skin, i.e., very soft to the touch while exhibiting a high degree of strain-stiffening. Linear chains are flexible and get rigid when they aggregate; on the other hand, stiff bottlebrush strands can provide a very soft matrix. The interplay of these distinct features provides LBBL copolymers with extraordinary mechanical and optical attributes.

5.4. Other types

Magnetorheological elastomers, also known as magnetosensitive elastomers, are smart elastomers embedded with ferromagnetic particles (T. Liu and Xu 2019). Magnetorheological plastomers can be used to make soft actuators (Qi et al. 2019).

6. Liquid crystal elastomers

The integration of liquid crystalline anisotropy with the elastic nature of rubbers is observed in liquid crystal elastomers (LCEs) (Ula et al. 2018). LCEs show exceptional physical and optical properties; they can be used to fabricate actuators and artificial muscle. LCEs can serve as dynamic platforms for 3D cell culture (Prévôt et al. 2018). These anisotropic scaffolds affect the attachment of cell, proliferation, and alignment.

7. Elastomers in biomedical applications

Linear elastomers are stiff and elastic polymers. They are only characterized by one degree of polymerization (n_x). On the other hand, brush elastomer architecture is defined by three degrees of polymerization (DP) (n_x, n_{sc} [DP of side chain], n_g [DP of spacers between side chains]). Brush elastomers are supersoft and very elastic polymers but they fail to replicate the intense strain-stiffening behavior of biological tissue (Sheiko et al. 2018). The stiffness of these elastomers ranges from 10^2 to 10^6 Pa, similar to that of synthetic gels. Molecular brushes behave as permanent softeners that reduce equilibrium shear modulus, resulting in supersoft elastomers (Pakula et al. 2006).

LBBL triblock copolymers are moldable plastomers and dynamic. These plastomers require ϕ_A and χ along with n_x, n_{sc}, n_g to be defined (see Fig. 2). LBBL plastomers are supersoft and can conform to the intense strain-stiffening attribute of soft tissue (Sheiko et al. 2018).

Unlike hydrogels and soft tissue, the softness of which relies on composition (i.e., water content), the mechanical properties of solvent-free plastomers depend on chemical structure and architecture, which remains constant upon deformation. These synthetic dry elastomers can closely mimic the biological stress-strain behavior of natural soft tissue (Vatankhah-Varnosfaderani et al. 2017a).

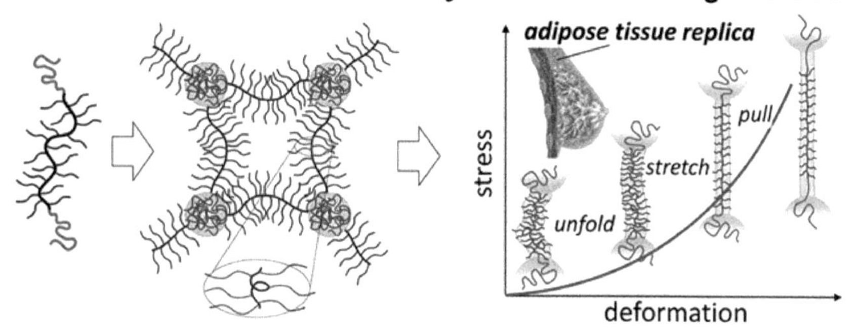

Fig. 2. Self-assembly and stress-strain behavior of linear–bottlebrush–linear (LBBL) plastomers (Keith et al. 2020)

8. Applications

8.1. Automotive and transportation

Various types of elastomers including SBR, natural rubber, polybutadiene, and nitrile rubber are used in automotive industries for tire and non-tire applications. Tire production is the predominant market for elastomers. SBR random copolymer and polybutadiene rubber are major polymeric constituents of tires. SBR is the most produced elastomer across the globe and there are two types of SBR based on the production process: emulsion SBR (E-SBR) and solution SBR (S-SBR). E-SBR is the conventional SBR production technology used worldwide; S-SBR has recently gained much attention in industrial scale because of fascinating characteristics that enable the manufacture of high-performance and green tires. New regulations for tire labeling in the European Union have resulted in increased demand for S-SBR. S-SBR enables tire manufacturers to fabricate energy-saving tires with decreased rolling resistance and enhanced wet grip (Hama and Inagaki). Decreased rolling resistance results in lower fuel consumption and diminished CO_2 emission (Grosch 1996). S-SBR is also used to manufacture high-performance shoes.

Furthermore, functionalized S-SBR improves filler incorporation into tire formulation, decreasing rolling resistance (Sun et al. 2018). For example, telechelic polymers based on S-SBR such as hydroxyl-terminated S-SBR (HTSSBR) can be manufactured (Qin et al. 2018b). HTSSBR can significantly decrease rolling, wear, and wet-skid resistance, which is highly interesting in the fabrication of next-generation energy-saving tires (Qin et al. 2018a).

Other studies are seeking a natural source to produce next-generation elastomers for tire industries (Lou et al. 2019).

8.2. Electric and electronic

The elastomer can be used as flexible electrical insulators for wires (Junian et al. 2015). Harder elastomers can be used as housing for electrical devices.

On the other hand, making conducting elastomers can further expand the applications of elastomers in stretchable conductor applications (Matsuhisa et al. 2019). Conducting elastomers can be made by the incorporation of conductive fillers such as carbon nanomaterials (Araby et al. 2014, Choi et al. 2019). These flexible conductive materials can be used in stretchable bioelectronics, wearable devices, and flexible solar cells (Noh 2016). For example, block copolymer elastic conductors based on polyaniline are used to fabricate highly conductive elastomers (Stoyanov et al. 2013). The soft triblock copolymer was based on maleic anhydride grafted to (poly(styrene-co-ethylenebutylene-co-styrene) (SEBS-g-MA). SEBS-g-MA was mixed with SEBS containing olefinic oil, which enables high loading of polyaniline without significant elasticity loss.

Various carbon-based nanomaterials such as carbon nanotubes and graphene can also be used to make conductive elastomers. It has been shown that incorporation of special MXene (Ti_3C_2 nanoplatelets) can remarkably improve the electrical and thermal conductivities of SBR to make it even more efficient than reduced graphene oxide (Q. Li et al. 2019).

8.3. Actuators and soft robotics

Soft robotics rely on soft actuators to make artificial muscles. Dielectric elastomer actuators (DEAs) are one of the major types of soft actuators that use electroactive polymers (Youn et al. 2020). Dielectricelastomer actuators are thin elastomer films sandwiched between two compliant electrodes similar to variable capacitors. Dielectric elastomers are a class of electroactive polymers that can produce large strains when an electric field is applied to compliant electrodes. These smart elastomers possess high elastic energy density and muscle-like mechanical properties.

Various elastomers have been used to make soft actuators; however, polyurethanes, acrylic elastomers, and silicones have been the most promising elastomers for DAE (Brochu and Pei 2010). Increasing the permittivity of elastomers, which can reduce the application potential and processing of elastomers to make thin films, is an important concern in research. Caspari et al. synthesized polar oligosiloxanes to make thin films using the doctor blade coating (Caspari et al. 2019).

Bottlebrush elastomers can be used as novel platforms for the fabrication of freestanding actuators with large stroke, even at low applied potential (Vatankhah-Varnoosfaderani et al. 2017b). On the other hand, inspired by morphogenesis in biological systems, Siéfert et al. manufactured shape-morphing elastomers or baromorphs, which are interesting for actuators (Siéfert et al. 2019). As shown in Fig. 3, Shian et al. fabricated a dielectric elastomer-based gripper that can be used as an artificial muscle for soft robotics (Shian et al. 2015).

8.4. Microfluidics

Microfluidics is the science of manipulating small quantities of a fluid using

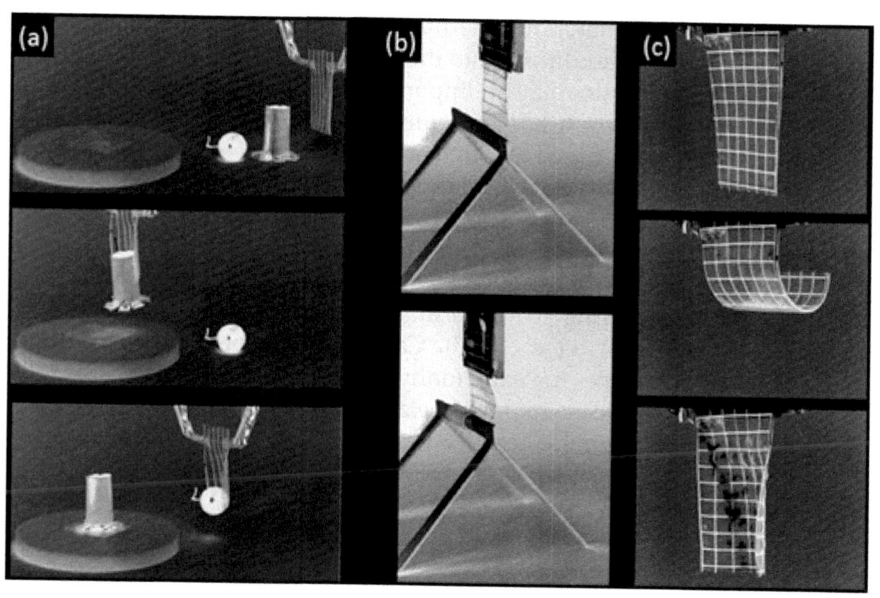

Fig. 3. Dielectric elastomer-based gripper for use in soft robotics (Shian et al. 2015)

miniaturized elements such as microchannels, microfilters, microvalves, micronozzles, and micropumps. The development of microfluidics along with microelectronics has enabled engineers to manufacture tiny devices (Cheng and Wu 2012). For example, it is involved in the making of microelectromechanical systems used in micro-propulsion systems of cube-satellite (Tummala and Dutta 2017). Besides, it can be combined with hydrogels for various purposes (Goy et al. 2019).

Soft, flexible, and extensible microfluidic devices affect the flow behavior of the fluids (Fallahi et al. 2019). Polydimethylsiloxane has been widely used to fabricate microfluidic chips. Flexible microfluidics can also be used to make sensors (Karthikeyan et al. 2019). Tsai et al. fabricated a sensitive pH sensor based on flexible microfluidics (Tsai et al. 2019). They combined aluminum-doped zinc oxide nanosheets with polydimethylsiloxane microfluidic chips. Furthermore, flexible microfluidics can be used to manufacture flexible biosensors for live cell analysis and point-of-care applications (Loo et al. 2019, Solis-Tinoco et al. 2019).

8.5. Tissue engineering

Elastomeric biomaterials are promising for soft tissue engineering, especially where tissues expand and contract, such as cardiac tissue. Various synthetic elastomers (e.g., polyesters) and natural proteins have been used in vascular tissue engineering (Hiob et al. 2016, Ye et al. 2018). For example, porous elastomeric scaffolds based on poly(glycerol sebacate) were used for skin tissue engineering (Zhang et al. 2016). The degradation rate of these scaffolds can be adjusted through the porous structure. *In vitro* cell culture investigation showed that mouse dermal fibroblasts can easily attach to the scaffolds, proliferate, and diffuse into the scaffolds (Zhang et al. 2016).

Biomimetic mechanics play an important role in soft robotics and tissue engineering. Sheiko Group has presented a roadmap to create tissue-like mechanical properties based on the manipulation of polymer networks (Sheiko et al. 2018). Bottlebrush plastomers are interesting platforms for tissue engineering applications. Bottlebrush plastomers can be used in skin-mimicking polymers (Vatankhah-Varnosfaderani et al. 2018).

Ahadian et al. used conductive elastomeric polyester scaffolds for cardiac tissue engineering to conform to the conductive and contractile cardiac tissue (Ahadian et al. 2017). Carbon nanotubes were embedded into poly(octamethylene maleate (anhydride) 1,2,4-butanetricarboxylate) (124 polymer) to fabricate moldable elastomers. Synthetic biodegradable elastomers were also used for heart valve tissue engineering (Xue et al. 2017).

8.6. Construction

Various admixtures are used to improve the mechanical and rheological properties of cement and concrete. For example, lignosulfonate plasticizers are used as water-reducing concrete admixture (Huang et al. 2018). Lignosulfonate is a derivative of natural lignin removed from wood during the Kraft pulping process. Various elastomers can be used to make modified concrete and mortar to enhance their workability, cement hydration, mechanical properties, and chemical resistance (Muhammad 2012, Weichold and Antons 2013).

Several elastomers have been used to make polymer-modified bitumen for waterproofing applications (Porto et al. 2019). Elastomers are also used to make construction adhesive (Steiner and Manville 2007). High-damping elastomeric materials and magnetorheological elastomers can be used to protect buildings against seismic events and winds (Lee et al. 2005, Yang et al. 2019).

9. Conclusion

The unique mechanical behavior of elastomers and plastomers makes them interesting polymers. They have been used for decades in many industrial applications such as tire production. Tire production industries have developed more efficient energy-saving tires based on novel and modified elastomers. On the other hand, natural forms and synthetic biological methods have drawn attention in the fabrication of novel bioelastomers such as recombinant elastomeric proteins for biomedical applications. Based on an engineering perspective involving modern chemistry, scientists have managed to fabricate solvent-free bottlebrush elastomers in which mechanical properties are independent of composition, unlike with polymer hydrogels. These unique architectural polymers demonstrate strain-stiffening behavior similar to animal tissues that seems highly promising for tissue engineering applications and soft robotics. Novel elastomer materials will gradually transform hard electronic, electric, and microfluidic devices to soft, flexible, and implantable devices. This large family of materials is still developing, and the future promises new grades of plastomers with new applications.

References

Abbasi, M., L. Faust and M. Wilhelm. 2019. Comb and bottlebrush polymers with superior rheological and mechanical properties. *Advanced Materials* 31(26): 1806484.

Aghaei-Ghareh-Bolagh, B., S.M. Mithieux and A.S. Weiss. 2016. Elastic proteins and elastomeric protein alloys. *Current Opinion in Biotechnology* 39: 56-60.

Aguilar, M.R. and J. San Román. 2019. Smart Polymers and Their Applications. Woodhead Publishing.

Ahadian, S., L.D. Huyer, M. Estili, B. Yee, N. Smith, Z. Xu, Y. Sun and M. Radisic. 2017. Moldable elastomeric polyester-carbon nanotube scaffolds for cardiac tissue engineering. *Acta Biomaterialia* 52: 81-91.

Annabi, N., S.M. Mithieux, G. Camci-Unal, M.R. Dokmeci, A.S. Weiss and A. Khademhosseini. 2013. Elastomeric recombinant protein-based biomaterials. *Biochemical Engineering Journal* 77, 110-118.

Araby, S., Q. Meng, L. Zhang, H. Kang, P. Majewski, Y. Tang and J. Ma. 2014. Electrically and thermally conductive elastomer/graphene nanocomposites by solution mixing. *Polymer* 55(1): 201-210.

Bhowmick, A.K. 2008. Current Topics in Elastomers Research. CRC Press. US.

Boczkowska, A. 2012. Advanced Elastomers: Technology, Properties and Applications. BoD – Books on Demand.

Brochu, P. and Q. Pei. 2010. Advances in dielectric elastomers for actuators and artificial muscles. *Macromolecular Rapid Communications* 31(1): 10-36.

Caspari, P., F.A. Nüesch and D.M. Opris. 2019. Synthesis of solvent-free processable and on-demand cross-linkable dielectric elastomers for actuators. *Journal of Materials Chemistry C* 7(39): 12139-12150.

Cheng, S. and Z. Wu. 2012. Microfluidic electronics. *Lab on a Chip* 12(16): 2782-2791.

Choi, S., S.I. Han, D. Kim, T. Hyeon and D.-H. Kim. 2019. High-performance stretchable conductive nanocomposites: Materials, processes, and device applications. *Chemical Society Reviews* 48(6): 1566-1595.

D'orazio, L., R. Greco, C. Mancarella, E. Martuscelli, G. Ragosta, C.J.P.E. Silvesrte and Science. 1982. Effect of the addition of ethylene-propylene random copolymers on the properties of high-density polyethylene/isotactic polypropylene blends: Part 1—Morphology and impact behavior of molded samples. *Polymer Engineering & Science* 22(9): 536-544.

Daniel, W.F., J. Burdyńska, M. Vatankhah-Varnoosfaderani, K. Matyjaszewski, J. Paturej, M. Rubinstein, A.V. Dobrynin and S.S. Sheiko. 2016. Solvent-free, supersoft and superelastic bottlebrush melts and networks. *Nature Materials* 15(2): 183-189.

De, S.K. and J.R. White. 2001. Rubber Technologist's Handbook. iSmithers Rapra Publishing. UK.

Duduta, M., E. Hajiesmaili, H. Zhao, R.J. Wood and D.R. Clarke. 2019. Realizing the potential of dielectric elastomer artificial muscles. *Proceedings of the National Academy of Sciences* 116(7): 2476-2481.

Fallahi, H., J. Zhang, H.-P. Phan and N.-T. Nguyen. 2019. Flexible microfluidics: Fundamentals, recent developments, and applications. *Micromachines* 10(12): 830.

Goy, C.B., R.E. Chaile and R.E. Madrid. 2019. Microfluidics and hydrogel: A powerful combination. *Reactive and Functional Polymers* 104314.

Grosch, K. 1996. The rolling resistance, wear and traction properties of tread compounds. *Rubber Chemistry and Technology* 69(3): 495-568.

Hama, H. and K. Inagaki. 2011. Development and Foresight of Solution SBR for Energy-Saving Tires. Sumitomo Chemical Co. Ltd. Petrochemicals Research Laboratory.

Hanhi, K., M. Poikelispaa and H.M. Tirila. 2007. Elastomeric materials. Tampere University of Technology, Tampere.

Hiob, M.A., G.W. Crouch and A.S. Weiss. 2016. Elastomers in vascular tissue engineering. *Current Opinion in Biotechnology* 40: 149-154.

Huang, C., J. Ma, W. Zhang, G. Huang and Q. Yong. 2018. Preparation of lignosulfonates from biorefinery lignins by sulfomethylation and their application as a water reducer for concrete. *Polymers* 10(8): 841.

Jenekhe, S.A. and D.J. Kiserow. 2004. Chromogenic Effects in Polymers: An Overview of the Diverse Ways of Tuning Optical Properties in Real Time. ACS Publications. US.

Jochum, F.D. and P. Theato. 2013. Temperature- and light-responsive smart polymer materials. *Chemical Society Reviews* 42(17): 7468-7483.

Junian, S., M. Makmud and J. Sahari. 2015. Natural rubber as electrical insulator: A review. *Journal of Advanced Review on Scientific Research* 6(1): 28-42.

Karthikeyan, M., J. Park and D.-W. Lee. 2019. Liquid metal based flexible microfluidic device for wireless sensor applications. *2019 International Conference on Optical MEMS and Nanophotonics (OMN)* pp. 68-69. IEEE.

Keith, A.N., M. Vatankhah-Varnosfaderani, C. Clair, F. Fahimipour, E, Dashtimoghadam, A. Lallam, M. Sztucki, D.A. Ivanov, H. Liang, A.V. Dobrynin and S.S. Sheiko. 2020. Bottlebrush bridge between soft gels and firm tissues. *ACS Central Science* 6(3): 413-419.

Khosravi, A., A. Fereidoon, M.M. Khorasani, G. Naderi, M.R. Ganjali, P. Zarrintaj, . . . S. Life. 2020. Soft and hard sections from cellulose-reinforced poly (lactic acid)-based food packaging films: A critical review. 23: 100429.

Lee, K.S., C.P. Fan, R. Sause and J. Ricles. 2005. Simplified design procedure for frame buildings with viscoelastic or elastomeric structural dampers. *Earthquake Engineering & Structural Dynamics* 34(10): 1271-1284.

Lendlein, A. and O.E. Gould. 2019. Reprogrammable recovery and actuation behaviour of shape-memory polymers. *Nature Reviews Materials* 4(2): 116-133.

Li, L. and K.L. Kiick. 2013. Resilin-based Materials for Biomedical Applications. ACS Publications. US.

Li, Q., B. Zhong, W. Zhang, Z. Jia, D. Jia, S. Qin, . . . X. Wang. 2019. Ti 3 C 2 MXene as a new nanofiller for robust and conductive elastomer composites. *Nanoscale* 11(31): 14712-14719.

Liu, C., H. Qin and P. Mather. 2007. Review of progress in shape-memory polymers. *Journal of Materials Chemistry* 17(16): 1543-1558.

Liu, Q., L. Jiang, R. Shi and L. Zhang. 2012. Synthesis, preparation, in vitro degradation, and application of novel degradable bioelastomers—A review. *Progress in Polymer Science* 37(5): 715-765.

Liu, T. and Y. Xu. 2019. Magnetorheological Elastomers: Materials and Applications. In Smart and Functional Soft Materials. IntechOpen. UK.

Liu, X., S. Zhao, X. Zhang, X. Li and Y.J.P. Bai. 2014. Preparation, structure, and properties of solution-polymerized styrene-butadiene rubber with functionalized end-groups and its silica-filled composites. *Polymer* 55(8): 1964-1976.

Loo, J.F., A.H. Ho, A.P. Turner and W.C. Mak. 2019. Integrated printed microfluidic biosensors. *Trends in Biotechnology*. 37(10): 1104-1120.

Lou, Z., S. Zhao, Q. Wang and H.J.A.C. Wei. 2019. N-doped carbon as peroxidase-like nanozymes for total antioxidant capacity assay. *Analytical Chemistry* 91(23): 15267-15274.

Lyons, R.E., D.C. Wong, M. Kim, N. Lekieffre, M.G. Huson, T. Vuocolo, . . . M.L. Colgrave. 2011. Molecular and functional characterisation of resilin across three insect orders. *Insect Biochemistry and Molecular Biology* 41(11): 881-890.

Maciejewska, M. and A. Krzywania-Kaliszewska. 2012. Thermo-shrinkable elastomers. *Advanced Elastomers: Technology, Properties and Applications* 181.

Mahmodi, G., P. Zarrintaj, A. Taghizadeh, M. Taghizadeh, S. Manouchehri, S. Dangwal, A. Ronte, S.-J.J.C.R. Kim and M.R. Saeb. 2020. From microporous to mesoporous mineral frameworks: An alliance between zeolite and chitosan. *Carbohydrate Research* 489: 107930.

Majidi, C. 2019. Soft-matter engineering for soft robotics. *Advanced Materials Technologies* 4(2): 1800477.

Matsuhisa, N., X. Chen, Z. Bao and T. Someya. 2019. Materials and structural designs of stretchable conductors. *Chemical Society Reviews* 48(11): 2946-2966.

Mineart, K.P., S.S. Tallury, T. Li, B. Lee and R.J. Spontak. 2016. Phase-change thermoplastic elastomer blends for tunable shape memory by physical design. *Industrial & Engineering Chemistry Research* 55(49): 12590-12597.

Mohammadi, Y., M. Ahmadi, M.R. Saeb, M.M. Khorasani, P. Yang and F.J.J.M. Stadler. 2014. A detailed model on kinetics and microstructure evolution during copolymerization of ethylene and 1-octene: From coordinative chain transfer to chain shuttling polymerization. *Macromolecules* 47(14): 4778-4789.

Muhammad, B. 2012. Technology, properties and application of NRL elastomers. *Advanced Elastomers –Technology, Properties and Applications* 266-288.

Noh, J.-S. 2016. Conductive elastomers for stretchable electronics, sensors and energy harvesters. *Polymers* 8(4): 123.

Pakula, T., Y. Zhang, K. Matyjaszewski, H.-i. Lee, H. Boerner, S. Qin and G.C. Berry. 2006. Molecular brushes as super-soft elastomers. *Polymer* 47(20): 7198-7206.

Palza, H., P. Zapata and C. Sagredo. 2018. Shape memory composites based on a thermoplastic elastomer polyethylene with carbon nanostructures stimulated by heat and solar radiation having piezoresistive behavior. *Polymer International* 67(8): 1046-1053.

Paran, S.M.R., H. Vahabi, M. Jouyandeh, F. Ducos, K. Formela and M.R. Saeb. 2019. Thermal decomposition kinetics of dynamically vulcanized polyamide 6-acrylonitrile butadiene rubber-halloysite nanotube nanocomposites. *Journal of Applied Polymer Science* 136(20): 47483.

Porto, M., P. Caputo, V. Loise, S. Eskandarsefat, B. Teltayev and C. Oliviero Rossi. 2019. Bitumen and bitumen modification: A review on latest advances. *Applied Sciences* 9(4): 742.

Prévôt, M.E., S. Ustunel and E. Hegmann. 2018. Liquid crystal elastomers—A path to biocompatible and biodegradable 3D-LCE scaffolds for tissue regeneration. *Materials* 11(3): 377.

Qi, S., J. Fu, Y. Xie, Y. Li, R. Gan and M. Yu. 2019. Versatile magnetorheological plastomer with 3D printability, switchable mechanics, shape memory, and self-healing capacity. *Composites Science and Technology* 183: 107817.

Qin, X., B. Han, J. Lu, Z. Wang, Z. Sun, D. Wang, T.P. Russell, L. Zhang and J. Liu. 2018. Rational design of advanced elastomer nanocomposites towards extremely energy-saving tires based on macromolecular assembly strategy. *Nano Energy* 48, 180-188.

Qin, X., J. Wang, B. Han, B. Wang, L. Mao and L. Zhang. 2018. Novel design of eco-friendly super elastomer materials with optimized hard segments micro-structure: Toward next-generation high-performance tires. *Frontiers in Chemistry* 6: 240.

Qiu, Y., E. Zhang, R. Plamthottam and Q. Pei. 2019. Dielectric elastomer artificial muscle: Materials innovations and device explorations. *Accounts of Chemical Research* 52(2): 316-325.

Ratna, D. and J. Karger-Kocsis. 2008. Recent advances in shape memory polymers and composites: A review. *Journal of Materials Science* 43(1): 254-269.

Reynolds, V.G., S. Mukherjee, R. Xie, A.E. Levi, A. Atassi, T. Uchiyama, H. Wang, M.L. Chabinyc and C.M. Bates. 2020. Super-soft solvent-free bottlebrush elastomers for touch sensing. *Materials Horizons* 7(1): 181-187.

Sadeghi, K., J.-Y. Yoon and J. Seo. 2019. Chromogenic polymers and their packaging applications: A review. *Polymer Reviews* 1-51.

Sheiko, S.S., M.H. Everhart, A.V. Dobrynin and M. Vatankhah-Varnosfaderani. 2018. Encoding tissue mechanics in silicone. *Science Robotics* 3(18): eaat7175.

Shian, S., K. Bertoldi and D.R. Clarke. 2015. Dielectric elastomer based "grippers" for soft robotics. *Advanced Materials* 27(43): 6814-6819.

Siéfert, E., E. Reyssat, J. Bico and B. Roman. 2019. Bio-inspired pneumatic shape-morphing elastomers. *Nature Materials* 18(1): 24-28.

Solis-Tinoco, V., S. Marquez, T. Quesada-Lopez, F. Villarroya, A. Homs-Corbera and L.M. Lechuga. 2019. Building of a flexible microfluidic plasmo-nanomechanical biosensor for live cell analysis. *Sensors and Actuators B: Chemical* 291: 48-57.

Steiner, P. and J. Manville. 2007. Elastomeric construction adhesives: Influence of solvent retention on strength. *Wood and Fiber Science* 10(3): 229-234.

Stoyanov, H., M. Kollosche, S. Risse, R. Waché and G. Kofod. 2013. Soft conductive elastomer materials for stretchable electronics and voltage controlled artificial muscles. *Advanced Materials* 25(4): 578-583.

Sun, C., S. Wen, H. Ma, Y. Li, L. Chen, Z. Wang, B. Yuan and L. Liu. 2018. Improvement of silica dispersion in solution polymerized styrene–butadiene rubber via introducing amino functional groups. *Industrial & Engineering Chemistry Research* 58(3): 1454-1461.

Tatham, A.S. and P.R. Shewry. 2000. Elastomeric proteins: Biological roles, structures and mechanisms. *Trends in Biochemical Sciences* 25(11): 567-571.

Tjin, M.S., P. Low and E. Fong. 2014. Recombinant elastomeric protein biopolymers: Progress and prospects. *Polymer Journal* 46(8): 444-451.

Tsai, Y.-T., S.-J. Chang, L.-W. Ji, Y.-J. Hsiao and I.-T. Tang. 2019. Fast detection and flexible microfluidic pH sensors based on Al-doped ZnO nanosheets with a novel morphology. *ACS Omega* 4(22): 19847-19855.

Tummala, A.R. and A. Dutta. 2017. An overview of cube-satellite propulsion technologies and trends. *Aerospace* 4(4): 58.

Ula, S.W., N.A. Traugutt, R.H. Volpe, R.R. Patel, K. Yu and C.M. Yakacki. 2018. Liquid crystal elastomers: An introduction and review of emerging technologies. *Liquid Crystals Reviews* 6(1): 78-107.

van Eldijk, M.B., C.L. McGann, K.L. Kiick and J.C. van Hest. 2011. Elastomeric polypeptides. *In*: Peptide-based Materials. pp. 71-116. Springer.

Vatankhah-Varnosfaderani, M., W.F. Daniel, M.H. Everhart, A.A. Pandya, H. Liang, K. Matyjaszewski, A.V. Dobrynin and S.S. Sheiko. 2017a. Mimicking biological stress-strain behaviour with synthetic elastomers. *Nature* 549(7673): 497-501.

Vatankhah-Varnoosfaderani, M., W.F. Daniel, A.P. Zhushma, Q. Li, B.J. Morgan, K. Matyjaszewski, D.P. Armstrong, R.J. Spontak, A.V. Dobrynin and S.S. Sheiko. 2017b. Bottlebrush elastomers: A new platform for freestanding electroactuation. *Advanced Materials* 29(2): 1604209.

Vatankhah-Varnosfaderani, M., A.N. Keith, Y. Cong, H. Liang, M. Rosenthal, M. Sztucki, . . . A.V. Dobrynin. 2018. Chameleon-like elastomers with molecularly encoded strain-adaptive stiffening and coloration. *Science* 359(6383): 1509-1513.

Wang, Y., U. Gupta, J. Zhu, P. Li, D. Du, L. Zhang, J. Ouyang, J. Liu and C.C. Foo. 2019. Bio-inspired soft robot driven by transparent artificial muscle. *2019 IEEE International Conference on Robotics and Biomimetics (ROBIO)* pp. 1959-1964. IEEE.

Wei, M., Y. Gao, X. Li and M.J. Serpe. 2017. Stimuli-responsive polymers and their applications. *Polymer Chemistry* 8(1): 127-143.

Weichold, O. and U. Antons. 2013. Hard and soft: Principles of polyner-impregnated concrete using elastomers. *Advanced Materials Research* 687: 118-123. Trans Tech Publ.

Xue, Y., V. Sant, J. Phillippi and S. Sant. 2017. Biodegradable and biomimetic elastomeric scaffolds for tissue-engineered heart valves. *Acta Biomaterialia* 48: 2-19.

Yang, J., S. Sun, S. Zhang and W. Li. 2019. Review of structural control technologies using magnetorheological elastomers. *Current Smart Materials* 4(1): 22-28.

Ye, H., K. Zhang, D. Kai, Z. Li and X.J. Loh. 2018. Polyester elastomers for soft tissue engineering. *Chemical Society Reviews* 47(12): 4545-4580.

Yeo, G.C., B. Aghaei-Ghareh-Bolagh, E.P. Brackenreg, M.A. Hiob, P. Lee and A.S. Weiss. 2015. Fabricated elastin. *Advanced Healthcare Materials* 4(16): 2530-2556.

Youn, J.-H., S.M. Jeong, G. Hwang, H. Kim, K. Hyeon, J. Park and K.-U. Kyung. 2020. Dielectric elastomer actuator for soft robotics applications and challenges. *Applied Sciences* 10(2): 640.

Zarrintaj, P., B. Bakhshandeh, M.R. Saeb, F. Sefat, I. Rezaeian, M.R. Ganjali, S. Ramakrishna and M. Mozafari. 2018a. Oligoaniline-based conductive biomaterials for tissue engineering. *Acta Biomaterialia* 72: 16-34.

Zarrintaj, P., M.R. Saeb, S. Ramakrishna and M. Mozafari. 2018b. Biomaterials selection for neuroprosthetics. *Current Opinion in Biomedical Engineering* 6: 99-109.

Zhang, X., C. Jia, X. Qiao, T. Liu and K. Sun. 2016. Porous poly(glycerol sebacate) (PGS) elastomer scaffolds for skin tissue engineering. *Polymer Testing* 54: 118-125.

CHAPTER

11

Polyurethane

Noushin Ezzati[1]*, Ebad Asadi[2], Majid Abdouss[2] and Elaheh Kowsari[2]
[1] Young Researchers and Elites Club, Science and Research Branch, Islamic Azad University, Tehran, Iran
[2] Department of Chemistry, Amirkabir University of Technology, Tehran, Iran

1. Introduction

Polyurethane (PU) was introduced in the 1930s by Otto Bayer for aerospace and military applications (Silva and Bordado 2004). It was first developed as rubber during World War II (Akindoyo et al. 2016). Combining the elasticity of rubber together with the toughness of metals, it is a suitable alternative for plastics or metals in many cases. Polyurethanes are among the most common, versatile and researched materials in the world today (Akindoyo et al. 2016).

Polyurethane polymers are defined according to the presence of –NH-CO-O- as the urethane link in macromolecular structure and formed by the reaction of polyols and isocyanate (Prisacariu and Scortanu 2011). The characteristics of PU depend on the chemical nature of macroglycol, the NCO/OH ratio, isocyanate type, chain extender, catalyst, and also the synthesis condition (Orgilés-Calpena et al. 2011). The PU polyurethane industry is heavily dependent on petroleum resources because the main compounds for preparation of PUs, including polyols and isocyanates, are derived from petroleum (Gama et al. 2015). In view of environmental concerns, the cost of fossil fuel, and future shortage of petroleum, many investigations have been conducted to find alternative sources for preparation of PUs. Non-petroleum-based polyols have been extensively studied for preparation of PUs (Li et al. 2017b).

Polyurethane is a block copolymer constructed of semi-crystalline hard and amorphous soft segments commonly derived from diisocyanates and long chain polyol component, respectively (Adak et al. 2018). From the reaction of –NCO groups of diisocyanates with –OH groups of polyol, PU polymer is achieved. The choice of raw material depends on the target application (Athanasiou et al. 1996, Penco et al. 1996).

*Corresponding authors: s416em@gmail.com

Polyurethanes are classified as flexible, rigid, thermoplastic, waterborne, binders, coating, sealant and elastomers (Szycher 1999).

2. Synthesis

Polyurethanes are composed of soft and hard segments. The soft segment that is responsible for flexibility of PU chains is composed of high-molecular-weight polyols. The hard segment that gives them strength and rigidity is formed by NCO-functionalized compound and low-molecular-weight chain extender (Kim et al. 2019). Polyurethane is commonly prepared via the reaction of isocyanate and polyol in the vicinity of catalyst or UV light. Isocyanate and polyol molecules should contain two or more isocyanate groups (R–(N=C=O)) and hydroxyl groups (R′–(OH)) (Rahman et al. 2019). Generally, long chain polymers provide stretchy structure and hard polymers are obtained from high crosslinking degree among short chains. Also, long chain with average crosslinking creates polymers suitable for preparation of foams. The incorporation of different additives such as fire retardants, blowing agents, pigments and crosslinkers in addition to some modification during polymer synthesis brings a wide range of improved characteristics to PU (Akindoyo et al. 2016).

Depending on the nature of reactants, including isocyanate, polyol and chain extender, their ratios, and preparation condition, various kinds of properties such as hardness and rheological characteristics are obtained (Tenorio-Alfonso et al. 2019), (Javaid et al. 2018a). The formation of PU from isocyanate and polyol is shown in Fig. 1. The optimized amount of NCO/OH molar ratio, the amount of catalyst, loading of crosslinking agent, and loading of blowing agent will influence the compression strength of PU (Li et al. 2018).

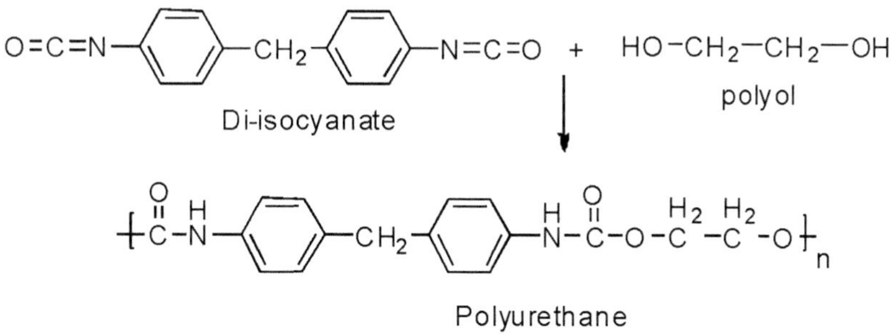

Fig. 1. Synthesis of polyurethane from isocanate and polyol

2.1. Polyol

Polyols used in preparation of PU often possess two or more OH groups. Different types of polyols are available to use in synthesis of PU, of which the main kinds are polyether polyols and polyester polyols (Ionescu 2005).

Polyether polyols provide the advantages of low cost, suitable viscosity, hydrolytic stability, and flexibility but their flammability and low thermal stability limit their use. Polyester polyols are prepared from raw ingredients and usually are obtained via direct polyesterication of diacids and glycols. Aliphatic polyesters have oxidative and hydrolytic stability. Aromatic polyesters possess flame-retardant characteristics. Polycarbonate polyols demonstrate oxidative and hydrolytic stability, but their high cost and undesirable viscosity restrict their applications. The low temperature flexibility and the solvent resistance of polybutadiene polyols are considered interesting properties, but these polyols have high cost, undesirable viscosity and thermal oxidizable characteristics (Sonnenschein 2014, Petrovic 2008).

One of the cost-efficient, industrial methods for fabrication of polyol is based on epoxidation of double bonds. This method is used for component containing an active hydrogen atom. This synthesis pathway provides the possibility of obtaining bio-polyol with promising features for a wide range of applications. The properties of the rigid foam, such as molecular weight, viscosity, and functionality, are directly influenced by the bio-polyol synthesis conditions (Narine et al. 2007).

In synthesis procedure, the target properties of PU specify the type of suitable polyol. For example, rigid PUs are prepared using low-molecular-weight polyol and flexible PUs are obtained from high-molecular-weight polyols (Petrovic 2008). Higher functionality of polyol results in more crosslinks in polymer network (Zain et al. 2016).

2.1.1. Green polyol preparation

The advantages of using accessible renewable, green, bio-based resources for preparation of polyol have directed many studies to investigate natural starting materials such as starch, cellulose, oil and sugar to eliminate the use of petroleum (Kurańska et al. 2016, Blache et al. 2018). Using these materials as polyol source can reduce the amounts of harmful and toxic reagents (Gama et al. 2015). Among these natural materials, vegetable oils that are low cost, non-toxic, biodegradable and easily available are preferable (Chaudhari et al. 2013). Several researchers investigate the preparation of oil-based polymeric polyols extracted from vegetables such as polyetheramides, alkyds, polyesteramides, piperazine-modified polyol, and epoxies. Since the fatty acid extracted from each kind of vegetable oil is different, the thermal tolerance, mechanical strength and chemical properties will be different (Patil et al. 2017). The chemical structure of bio-polyols plays an important role in the foaming process and the final product properties. The bio-polyols based on triethanamine have significant desirable effects for the foaming process because they avoid use of catalyst and provide high reactivity of the reaction mixture. Also, PU produced by polyols from epoxidation method has better mechanical strength due to less reactive system (Prociak et al. 2018).

A process similar to oxypropylation is used for modification of castor oil, rapeseed oil, and linseed oil (Zhang et al. 2014, Hu et al. 2002) or acid liquefaction of biomass residues such as soybean, alginic acid, sugarcane bagasse, palm, coffee ground wastes, cork, lignin, protein and carbohydrates (Patil et al. 2017, Hilmi and Zainuddin 2019, Kwon et al. 2007, Pechar et al. 2006, Yang et al. 2007, Soares et al. 2014).

Bio-polyols commonly are produced through liquefaction of lignocellulosic biomass using polyhydric alcohols as liquefying agent. In the absence of oxygen and through fast pyrolysis, lignocellulosic biomass can be transformed to liquid oil. This kind of biomass possesses more eco-friendly features (Torri et al. 2016). Hydrothermal liquefaction and fast pyrolysis have been widely employed to prepare bio-chemicals such as bio-phenols and bio-polyols (Li et al. 2017a). In general, considering the high temperature required for pyrolysis, liquefaction is more desirable (Cheng et al. 2010, Ye et al. 2014). The liquefaction procedure is mainly influenced by reaction time, temperature, solid/liquid ratio, catalysts and the nature of solvents used (Li et al. 2017a). The influence of solvents has been investigated in some studies. It was found that in some cases a mixture of various solvents increases the efficiency of the process (Cheng et al. 2010).

Additionally, it has been proved that biomass liquefaction occurs through three main reactions: hydrolysis, decomposition and poly-condensation. The type of material and liquefaction condition directly influence the composition and structure of finished product (Li et al. 2017a). For example, bio-oils obtained from lignocellulose biomass liquefied in alcohol, water and phenol produce substitute phenol for the synthesis of bisphenol F resins/adhesives (Ye et al. 2014). In contrast, bio-oil obtained from biomass liquefaction in polyhydric alcohols are rich in hydroxyl groups and thus suitable as bio-polyols for bio-based PU fabrication (Mahmood et al. 2015). It has been accepted that the mixture of ethanol and water could improve liquefaction procedure. Additionally, a base catalyst and suitable medium temperature promote the cost efficiency of the reaction (Cheng et al. 2010, Feng 2013). Lignocellulosic biomass has many advantages over petroleum, such as abundance carbon neutrality, variety and low contents of S and N (Li et al. 2017a). These kinds of polyols can be prepared by conversion of some biomass into bio-oils through thermochemical, biological or chemical methods. Chemical methods require specific facilities and are generally based on acid hydrolysis. Biological methods are less time consuming. Thermochemical methods such as liquefaction or pyrolysis are other ways to prepare PU foams with features comparable with petroleum analogs (Li et al. 2017b).

Three types of bio-polyols with various OH group position have been reported. The first bio-polyol with terminal primary OH group extracted from canola oil, the second one obtained from soybean, and the third one produced by crude castor oil. The results revealed the reactivity of PU from castor oil and soybean on the cream, rising and gel time were lower than those of PU obtained from canola oil-based polyol. As the molar mass decreases, the brittleness and compressive strength increase (Kurańska et al. 2015).

Lignin has a rigid 3D structure composed of various kinds of phenylpropane. Lignin is a byproduct of the paper and pulping industry. Its low cost and availability make it a suitable alternative for petroleum feedstock (Pettersen 1984, Li et al. 1997, El Mansouri and Salvadó 2007).

Owing to a large amount of phenolic and aliphatic hydroxyl groups, lignin has been a suitable candidate for preparation of thermoplastic PU films. Lignin obtained from kraft pulping and organosolv pretreatment is investigated for production of rigid polyols (Mittal 2012). Lignin is a non-covalent filler with large numbers of aromatic and aliphatic hydroxyl groups that provide high polarity structure. Considering the fact that lignin is non-soluble in thermoset resins, chemical modification of lignin is vital for preparation of homogenous composites (Zhang et al. 2015).

One of the most common pathways for preparation of polyols is oxirane ring opening techniques as an equilibrium exchange reaction. During this procedure, the oxirane ring opens by another fatty acid moiety. For epoxidation of vegetable oil, the triglyceride reacts with methanoic acid and then peroxide is added to the reaction pot. The preparation procedure of polyol, the reaction temperature, and carboxyl-to-epoxy ratio influence the efficiency of reaction and quality of products (Venkatesh and Jaisankar 2019).

Castor oil is easily available, low in cost, and environment-friendly. It has become one of the common materials for fabrication of bio-based PUs. Compared to other polyols, polyol fabricated from castor oil results in low crosslinking density and undesirable mechanical properties in the final PU, but by increasing the hydroxyl number, the mechanical properties can be improved. For this purpose the reaction of glycerol with castor oil is proposed. This technique first increases the hydroxyl number of castor oil, and then the obtained glyceride (soft segment) is polymerized with aliphatic isophorone diisocyanate (hard segment) to fabricate PU (Du et al. 2018, Tenorio-Alfonso et al. 2019).

There is another specific category of polyols with superior qualities known as specialty polyols that are in demand for preparation of adhesives, sealants, elastomers, and adhesives. Some of these polyols are polycarbonate polyols, polycaprolactone polyols, polysulfide polyols and butadiene polyols (Akindoyo et al. 2016).

2.1.2. Disadvantages of bio-based polyol

Prociak et al. reported that the replacement of petroleum-based polyol with polyols obtained from rapeseed enhances the brittleness of foams. Also, compared with the reference formulation, bio-based polyols have a discernibly lower reaction rate than petrochemical polyols.

2.2. Isocyanate

Isocyanates are known as the NCO chemical group and belong to hard segment of polymer (de Oliveira et al. 2013). They are categorized as heterofunctional and difunctional or aliphatic and aromatic. Generally, the aromatic isocyanates are more reactive than aliphatic ones (Gabriel et al. 2017). The composites based on aromatic diisocyanate present highest molecular weight and provide higher rate of reaction. The mechanical characteristic investigation revealed that PU with higher molecular weight exhibited higher mechanical strength (Javaid et al. 2018b).

Aliphatic diisocyanate, methylene diphenyl diisocyanate (MDI) and toluene diisocyanate (TDI) are the most common isocyanates (Tachibana and Abe 2019). MDI and TDI are petroleum-based isocyanates used to improve mechanical and thermal strength and are more cost efficient. One of the challenges of using such diisocyanates is their toxic nature. For medical use, aromatic diisocyanates such as MDI (4, 4′-methylenebis (phenyl isocyanate)) and TDI (tolylene-2,4-diisocyanate) are not proposed because they can degrade to aromatic amines, which are mutagenic or carcinogenic materials. Instead, aliphatic or cyclic diisocyanates such as HDI (1,6 hexamethylene diisocyanate), HMDI (4,4′-methylenebis(cyclohexylisocyanate)), IPDI (isophorone diisocyanate), LDI (lysine methyl esterdiisocyanate), and BDI (1, 4-diisocyanatobutane) are better choices in light of the non-toxic nature of the degradation products (Zdrahala and Zdrahala 1999, Hu et al. 2015, Han et al. 2011, Wang et al. 2013). 1,4-butanediisocyanate has been used for preparation of thermo-responsive PU for biomedical applications. This polymer is biodegradable and hydrolysable in nature. Bio-based aliphatic isocyanate has been widely investigated for use in PU preparation. Also, modification in polyol reaction procedure or incorporation of additives has been proposed to reduce the volatility of toxic gas (Tachibana and Abe 2019, Hao et al. 2016, Calvo-Correas et al. 2015).

2.3. Catalyst

Catalysts perform an important role in the kinetics of PU reactions. The type and amount of catalyst have direct impact on the molecular weight of the fabricated polymer (Parcheta and Datta 2018). Also, the optimum amount of catalyst must be considered because higher amount of catalysts results in higher density and undesirable features of foam (Zhao et al. 2014, Oertel and Abele 1994, Saleh 1990).

Amines are the most common organic catalysts used in PU synthesis reactions (Muuronen et al. 2019). Tertiary amines catalyze both isocyanate–hydroxyl and isocyanate–water reactions. In view of environmental issues and other drawbacks generated by using tertiary amine as catalyst, the necessity of other types of catalysts such as non-fugitive catalysts is clear. For this purpose, another kind of catalyst has been designed to chemically bind into PU matrix through its functionalization with isocyanate reactive groups such as urea, amino/amido or hydroxyl groups (Silva and Bordado 2004). The most common mechanism of this catalyst is the complexation of isocyanate with the tertiary amine followed by the attack of the nucleophilic reagent and the formation of a hydrogen-bonded complex between the nucleophilic reagent and the tertiary amine, which in the second step attacks the isocyanate (Silva and Bordado 2004, Muuronen et al. 2019). Other important group of catalysts for PU reaction is organometallic catalysts, which are more effective in isocyanate–hydroxyl reactions (Squiller 1987).

2.4. Chain extender

Chain extenders are commonly low-molecular-weight ingredients such as glycerol, hydroxyl amines or diamines that are applied to influence the end of PU chains during the preparation procedure. At the final stage the chain extender reacts with isocyanate and influences the mechanical response of PU. In a study developed by Prisacariu, it was found that with the chain extender with even numbers of CH_2 group such as 2, 4, 6, the smallest compression set values are obtained, which results in better elastic mechanical behavior. All in all, the length and type of chain extender directly influences the mechanical response of PU (Prisacariu and Scortanu 2011). Additionally, the chemical structure of chain extender influences the properties of PU; for example, a diol as chain extender results in higher mean particle size of PU. The glass transition temperature decreases along with increase in the chain extender length. The decrease of glass transition temperature is attributed to easy mobility of macromolecules. This mobility is related to two main mechanisms. The first mechanism is the separation of hard and soft segments, which results in greater flexibility. The second mechanism concerns the vibration of the main chain (Orgilés-Calpena et al. 2011, Chanda 2006).

Cakić et al. reported preparation of PU-silica nanocomposite modified by (3-aminopropyl) triethoxysilane (APTES) as chain extender in PU backbone. Sol-gel method was used in this process. The terminal amino group of APTES reacted with the functional groups of diisocyanate. According to TGA investigation it was revealed that APTES improved the thermal stability of composite by increasing the crosslink density (Cakić et al. 2019). It was also demonstrated that the morphology of polymer is affected by APTES and larger particles are obtained via this reagent (Cakić et al. 2019). Sadeghi reported using polyhedral oligomeric silsesquioxane (POSS) and 1,4-Butanediol (BDO) as two chain extenders simultaneously for better phase separation. It was revealed that POSS nanoparticles improve the recovery stress and responsiveness in shape memory PU. Also, it was observed that combined use of POSS and BDO improves hard and soft phase separation (Kazemi et al. 2019).

2.5. NCO/OH molar ratios

Some studies proved that a suitable value of NCO/OH improves the elastomeric properties of the foams, whereas with high ratio of NCO/OH, the obtained foam will be brittle (Zhao et al. 2014, Oertel and Abele 1994, Saleh 1990). The NCO/OH ratio is the equivalent ratio of components containing NCO and OH groups (Gogoi et al. 2013). As a result, specific isocyanate and polyol used in the synthesis process determine the properties of the final product (de Oliveira et al. 2013). Increase in NCO/OH ratio increases the available isocyanate amount of pre-polymer and provides more reaction sites, which results in the increment of crosslinking degree (Gogoi et al. 2013). Furthermore, it has been demonstrated that hydroxyl-to-isocyanate (OH/NCO) ratio is an important factor influencing thermal properties. With increasing NCO/OH ratio, higher content of hard segments is obtained, which results in improved thermal stability (García-Pacios et al. 2011). By increasing the NCO/OH molar ratio, the tensile strength, tear strength and hardness of PU increases, while the viscosity and elongation at break decreases (Petrović et al. 2002). The increment in OH/NCO ratio provides the formation of side products that change the decomposition temperature. For example, with excessive amount of isocyanate, allophanate and biuret are formed (Javaid et al. 2018b, Gogoi et al. 2013, de Oliveira et al. 2013).

As described above, the performance of PU directly depends on the NCO/OH molar ratio. For coating applications it was reported by Lopes de Oliveira that the NCO/OH molar ratio of 1.2 is the ratio that provides the best anticorrosion performance (de Oliveira et al. 2013). In cast resin, the NCO/OH ratio performs an important role because at low ratio the cast resin provides good flexibility and with NCO/OH ratio below 0.7 the obtained polymer is too flexible. Also, the density is affected by this ratio; with about 4% decrease in NCO/OH ratio the density changes from 1.0 to 4.0 (Petrović et al. 2002). By controlling the NCO/OH ratio, the morphology and other characteristics of the final product can easily be regulated. The weight average, molecular weight and viscosity are affected by this ratio. The Tg value also decreases with increasing NCO/OH ratio as a result of decrease in average molecular weight of pre-polymer. More short chain segments improve the free volume within the polymer and cause increment in polymer segment mobility and decrease the pre-polymer Tg (Gogoi et al. 2014).

2.6. Phase separation

The phase separation in PU structure is attributed to the synthesis method, crystallization ability, the content of hydrogen bonding, the type of hard and soft segments, the soft segment molecular structure, the molar ratio of hard to soft segment, the chemical nature of hard and soft segments, and other factors. The H-bond is the main factor that affects the crystallinity of the PU sample. The small diisocyanate PUs exhibit the highest H-bonding and limited hard and soft segment mobility (Tatai et al. 2007). The hard phase in PU foam could be considered a crosslinking site that collaborates with the cohesive force and stiffness. As a result, any increment in cohesive force is attributed to high level of micro-phase separation. With increment in hydrogen bonding, higher resistance against deformation is produced, and with increase in the separated hard domains, compressive strength increases (Abdollahi Baghban et al. 2019, Garrett et al. 2003).

2.7. Degradation

Generally, the degradation of PUs follows decomposition of hard segments. As a result, the physicochemical structure, length, symmetry and composition of hard segments directly influence the thermal stability (Javni et al. 2004, Liu and Ma 2002, Awad and Wilkie 2010). Also, the composition and molecular weight of soft segment can change the PU thermal degradation behavior (Javaid et al. 2018b). The first stage of degradation of PUs is cleavage of urethane linkage, which is related to the type of substituent that exists in the polyols and toluene diisocyanate. Following the degradation of hard segments, thermal transition is carried out that produces carbon dioxide, primary and secondary amines, alcohol, isocyanate and olefins. The next stage is due to the breakdown of aliphatic alkyl group of fatty acid. The third stage of thermal degradation is related to complete decomposition of crosslinker. The decomposition of soft segment occurs in the second or third stage. The structure of diol and diisocyanate influences the temperature of initial decomposition process (Venkatesh and Jaisankar 2019). The higher amount of oxygen in polyol causes more weight loss during thermal degradation. On the other hand, increase in hydroxyl-to-isocyanate ratio influences decomposition temperature, which associated with formation of side materials (Javaid et al. 2018b). The degradation properties are commonly studied by DLS, GPC and H NMR (Luan et al. 2019).

Harsh environmental conditions saturated by oxygen and water molecules in addition to UV radiation accelerate polymer degradation. This undesirable process includes the breaking of unsaturated bonds and production of ketone, aldehyde, peroxides and hydroperoxide groups (Yang et al. 2001, Martin et al. 2012). This degradation causes some changes in mechanical, physical and chemical structure of polymer, exhibited by cracking, blistering, loss of gloss and color change (Yang et al. 2001). Aromatic PU is more sensitive to UV radiation exposure because of production of mono- and di-quinone imides. Aliphatic-grade PU is therefore a better choice than aromatic-grade PU in a harsh environment (Rashvand and Ranjbar 2014).

3. Types of polyurethane

3.1. Thermoset polyurethane

Thermoset polymers are known for their promising thermo-mechanical characteristics. This class of PU polymers, because of its versatile chemistry and robustness, is widely used in industry. The essential carbamate bonds of thermoset PU are exchangeable, while thermoset polymers cannot be reprocessed (Zheng et al. 2016), (Fortman et al. 2015). Because of its excellent mechanical characteristics, rubber-like elasticity, tear resistance, and promising abrasion resistance, thermoset PU is considered a potential material for bio-composite preparation (Al-Oqla and El-Shekeil 2019).

Zheng et al. demonstrated that classical thermoset PUs with shape memory properties have thermal plasticity, which is attributed to carbamate-only bond exchange. On the other hand, the thermal plasticity is broadly applicable to all thermoset PUs. The combination of plasticity and elasticity opens a wide range of application for this kind of PU (Zheng et al. 2016). Also, it should be considered that, because of the nature of connections in the polymer chain, thermoset polymers cannot be reprocessed through temperature and pressure (de Avila Delucis et al. 2019).

3.2. Thermoplastic polyurethanes

Like other PUs, thermoplastic PUs are built of alternating hard and soft segments. Flexible oligomer diols with estimated molar mass of 500 to 3000 g/mol are soft segments. The soft segment is copolymerized with hard segments composed of short diols and diisocyanates (Al-Oqla and El-Shekeil 2019). Isosorbide (ISO) is a green monomer for preparation of thermoplastic PU. Because it is non-toxic and has aliphatic, chiral and rigid structure, higher hardness and higher glass transition are expected from the polymers based on ISO. Moreover, isosorbide can be used as chain extender for preparation of thermoplastic PU, resulting in outstanding thermal and mechanical properties in comparison with common petroleum-based monomers (Blache et al. 2018). Thermoplastic PU is versatile and exhibits characteristics between rubber and plastic (Arenas et al. 2019). Thermoplastic PU is also flexible and easy to process. The possibility of melt processability makes it a great candidate where high flexibility is demanded. Thermoplastic PUs have a wide range of applications in coatings, adhesives, medical devices, and sealants (Li et al. 2019).

3.3. Waterborne polyurethane

Conventional PUs are prepared using organic solvents. In view of environmental concerns, waterborne structures have received tremendous attention in recent years. During preparation procedure of waterborne PU, only water evolves from polymer, so the waterborne PU is considered a green material in the coating and adhesive industry (Rahman et al. 2013). The incorporation of some filler such as carbon nanotubes, silica particles, nanoclay, or graphene has been widely recommended in many studies to overcome the unfavorable mechanical and thermal properties of waterborne PU (Kim et al. 2019).

3.4. Polyurethane foams

Foams are among the most versatile polymeric materials based on PU. They are used in the furniture industry and in automotive applications as sound insulation, bumpers and seat cushioning (Carriço et al. 2016). Rigid PU foams based on bio-polyols and petrochemical materials are intrinsically flammable (Kurańska et al. 2016), but because of their insulation properties, low pollution impact, affordable cost, low fragility and easy adhesion for coating, rigid foams are widely used (de Avila Delucis et al. 2019, Hilmi and Zainuddin 2019).

Polyurethane foams are composed of polyols, chain extenders, blowing agents, gelling and blowing catalysts, surfactants, and isocyanates as raw materials. The final characteristics of PU foams depend on the components and the preparation method. The two important chemical reactions occurring simultaneously are the gelling reaction between isocyanates and hydroxyl group, which produces urethane group, and the blowing reaction between amine and isocyanate groups, which results in urea group. The amount and type of materials used in preparation of foam directly influence the properties of the final product. Additionally, by controlling the rate of blowing and gelling reactions, the properties of foam can be optimized (Abdollahi Baghban et al. 2018, Oliviero et al. 2017).

One of the challenges bio-polyols present in the manufacture of PU foams is lower mechanical strength. The mechanical characteristics of bio-based foams are attributed to the architecture of cellular structure such as edge length, cell wall thickness, cell size and anisotropy (Hawkins et al. 2005). To resolve this issue, various types of fillers have been proposed to reinforce the bio-foam properties (Hilmi and Zainuddin 2019).

3.5. Supramolecular polyurethane

Polyurethanes that are easy to prepare and have other special characteristics are suitable for preparation of supramolecular structures (Houton et al. 2015). Preparation of supramolecular polymer networks was proposed according to non-covalent interactions (Voorhaar and Hoogenboom 2016). In such supramolecular polymers, the low-energy bonding of non-covalent bonds such as metal coordination, host-guest complexation, hydrogen bonding, π-π staking, and electrostatic interaction can provide a wide range of unique characteristics such as better processability and self-healing behavior (Comí et al. 2017).

3.6. Flame retardant

Flame retardant composites based on polymers are used to reduce economic losses and fire hazards in practical applications. Some fillers such as Go can improve the fire-retardant property (Du et al. 2020). Addition of isocyanurate trimerization structures in PU structure or introduction of flame retardant additives such as nitrogen, phosphorus and halogen can improve fire-resistant properties (Kurańska et al. 2016, Bian et al. 2007). The use of halogenated flame retardants has declined in view of their potential health risks. Instead, expandable graphite filler was proposed to reduce flammability (Kurańska et al. 2016).

4. Polyurethane properties

4.1. Thermal properties

One of the most useful characteristics of PUs is their heat insulating performance; they have the lowest thermal conductivity coefficient among similar materials. Incorporation of isocyanurate rings in polymer structure can improve some characteristics such as dimensional stability, thermal tolerance and fire-retardant properties (Prociak et al. 2017). Since the thermal decomposition of hard segments initiates the degradation of PUs, the chemical structure, composition, length, content, distribution, and symmetry of hard segments perform an important role in improving the thermal stability of PUs. The molecular weight distribution and chemical structure of soft segment also influence the thermal behavior of PU. Moreover, the higher amount of oxygen in polyol provides more weight loss during thermal degradation (Javaid et al. 2018b). Chemical modification of PU by heterocyclic imide has been proposed to improve thermal stability (Yeganeh et al. 2007).

4.2. Mechanical characteristics

Polyurethane inherently possesses low mechanical performance and stiffness. This undesirable characteristic limits its application in many fields. One of the proposed solutions to overcome this deficiency is incorporation of high module nanoparticles in polymer matrix (Babaie et al. 2019, Bera and Maji 2017). The mechanical properties of PU can be modified using toluene diisocyanate and methylene diphenyl diisocyanate. Such components improve the mechanical performance of PU, but because of the carcinogenic nature of the degradation products, their medical use is limited (Kalajahi et al. 2016). Recently, many researchers have studied ways to reinforce the mechanical characteristics of PU. Carbon nanofibers (Burgaz and Kendirlioglu 2019), silicon carbid (Burgaz and Kendirlioglu 2019), carbon nanotubes (Han and Cho 2018), and graphene nano-sheets (Song et al. 2019) are some common candidates for improvement in the mechanical performance of PU.

4.3. Shape memory

Shape memory polymers belong to the category of smart materials that can adopt programmed shapes and recover their original shape upon external stimulation (Zheng et al. 2016). Polyurethane is one such polymer. The origin of its shape memory is its heterogeneous structure and the alternately repeating soft and hard segments. Generally, the soft segment sustains the original shape of the polymer, whereas the hard segment, composed of urethane group, chain extender and the remains of diisocyanate, controls the original shape of the polymer (Babaie et al. 2019, Ha et al. 2019). It has been demonstrated that phase separation significantly influences the morphology (Babaie et al. 2019). The hydrogen-bonding content among soft and hard segments directly affects the phase separation degree such that the hydrogen bond between hard segments enhances the phase separation degree and the hydrogen bond among soft and hard segments increases phase mixing degree (Sami et al. 2014). Shape memory properties can be improved by incorporation of fillers such as carbon nanofibers, graphene, nano-sheets, or multi-wall carbon nanotubes (Babaie et al. 2019). The shape memory PUs possess reversible elongation and recovery ratio, outstanding flexibility, and easy processability compared to shape memory ceramics (Sofla et al. 2019).

4.4. Electrical conductivity

The undesirable electrical characteristics of PU have limited its biomedical applications, while this weak electrical conductivity allows cells or cultured tissue upon biomaterials to be stimulated. Metal nanostructures carbon black or carbon-based materials can improve the PU electrical performance (Bahrami et al. 2019). Graphene particles between polymer chains act as electron transfer pathway and provide the fast charge of electrons through polymer (Bahrami et al. 2019).

4.5. Acoustic properties

One of the main forms of environmental pollution, especially in cities, is noise pollution. Most air-borne and structural noise is attributed to low frequencies between 30 and 500 Hz and medium to high frequencies between 500 and 8000 Hz (Sung and Kim 2017). Different kinds of polymeric foams, glass wools and other similar materials are commonly used to reduce annoying noises (Park et al. 2017). The most common insulating material used as sound damping is porous viscoelastic PU foam (Abdollahi Baghban et al. 2019). The energy dissipation of acoustic waves is believed to occur through viscoinertial and thermal damping. The viscoelastic frame damping mechanisms and the specifications of foams include morphology, porosity, average cell size, cell size distribution, thickness of cell walls, macroscopic density, thermal characteristic lengths, airflow resistivity and stiffness (Abdollahi Baghban et al. 2018). It has been revealed that acoustic specification of PU foams can be improved by optimizing the chemical structure, molecular weight, hydroxyl values of polyols, type of isocyanate, water content, composition along the chain extenders, hybridization of organic and/or inorganic fillers (micron or nanosize) or fibers during manufacturing, and also experimental foaming processes such as one-step, two-step, or pre-polymer methods (Abdollahi Baghban et al. 2018, Gama et al. 2017). It has been proposed that the phase separation in polymer matrix could improve the sound damping behavior. Also, the micro-phase separation increases open pores and cell content and influences the viscoelectric behavior of foam. Cell rupture increases the open-cell content percentage, which can increase the collisions of sound waves with gas molecules, struts and walls (Abdollahi Baghban et al. 2019, Baghban et al. 2019).

Furthermore, considering the hydrogen bonding between urea-urea groups in hard segments, it seems that energy dissipation mechanism is carried out by hard segments. When the hard segment vibrates, the soft segment begins to vibrate. It is concluded that in addition to hard segment structure, the sound dissipation in PU is influenced by the dynamic of the polymer chain segment and dipole polarization made by micro-phase separation (Abdollahi Baghban et al. 2019).

4.6. Polyurethane adhesive properties

Many studies have focused on bio-secured adhesives used for bonding wood. For this purpose, wood adhesives based on kraft lignin, coffee bean shell or soy protein have been investigated. Tenorio-Alfonso and co-workers reported bio-based PU adhesive as an alternative to common traditional petroleum-based ones. They proposed the reaction of diisocyanate compound with the –OH group castor oil, which is part of the polymer network and does not vaporize out of the adhesive, as solvents do. Additionally, it has been reported that when the NCO/OH molar ratio is increased to 2.0 the chain crosslinking density increases and improves the adhesive bond strength (Tenorio-Alfonso et al. 2019, Somani et al. 2003).

4.7. UV-resistant PU foams

Delucis et al. reported preparation and photodegradation characteristic of PU foams filled with forest-based wastes such as paper sludge, wood and bark. The highest UV radiation resistance was demonstrated by composites filled with bark and kraft lignin. Also, the incorporation of UV-resistant nano and micro inorganic materials such as ZnO, GeO_2 or TiO_2 can improve UV resistance. However, some of these materials have photocatalytic effects too (Adak et al. 2019). Additionally, recent studies have revealed the promising characteristics of polymer-graphene composite for UV resistance applications (Xie et al. 2015).

5. Polyurethane composites

The most challenging method of preparation of polymer nanocomposite is interfacial engineering between the polymer chains and the filler. When this interaction is strong, due to efficient load transfer upon tensile strength, considerable reinforcement will be obtained. In contrast, the weak interactions of polymer and fillers including hydrogen bonds, π-π interfacial interactions, metal ligand co-orientation and ionic bonding obviously weaken the reinforcement effect. Surface modification is the most common method to make high dispersion polymer-nano filler composites (Zare et al. 2017). The bio-composite strength is influenced by the conditions in the design processes, such as the interaction and compatibility of filler with polymer and the volume of filler in polymer matrix. The optimum content of filler should be experimentally calculated in view of the fact that with lower or higher amount of filler, there will be a decrease in the composite strength. On the other hand, good interfacial bonding between matrix and filler allow good stress transfer through the composite and improve mechanical strength (Al-Oqla and El-Shekeil 2019).

Because of the hygroscopic nature of PU matrix in comparison with other polymers such as polypropylene, polyethylene, or polyvinylchloride, it is a suitable choice for natural fiber composites.

Nano- and microcomposites based on PU with the thermoset and thermoplastic characterization have been proposed in many studies

(Al-Oqla and El-Shekeil 2019). Considering the high tendency of nanofillers to aggregate, fabrication of high performance nanocomposites with uniform dispersion of nanofiller through polymer chains and efficient interfacial interaction of polymer and filler have been the focus of many studies (Li et al. 2019).

The method of preparation in addition to compatibility of filler with polymer and polymer-filler interactions can directly influence the properties of the final product (Adak et al. 2019).

In the preparation procedure, compared with solution mixing or *in situ* polymerization, melt blending is preferable because it requires less organic solution. The drawback of this method is the agglomeration of filler particles through matrix, since the mechanical force during melt blending pathway is not suitable to destroy intramolecular forces between filler particles (Sheng et al. 2019).

Although *in situ* polymerization is known as a facile composite preparation providing more homogenous composite through suitable dispersion of filler in polymer matrix, the masking effect of polymer chains diminishes its surface properties in some applications, such as antimicrobial products (Ahmadi et al. 2019).

Solution mixing is an outstanding pathway for dispersion of nanomaterials such as clay in polymer matrix, but the environmental concerns and high cost of this method have limited its use, especially for industrial application (Kim et al. 2010). Sonication and high-speed stirring are commonly used in solution mixing method for better dispersion of nanoclay particles in polymer matrix. Also, high shear melt mixing has been proposed for delamination of clay particles, which results in homogeneous polymer (Adak et al. 2018).

Polyurethane demonstrates outstanding performance, but, depending on the target application, addition of some filler can improve the efficiency of any composite obtained.

Cellulose as the most abundant natural biopolymer with hydroxyl group on its chain backbone provides strong hydrogen bond with other hydroxyl groups on neighboring chains. This intramolecular interaction causes the cellulose chains to assemble into an ordered structure that demonstrates high tensile strength. Various kinds of cellulose-based materials such as cellulose nanofibers, cellulose nanocrystal, and bacterial cellulose have been incorporated in PU matrix. Incorporation of cellulose into the polymer can change some of the properties of polymer such as glass transition temperature, owing to the interaction of PU with cellulose chains (Kim et al. 2019, Simón et al. 2014). Also, lower contact angle, improved specular gloss and hardness are observed in reinforced rigid PU foams filled with cellulose nanofibers (de Avila Delucis et al. 2019).

Chitosan as a naturally occurring analog of chitin is linear amino polysaccharides of randomly substituted N-acetyl-D-glucosamine and d-glucosamine. Its suitable hydrophobicity and crystallinity make it a good material for surface coating. Also, chitosan possesses antimicrobial performance attributed to deprotonated amino group on its surface, which makes it a strong nucleophile (Hilmi and Zainuddin 2019, Javaid et al. 2018b). Chitosan-based PU composites are a promising scaffold with outstanding flexibility in addition to biostability, which makes them a good candidate for medical applications (Hilmi and Zainuddin 2019). Using chitosan as chain extender in preparation of PUs gives the final composite suitable thermo-mechanical and structural characteristics. It has also been proved that more chitosan content in PU matrix improves the adhesive power and biocompatibility (Takahashi et al. 1996). Also, the thermal resistance property of PU improves with chitosan loading (Javaid et al. 2018a). PU-chitosan composites are used as surgical thread owing to their good elasticity, ideal mechanical properties, tensile strength and non-toxic nature (Zia et al. 2008). Considering the presence of hydroxyl and amino group, chitosan could form intramolecular hydrogen bonds that increase its crosslinking capability and viscosity (Javaid et al. 2018a). PU-chitosan composites are known to show better adhesive power and biocompatibility in addition to suitable mechanical properties, elasticity, good tensile strength and non-toxic nature.

Thermal and mechanical properties of PU-carbon nanotube were investigated and the 10°C increment in glass transition temperature was observed. Also, the module and tensile strength of PU was improved after addition of only 2 wt% carbon nanotubes to PU matrix. Graphene is one atom packed in a honeycomb crystalline lattice, with the hybrid of sp^2 and 2D planar sheet of carbon. Because graphene has outstanding optical, chemical, electrical conductivity (electro-conductivity of 10,000 $s.cm^{-1}$) and mechanical properties (tensile module of 1100 GPa), it has attracted attention as filler in polymer matrix (Bahrami et al. 2019). In view of its great mechanical strength, large theoretical specific surface area and low cost, it has been extensively used as filler in many composites such as electronic devices, supercapacitors and sensors. Also, it has been reported (Qu et al. 2014) that graphene-based composites have outstanding application as UV protection material (Song et al. 2019, Du et al. 2020). For better incorporation of graphene in polymer, functionalized graphene has been proposed (Adak et al. 2019). Also, graphene nanoparticles are used for modification of electrical performance of PU (Bahrami et al. 2019). Because of few hydrophobic groups in addition to van der Waals forces on graphene sheets, the water solubility of graphene is low. To overcome this drawback, GO is used instead. Thanks to large numbers of oxygen-containing functional groups, GO demonstrates suitable dispersion in aqueous solution such as carboxyl, carbonyl, hydroxyl, and epoxy groups (Song et al. 2019).

Carbon nanotubes, owing to their promising conductivity, high strength, outstanding temperature resistance, and light weight, have been extensively used as filler in many polymers (Burgaz and Kendirlioglu 2019). Aggregation and poor dispersion are the most important challenges in using carbon nanotubes as filler in polymer matrix. For this reason, different pathways to functionalize carbon nanotubes have been proposed for better dispersion of carbon nanotubes in polymer matrix (Moniruzzaman and Winey 2006).

Multi-wall carbon nanotubes with lower price are used extensively in many composites, but their zero band gap and metallic properties have limited their applications in some fields such as supercapacitors and electrodes. In contrast, single-wall carbon nanotubes possess different band gap energies based on chirality. Through functionalization of single-wall carbon nanotubes the band gap can be tuned to serve the target properties (Ha et al. 2019).

It has been observed that using nanosilica as filler improves thermal and mechanical properties. The incorporation of 2.5% nanosilica to rigid PU foams increases the softening point and the temperature of extrapolated beginning of first loss in mass by 2 °C. Also, addition of silica decreases the compression strength of PU foams and inhibits foam growth because it increases the viscosity (Burgaz and Kendirlioglu 2019).

6. Advanced polyurethane composites

6.1. Smart polyurethane

Over the last decades, advanced polymers have attracted copious attention owing to their promising applications in new fields of research. They have demonstrated capabilities for reducing resource use to minimize energy consumption (Bayan and Karak 2018). Self-healing polymers are a new class of advanced materials with the ability to repair sites damaged by mechanical force (Bayan and Karak 2018, Bayan and Karak 2017). Healing the damaged sites in polymer structure is an important subject to extend polymer lifespan. This capability is important in automotives, electronic devices, medical apparatus, packaging and other applications (Wong et al. 2019). High polymeric chain mobility and chain diffusion are the keys to this self-healing characteristic. Polyurethane based on aromatic diisocyanate is susceptible to UV radiation because of auto-oxidation degradation. Also, the presence of Si-GO particles in polymer matrix improves the self-healing process, which results in dissipation of income energy in polymer matrix.

Self-cleaning polymers are another class of smart polymers with special characteristics to keep the surface free from dust and dirt. The self-cleaning characteristics of polymers require a hydrophobic/hydrophilic surface. Hyper-branched PU possesses properties such as high reactivity, low melt and solution viscosity and suitable compatibility with other polymers. Owing to aforementioned characteristics PUs are considered suitable for preparation of smart polymers (Bayan and Karak 2018).

6.2. Biomedical applications

Polyurethane is known as the most common group of synthetic biomaterials in the healthcare industry. A wide range of biodegradable PU with promising mechanical, biological and physicochemical characteristics has been used in regenerative medicine (Torres 2019). Considering the advantages of sponge-like PUs for biomedical applications such as tailorable physicochemical properties, it has been the base of many composites especially for biomedical applications (Marzec et al. 2017). This class of polymers has a wide range of uses in biomaterials, for example, those involved in bone forming. High affinity to bone-forming cells and ideal characteristics for bone repair were found in these renewable materials (Mi et al. 2014). Hydroxyapatite fillers exhibited improvement in elastic modules and compressive strength of PU composite scaffolds. The mechanical characteristics are enhanced with increase of filler loading to 40%. Composites with hydroxyapatite loading demonstrated high affinity to osteoplastic cells and were suitable templates for cell growth and proliferation. They are also compatible with the tissue host, with no immune rejection. This kind of composite could be applied for bone repair (Du et al. 2018). The use of hydroxyapatite as nanoparticles in PU matrix has been extensively studied (Santerre et al. 1994). It has been demonstrated that with 30% loading of hyaluronic acid nanoparticles in PU matrix, due to the formation of hydrogen-bonding interactions of amide groups of PU with hydroxyl groups of hyaluronic acid, the thermal stability improves (Navarro-Baena et al. 2015). Javaid et al. focused on preparation and characterization of PU composites with chitosan and montmorillonite clay as filler. Step growth polymerization method was employed for reacting hydroxyl-terminated polybutadiene and toluene diisocyanate to increase antimicrobial and thermal characteristics. The results indicated that this composite owing to suitable mechanical strength and antibacterial performance could be considered for biological applications (Prociak et al. 2017).

7. Conclusion

Polyurethane, with the elasticity of rubber and the toughness of metals, has attracted extensive attention in recent years. Polyurethane is composed of hard and soft segments that are responsible for all the characteristics of polymer. Variations in preparation methods, ingredients and reagents, as well as the use of fillers, can overcome the defects of PU. The new outstanding aspect of PU application is its intrinsic capability to be used as a smart material in many fields.

The demand for PU has gone up in various fields. The shortage of petroleum-based resources and environmental issues have become a main concern in preparation of such polymers. Recently, many studies have been conducted on bio-based materials for preparation of PU. Many outstanding results have been reported in such studies, but industrial applications of PU will have to be considered more precisely.

References

Abdollahi Baghban, Sahar, Manouchehr Khorasani and Gity Mir Mohamad Sadeghi. 2019. Soundproofing flexible polyurethane foams: Effect of chemical structure of chain extenders on micro-phase separation and acoustic damping. *Journal of Cellular Plastics* 0021955X19864387.

Abdollahi Baghban, Sahar, Manouchehr Khorasani and Gity Mir Mohamad Sadeghi. 2018. Soundproofing flexible polyurethane foams: The impact of polyester chemical structure on the microphase separation and acoustic damping. *Journal of Applied Polymer Science* 135(46): 46744.

Adak, Bapan, Bhupendra Singh Butola and Mangala Joshi. 2018. Effect of organoclay-type and clay-polyurethane interaction chemistry for tuning the morphology, gas barrier and mechanical properties of clay/polyurethane nanocomposites. *Applied Clay Science* 161: 343-353.

Adak, Bapan, Mangala Joshi and B.S. Butola. 2019. Polyurethane/functionalized-graphene nanocomposite films with enhanced weather resistance and gas barrier properties. *Composites Part B: Engineering* 176: 107303.

Ahmadi, Younes, Mithilesh Yadav and Sharif Ahmad. 2019. Oleo-polyurethane-carbon nanocomposites: Effects of in-situ polymerization and sustainable precursor on structure, mechanical, thermal and antimicrobial surface-activity. *Composites Part B: Engineering* 164: 683-692.

Akindoyo, John O., Md.D.H. Beg, Suriati Ghazali, M.R. Islam, Nitthiyah Jeyaratnam and A.R. Yuvaraj. 2016. Polyurethane types, synthesis and applications – A review. *RSC Advances* 6(115): 114453-114482.

Al-Oqla, Faris M. and Y.A. El-Shekeil. 2019. Investigating and predicting the performance deteriorations and trends of polyurethane bio-composites for more realistic sustainable design possibilities. *Journal of Cleaner Production* 222: 865-870.

Arenas, Jorge P., Jose L. Castaño, Loreto Troncoso and Maria L. Auad. 2019. Thermoplastic polyurethane/laponite nanocomposite for reducing impact sound in a floating floor. *Applied Acoustics* 155: 401-406.

Athanasiou, Kyriacos A., Gabriele G. Niederauer and C. Mauli Agrawal. 1996. Sterilization, toxicity, biocompatibility and clinical applications of polylactic acid/polyglycolic acid copolymers. *Biomaterials* 17(2): 93-102.

Awad, Walid H. and Charles A. Wilkie. 2010. Investigation of the thermal degradation of polyurea: The effect of ammonium polyphosphate and expandable graphite. *Polymer* 51(11): 2277-2285.

Babaie, Amin, Mostafa Rezaei and Reza Lotfi Mayan Sofla. 2019. Investigation of the effects of polycaprolactone molecular weight and graphene content on crystallinity, mechanical properties and shape memory behavior of polyurethane/graphene nanocomposites. *Journal of the Mechanical Behavior of Biomedical Materials* 96: 53-68.

Baghban, Sahar Abdollahi, Manouchehr Khorasani and Gity Mir Mohamad Sadeghi. 2019. Acoustic damping flexible polyurethane foams: Effect of isocyanate index and water content on the soundproofing. *Journal of Applied Polymer Science* 136(15): 47363.

Bahrami, Saeid, Atefeh Solouk, Hamid Mirzadeh and Alexander M. Seifalian. 2019. Electroconductive polyurethane/graphene nanocomposite for biomedical applications. *Composites Part B: Engineering* 168: 421-431.

Bayan, Rajarshi and Niranjan Karak. 2017. Renewable resource derived aliphatic hyperbranched polyurethane/aluminium hydroxide-reduced graphene oxide nanocomposites as robust, thermostable material with multi-stimuli responsive shape memory features. *New Journal of Chemistry* 41(17): 8781-8790.

Bayan, Rajarshi and Niranjan Karak. 2018. Bio-derived aliphatic hyperbranched polyurethane nanocomposites with inherent self healing tendency and surface hydrophobicity: Towards creating high performance smart materials. *Composites Part A: Applied Science and Manufacturing* 110: 142-153.

Bera, Madhab and Pradip K. Maji. 2017. Effect of structural disparity of graphene-based materials on thermo-mechanical and surface properties of thermoplastic polyurethane nanocomposites. *Polymer* 119: 118-133.

Bian, Xiang-Cheng, Jian-Hua Tang, Zhong-Ming Li, Zhong-Yuan Lu and Ai Lu. 2007. Dependence of flame-retardant properties on density of expandable graphite filled rigid polyurethane foam. *Journal of Applied Polymer Science* 104(5): 3347-3355.

Blache Héloïse, Françoise Méchin, Alain Rousseau, Étienne Fleury Jean-Pierre Pascault, Pierre Alcouffe, Nicolas Jacquel and René Saint-Loup. 2018. New bio-based thermoplastic polyurethane elastomers from isosorbide and rapeseed oil derivatives. *Industrial Crops and Products* 121: 303-312.

Burgaz, Engin and Caner Kendirlioglu. 2019. Thermomechanical behavior and thermal stability of polyurethane rigid nanocomposite foams containing binary nanoparticle mixtures. *Polymer Testing*: 105930.

Cakić, Suzana M., Maja D. Valcic, Ivan S. Ristić, Tanja Radusin, Miroslav J. Cvetinov and Jaroslava Budinski-Simendić. 2019. Waterborne polyurethane-silica nanocomposite adhesives based on castor oil-recycled polyols: Effects of (3-aminopropyl) triethoxysilane (APTES) content on properties. *International Journal of Adhesion and Adhesives* 90: 22-31.

Calvo-Correas, Tamara, Arantzazu Santamaria-Echart, Ainara Saralegi, Loli Martin, Ángel Valea, M. Angeles Corcuera and Arantxa Eceizaa. 2015. Thermally-responsive biopolyurethanes from a biobased diisocyanate. *European Polymer Journal* 70: 173-185.

Carriço, Camila S., Thaís Fraga and Vânya M.D. Pasa. 2016. Production and characterization of polyurethane foams from a simple mixture of castor oil, crude glycerol and untreated lignin as bio-based polyols. *European Polymer Journal* 85: 53-61.

Chanda, Manas. 2006. Introduction to Polymer Science and Chemistry: A Problem-solving Approach: CRC Press. Taylor & Francis Group, Boca Raton, Fla.

Chaudhari, Ashok B., Pyus D. Tatiya, Rahul K. Hedaoo, Ravindra D. Kulkarni and Vikas V. Gite. 2013. Polyurethane prepared from neem oil polyesteramides for self-healing anticorrosive coatings. *Industrial & Engineering Chemistry Research* 52(30): 10189-10197.

Cheng, Shuna, Ian D'cruz, Mingcun Wang, Mathew Leitch and Chunbao Xu. 2010. Highly efficient liquefaction of woody biomass in hot-compressed alcohol-water co-solvents. *Energy & Fuels* 24(9): 4659-4667.

Comí, Marc, Gerard Lligadas, Juan C. Ronda, Marina Galià and Virginia Cádiz. 2017. Adaptive bio-based polyurethane elastomers engineered by ionic hydrogen bonding interactions. *European Polymer Journal* 91: 408-419.

de Avila Delucis, Rafael, Eduardo Fischer Kerche, Darci Alberto Gatto, Washington Luiz Magalhães Esteves, Cesar Liberato Petzhold and Sandro Campos Amico. 2019. Surface response and photodegradation performance of bio-based polyurethane-forest derivatives foam composites. *Polymer Testing* 80: 106102.

de Oliveira, Mara Cristina Lopes, Renato Altobelli Antunes and Isolda Costa. 2013. Effect of the NCO/OH molar ratio on the physical aging and on the electrochemical behavior of polyurethane-urea hybrid coatings. *Int. J. Electrochem. Sci* 8: 4679-4689.

Du, Jingjing, Yi Zuo, Lili Lin, Di Huang, Lulu Niu, Yan Wei, Kaiqun Wang, Qiaoxia Lin, Qin Zou and Yubao Li. 2018. Effect of hydroxyapatite fillers on the mechanical properties and osteogenesis capacity of bio-based polyurethane composite scaffolds. *Journal of the Mechanical Behavior of Biomedical Materials* 88: 150-159.

Du, Weining, Yong Jin, Shuangquan Lai, Liangjie Shi, Yichao Shen and Heng Yang. 2020. Multifunctional light-responsive graphene-based polyurethane composites with shape memory, self-healing, and flame retardancy properties. *Composites Part A: Applied Science and Manufacturing* 128: 105686.

El Mansouri, Nour-Eddine and Joan Salvadó. 2007. Analytical methods for determining functional groups in various technical lignins. *Industrial Crops and Products* 26(2): 116-124.

Feng, Z.Q., H.P. Kang and G.Q. Han. 2013. Polyurethane grouting material modified by inorganic salts in coal mine. *Chinese Journal of Geotechnical Engineering* 35(8): 1559-1564.

Fortman, David J., Jacob P. Brutman, Christopher J. Cramer, Marc A. Hillmyer and William R. Dichtel. 2015. Mechanically activated, catalyst-free polyhydroxyurethane vitrimers. *Journal of the American Chemical Society* 137(44): 14019-14022.

Gabriel, Lais P., Maria Elizabeth M. dos Santos, André L. Jardini, Gilmara N.T. Bastos, Carmen G.B.T. Dias, Thomas J. Webster and Rubens Maciel Filho. 2017. Bio-based polyurethane for tissue engineering applications: How hydroxyapatite nanoparticles influence the structure, thermal and biological behavior of polyurethane composites. *Nanomedicine: Nanotechnology, Biology and Medicine* 13(1): 201-208.

Gama, Nuno, Rui Silva, António P.O. Carvalho, Artur Ferreira and Ana Barros-Timmons. 2017. Sound absorption properties of polyurethane foams derived from crude glycerol and liquefied coffee grounds polyol. *Polymer Testing* 62: 13-22.

Gama, Nuno V., Belinda Soares, Carmen S.R. Freire, Rui Silva, Carlos P. Neto, Ana Barros-Timmons and Artur Ferreira. 2015. Bio-based polyurethane foams toward applications beyond thermal insulation. *Materials & Design* 76: 77-85.

García-Pacios, Vanesa, Víctor Costa, Manuel Colera and José Miguel Martín-Martínez. 2011. Waterborne polyurethane dispersions obtained with polycarbonate of hexanediol intended for use as coatings. *Progress in Organic Coatings* 71(2): 136-146.

Garrett, James T., Ruijian Xu, Jaedong Cho and James Runt. 2003. Phase separation of diamine chain-extended poly (urethane) copolymers: FTIR spectroscopy and phase transitions. *Polymer* 44(9): 2711-2719.

Gogoi, Runumi, M. Sarwar Alam and Rakesh Kumar Khandal. 2014. Effect of increasing NCO/OH molar ratio on the physicomechanical and thermal properties of isocyanate terminated polyurethane prepolymer. *International Journal of Basic and Applied Sciences* 3(2): 118.

Gogoi, Runumi, Utpal Kumar Niyogi, M. Sarwar Alam and Dayal Singh Mehra. 2013. Study of effect of NCO/OH molar ratio and molecular weight of polyol on the physico-mechanical properties of polyurethane plaster cast. *World Applied Sciences Journal* 21(2): 276-283.

Ha, Yu-Mi, Young-O. Kim, Young-Nam Kim, Jaewoo Kim, Jae-Suk Lee, Jae Whan Cho, Morinobu Endo, Hiroyuki Muramatsu, Yoong Ahm Kim and Yong Chae Jung. 2019. Rapidly self-heating shape memory polyurethane nanocomposite with boron-doped single-walled carbon nanotubes using near-infrared laser. *Composites Part B: Engineering* 175: 107065.

Han, J., R.W. Cao, B. Chen, L. Ye, A.Y. Zhang, J. Zhang and Z.G. Feng. 2011. Electrospinning and biocompatibility evaluation of biodegradable polyurethanes based on L-lysine diisocyanate and L-lysine chain extender. *Journal of Biomedical Materials Research: Part A* 96: 705-714.

Han, Na Rae and Jae Whan Cho. 2018. Effect of click coupled hybrids of graphene oxide and thin-walled carbon nanotubes on the mechanical properties of polyurethane nanocomposites. *Composites Part A: Applied Science and Manufacturing* 109: 376-381.

Hao, Hongye, Jingyu Shao, Ya Deng, Shan He, Feng Luo, Yingke Wu, Jiehua Li, Hong Tan, Jianshu Lia and Qiang Fua. 2016. Synthesis and characterization of biodegradable lysine-based waterborne polyurethane for soft tissue engineering applications. *Biomaterials Science* 4(11): 1682-1690.

Hawkins, Michelle Cameron, Brendan O'Toole and Dacia Jackovich. 2005. Cell morphology and mechanical properties of rigid polyurethane foam. *Journal of Cellular Plastics* 41(3): 267-285.

Hilmi, Hazmi and Firuz Zainuddin. 2019. Mechanical properties of polytetrafluoroethylene (PTFE) powder reinforced bio-based palm oil polyurethane (POPU) composite foam. *Materials Today: Proceedings* 16: 1708-1714.

Houton, Kelly A., George M. Burslem and Andrew J. Wilson. 2015. Development of solvent-free synthesis of hydrogen-bonded supramolecular polyurethanes. *Chemical Science* 6(4): 2382-2388.

Hu, Bin, Chen Ye and Changyou Gao. 2015. Synthesis and characterization of biodegradable polyurethanes with unsaturated carbon bonds based on poly (propylene fumarate). *Journal of Applied Polymer Science* 132(24).

Hu, Yan Hong, Yun Gao, De Ning Wang, Chun Pu Hu, Stella Zu, Lieve Vanoverloop and David Randall. 2002. Rigid polyurethane foam prepared from a rape seed oil based polyol. *Journal of Applied Polymer Science* 84(3): 591-597.

Ionescu, Mihail. 2005. Chemistry and Technology of Polyols for Polyurethanes. iSmithers Rapra Publishing. Shawbury, Shrewsburty, Shropshire, UK: Rapra Technology, Ltd.

Javaid, Muhammad Asif, Rasheed Ahmad Khera, Khalid Mahmood Zia, Kei Saito, Ijaz Ahmad Bhatti and Muhammad Asghar. 2018. Synthesis and characterization of chitosan modified polyurethane bio-nanocomposites with biomedical potential. *International Journal of Biological Macromolecules* 115: 375-384.

Javaid, Muhammad Asif, Muhammad Rizwan, Rasheed Ahmad Khera, Khalid Mahmood Zia, Kei Saito, Muhammad Zuber, Javed Iqbal and Peter Langer. 2018. Thermal degradation behavior and X-ray diffraction studies of chitosan based polyurethane bio-nanocomposites using different diisocyanates. *International Journal of Biological Macromolecules* 117: 762-772.

Javni, I., W. Zhang and Z.S. Petrović. 2004. Soybean-oil-based polyisocyanurate rigid foams. *Journal of Polymers and the Environment* 12(3): 123-129.

Kalajahi, Alireza Eyvazzadeh, Mostafa Rezaei and Farhang Abbasi. 2016. Preparation, characterization, and thermo-mechanical properties of poly (ε-caprolactone)-piperazine-based polyurethane-urea shape memory polymers. *Journal of Materials Science* 51(9): 4379-4389.

Kazemi, Forouzan, Gity Mir Mohamad Sadeghi and Hamid Reza Kazemi. 2019. Synthesis and evaluation of the effect of structural parameters on recovery rate of shape memory polyurethane-POSS nanocomposites. *European Polymer Journal* 114: 446-451.

Kim, Hyunwoo, Yutaka Miura and Christopher W. Macosko. 2010. Graphene/polyurethane nanocomposites for improved gas barrier and electrical conductivity. *Chemistry of Materials* 22(11): 3441-3450.

Kim, Min Su, Kyoung Moon Ryu, Sang Hun Lee, Young Chul Choi and Young Gyu Jeong. 2019. Influences of cellulose nanofibril on microstructures and physical properties of waterborne polyurethane-based nanocomposite films. *Carbohydrate Polymers* 225: 115233.

Kurańska, Maria, Ugis Cabulis, Monika Auguścik, Aleksander Prociak, Joanna Ryszkowska and Mikelis Kirpluks. 2016. Bio-based polyurethane-polyisocyanurate composites with an intumescent flame retardant. *Polymer Degradation and Stability* 127: 11-19.

Kurańska, Maria, Aleksander Prociak, Ugis Cabulis and Mikelis Kirpluks. 2015. Water-blown polyurethane-polyisocyanurate foams based on bio-polyols with wood fibers. *Polimery* 60(11-12): 705-712.

Kwon, Oh-Jin, Seong-Ryul Yang, Dae-Hyun Kim and Jong-Shin Park. 2007. Characterization of polyurethane foam prepared by using starch as polyol. *Journal of Applied Polymer Science* 103(3): 1544-1553.

Li, Hongwei, Shanghuan Feng, Zhongshun Yuan, Qin Wei and Chunbao Charles Xu. 2017. Highly efficient liquefaction of wheat straw for the production of bio-polyols and bio-based polyurethane foams. *Industrial Crops and Products* 109: 426-433.

Li, Hongwei, Nubla Mahmood, Zhen Ma, Mingqiang Zhu, Junqi Wanga, Jilu Zheng, Zhongshun Yuan, Qin Wei and Charles (Chunbao) Xu. 2017. Preparation and characterization of bio-polyol and bio-based flexible polyurethane foams from fast pyrolysis of wheat straw. *Industrial Crops and Products* 103: 64-72.

Li, Hongwei, Chunbao Charles Xu, Zhongshun Yuan and Qin Wei. 2018. Synthesis of bio-based polyurethane foams with liquefied wheat straw: Process optimization. *Biomass and Bioenergy* 111: 134-140.

Li, Le, Lin Xu, Wei Ding, Hengyi Lu, Chao Zhang and Tianxi Liu. 2019. Molecular-engineered hybrid carbon nanofillers for thermoplastic polyurethane nanocomposites with high mechanical strength and toughness. *Composites Part B: Engineering* 177: 107381.

Li, Yan, Juraj Mlynar and Simo Sarkanen. 1997. The first 85% kraft lignin-based thermoplastics. *Journal of Polymer Science Part B: Polymer Physics* 35(12): 1899-1910.

Liu, Jin and Dezhu Ma. 2002. Study on synthesis and thermal properties of polyurethane-imide copolymers with multiple hard segments. *Journal of Applied Polymer Science* 84(12): 2206-2215.

Luan, Huacheng, Yun Zhu and Guiyou Wang. 2019. Synthesis, self-assembly, biodegradation and drug delivery of polyurethane copolymers from bio-based poly(1,3-propylene succinate). *Reactive and Functional Polymers* 141: 9-20.

Mahmood, Nubla, Zhongshun Yuan, John Schmidt and Chunbao Charles Xu. 2015. Preparation of bio-based rigid polyurethane foam using hydrolytically depolymerized kraft lignin via direct replacement or oxypropylation. *European Polymer Journal* 68: 1-9.

Martin, D.J., A.F. Osman, Y. Andriani and G.A. Edwards. 2012. Thermoplastic polyurethane (TPU)-based polymer nanocomposites. *In*: Gao, F. (ed.). Advances in Polymer Nanocomposites: Types and Applications, 1st edn. pp. 321-350. Woodhead Publishing, Cambridge.

Marzec, M., Justyna Kucińska-Lipka, Ilona Kalaszczyńska and Helena Janik. 2017. Development of polyurethanes for bone repair. *Materials Science and Engineering C* 80: 736-747.

Mi, Hao-Yang, Xin Jing, Max R. Salick, Travis M. Cordie, Xiang-Fang Peng and Lih-Sheng Turng. 2014. Morphology, mechanical properties, and mineralization of rigid thermoplastic polyurethane/hydroxyapatite scaffolds for bone tissue applications: Effects of fabrication approaches and hydroxyapatite size. *Journal of Materials Science* 49(5): 2324-2337.

Moniruzzaman, Mohammad and Karen I. Winey. 2006. Polymer nanocomposites containing carbon nanotubes. *Macromolecules* 39(16): 5194-5205.

Muuronen, Mikko, Peter Deglmann and Željko Tomović. 2019. Design principles for rational polyurethane catalyst development. *The Journal of Organic Chemistry*. 84(12): 8202-8209.

Narine, Suresh S., Xiaohua Kong, Laziz Bouzidi and Peter Sporns. 2007. Physical properties of polyurethanes produced from polyols from seed oils: II. Foams. *Journal of the American Oil Chemists' Society* 84(1): 65-72.

Navarro-Baena, Iván, Marina P. Arrieta, Agueda Sonseca, Luigi Torre, Daniel López, Enrique Giménez, José M. Kenny and Laura Peponi. 2015. Biodegradable nanocomposites based on poly (ester-urethane) and nanosized hydroxyapatite: Plastificant and reinforcement effects. *Polymer Degradation and Stability* 121: 171-179.

Oertel, Günter and Lothar Abele. 1994. Polyurethane handbook: chemistry, raw materials, processing, application, properties. Munich. New York: Hanser Publishers; New York: Distributed in the USA by Macmillan Pub. Co.

Oliviero, Maria, Letizia Verdolotti, Mariamelia Stanzione, Marino Lavorgna, Salvatore Iannace, Maurizio Tarello and Andrea Sorrentino. 2017. Bio-based flexible polyurethane foams derived from succinic polyol: Mechanical and acoustic performances. *Journal of Applied Polymer Science* 134(45): 45113.

Orgilés-Calpena, Elena, Francisca Arán-Aís, Ana M. Torró-Palau and César Orgilés-Barceló. 2011. Characterization of polyurethanes containing different chain extenders. *Progress in Rubber Plastics and Recycling Technology* 27(3): 145-160.

Parcheta, Paulina and Janusz Datta. 2018. Structure-rheology relationship of fully bio-based linear polyester polyols for polyurethanes – Synthesis and investigation. *Polymer Testing* 67: 110-121.

Park, Ju Hyuk, Kyung Suh Minn, Hyeong Rae Lee, Sei Hyun Yang, Cheng Bin Yu, Seong Yeol Pak, Chi Sung Oh, Young Seok Song, Yeon June Kang and Jae Ryoun Youn. 2017. Cell openness manipulation of low density polyurethane foam for efficient sound absorption. *Journal of Sound and Vibration* 406: 224-236.

Patil, Chandrashekhar K., Sandip D. Rajput, Ravindra J. Marathe, Ravindra D. Kulkarni, Hemant Phadnis, Daewon Sohn, Pramod P. Mahulikar and Vikas V. Gite. 2017. Synthesis of bio-based polyurethane coatings from vegetable oil and dicarboxylic acids. *Progress in Organic Coatings* 106: 87-95.

Pechar, T.W., S. Sohn, G.L. Wilkes, S. Ghosh, C.E. Frazier, A. Fornof and Timothy Edward Long. 2006. Characterization and comparison of polyurethane networks prepared using soybean-based polyols with varying hydroxyl content and their blends with petroleum-based polyols. *Journal of Applied Polymer Science* 101 (3):1432-1443.

Penco, Maurizio, Silvia Marcioni, Paolo Ferruti, Salvatore D'Antone and Romano Deghenghi. 1996. Degradation behaviour of block copolymers containing poly (lactic-glycolic acid) and poly (ethylene glycol) segments. *Biomaterials* 17(16): 1583-1590.

Petrovic, Z. 2008. Polyurethanes from vegetable oils. *Polym Rev* 48: 109-155.

Petrović, Zoran S., Wei Zhang, Alisa Zlatanić, Charlene C. Lava and Michal Ilavskyý. 2002. Effect of OH/NCO molar ratio on properties of soy-based polyurethane networks. *Journal of Polymers and the Environment* 10(1-2): 5-12.

Pettersen, Roger C. 1984. The chemical composition of wood. *The Chemistry of Solid Wood* 207: 57-126.

Prisacariu, Cristina and Elena Scortanu. 2011. Influence of the type of chain extender and urethane group content on the mechanical properties of polyurethane elastomers with flexible hard segments. *High Performance Polymers* 23(4): 308-313.

Prociak, Aleksander, Maria Kurańska, Ugis Cabulis and Mikelis Kirpluks. 2017. Rapeseed oil as main component in synthesis of bio-polyurethane-polyisocyanurate porous materials modified with carbon fibers. *Polymer Testing* 59: 478-486.

Prociak, Aleksander, Maria Kurańska, Ugis Cabulis, Joanna Ryszkowska, Milena Leszczyńska, Katarzyna Uram and Mikelis Kirpluks. 2018. Effect of bio-polyols with different chemical structures on foaming of polyurethane systems and foam properties. *Industrial Crops and Products* 120: 262-270.

Qu, Lijun, Mingwei Tian, Xili Hu, Yujiao Wang, Shifeng Zhu, Xiaoqing Guo, Guangting Han, Xiansheng Zhang, Kaikai Sun and Xiaoning Tang. 2014. Functionalization of cotton fabric at low graphene nanoplate content for ultrastrong ultraviolet blocking. *Carbon* 80: 565-574.

Rahman, Mohammad Mizanur, Aleya Hasneen, Won-Ki Lee and Kwon Taek Lim. 2013. Preparation and properties of sol-gel waterborne polyurethane adhesive. *Journal of Sol-gel Science and Technology* 67(3): 473-479.

Rahman, Mohammad Mizanur, Mohammad Mahbub Rabbani and Joyanta Kumar Saha. 2019. Polyurethane and its derivatives. *Functional Polymers* 225-240.

Rashvand, M. and Z. Ranjbar. 2014. Degradation and stabilization of an aromatic polyurethane coating during an artificial aging test via FTIR spectroscopy. *Materials and Corrosion* 65(1): 76-81.

Roche, A.A., M.J. Romand and F. Sidoroff. 1984. Adhesive joints: Formation, characteristics and testing. pp. 19–30. Edited by Mittal, K.L. New York: Plenum Press.

Saleh, Nabil A. 1990. Structure-property Relationships in Polymer-polyols-based Polyurethane Elastomers. © Nabil Abdulkader Saleh.

Sami, Selim, Erol Yildirim, Mine Yurtsever, Ersin Yurtsever, Emel Yilgor, Iskender Yilgor and Garth L. Wilkes. 2014. Understanding the influence of hydrogen bonding and diisocyanate symmetry on the morphology and properties of segmented polyurethanes and polyureas: Computational and experimental study. *Polymer* 55(18): 4563-4576.

Santerre, J.P., R.S. Labow, D.G. Duguay, D. Erfle and G.A. Adams. 1994. Biodegradation evaluation of polyether and polyester-urethanes with oxidative and hydrolytic enzymes. *Journal of Biomedical Materials Research* 28(10): 1187-1199.

Sheng, Xinxin, Yanfeng Zhao, Li Zhang and Xiang Lu. 2019. Properties of two-dimensional Ti_3C_2 MXene/thermoplastic polyurethane nanocomposites with effective reinforcement via melt blending. *Composites Science and Technology* 181: 107710.

Silva, Ana L. and João C. Bordado. 2004. Recent developments in polyurethane catalysis: Catalytic mechanisms review. *Catalysis Reviews* 46(1): 31-51.

Simón, D., A.M. Borreguero, A. De Lucas and J.F. Rodríguez. 2014. Glycolysis of flexible polyurethane wastes containing polymeric polyols. *Polymer Degradation and Stability* 109: 115-121.

Soares, Belinda, Nuno Gama, Carmen Freire, Ana Barros-Timmons, Inês Brandão, Rui Silva, Carlos Pascoal Neto and Artur Ferreira. 2014. Ecopolyol production from industrial cork powder via acid liquefaction using polyhydric alcohols. *ACS Sustainable Chemistry & Engineering* 2(4): 846-854.

Sofla, Reza Lotfi Mayan, Mostafa Rezaei, Amin Babaie and Mortaza Nasiri. 2019. Preparation of electroactive shape memory polyurethane/graphene nanocomposites and investigation of relationship between rheology, morphology and electrical properties. *Composites Part B: Engineering* 175: 107090.

Somani, Keyur P., Sujata S. Kansara, Natvar K. Patel and Animesh K. Rakshit. 2003. Castor oil based polyurethane adhesives for wood-to-wood bonding. *International Journal of Adhesion and Adhesives* 23(4): 269-275.

Song, Weihua, Bo Wang, Lihua Fan, Fangqing Ge and Chaoxia Wang. 2019. Graphene oxide/waterborne polyurethane composites for fine pattern fabrication and ultrastrong ultraviolet protection cotton fabric via screen printing. *Applied Surface Science* 463: 403-411.

Sonnenschein, Mark F. 2014. Polyurethanes: Science, Technology, Markets, and Trends. Vol. 11. John Wiley & Sons. Hoboken, NJ.

Squiller, Edward P. 1987. Catalysis in aliphatic isocyanate-alcohol reactions. Paper read at Miles Water-borne and higher-solids coatings symposium.

Sung, Giwook and Jung Hyeun Kim. 2017. Effect of high molecular weight isocyanate contents on manufacturing polyurethane foams for improved sound absorption coefficient. *Korean Journal of Chemical Engineering* 34(4): 1222-1228.

Szycher, Michael. 1999. Szycher's Handbook of Polyurethanes. 4–6. Boca Raton, FL: CRC Press.

Tachibana, Koichiro and Hideki Abe. 2019. Studies on thermo-mechanical and thermal degradation properties of bio-based polyurethanes synthesized from vanillin-derived diol and lysine diisocyanate. *Polymer Degradation and Stability* 167: 283-291.

Takahashi, Toshisada, Noriya Hayashi and Shunichi Hayashi. 1996. Structure and properties of shape-memory polyurethane block copolymers. *Journal of Applied Polymer Science* 60(7): 1061-1069.

Tatai, Lisa, Tim G. Moore, Raju Adhikari, François Malherbe, Ranjith Jayasekara, Ian Griffiths and Pathiraja A. Gunatillake. 2007. Thermoplastic biodegradable polyurethanes: The effect of chain extender structure on properties and in-vitro degradation. *Biomaterials* 28(36): 5407-5417.

Tenorio-Alfonso, Adrián, M. Carmen Sánchez and José M. Franco. 2019. Synthesis and mechanical properties of bio-sourced polyurethane adhesives obtained from castor oil and MDI-modified cellulose acetate: Influence of cellulose acetate modification. *International Journal of Adhesion and Adhesives* 102404.

Torres, Maykel González. 2019. Polyurethane/urea composite scaffolds based on poly (3-hydroxybutyrate-g-2-amino-ethyl methacrylate). *Composites Part B: Engineering* 160: 362-368.

Torri, Isadora Dalla Vecchia, Ville Paasikallio, Candice Schmitt Faccini, Rafael Huff, Elina Bastos Caramão, Vera Sacon, Anja Oasmaa and Claudia Alcaraz Zini. 2016. Bio-oil production of softwood and hardwood forest industry residues through fast and intermediate pyrolysis and its chromatographic characterization. *Bioresource Technology* 200: 680-690.

Venkatesh, D. and V. Jaisankar. 2019. Synthesis and characterization of bio-polyurethanes prepared using certain bio-based polyols. *Materials Today: Proceedings* 14: 482-491.

Voorhaar, Lenny and Richard Hoogenboom. 2016. Supramolecular polymer networks: Hydrogels and bulk materials. *Chemical Society Reviews* 45(14): 4013-4031.

Wang, J., Z. Zheng, Q. Wang, P. Du, J. Shi and X. Wang. 2013. Synthesis and characterization of biodegradable polyurethanes based on L-cystine/cysteine and Pol(e-caprolactone). *Journal of Applied Polymer Science* 128: 4047-4057.

Wong, Chee Sien, Nurul Izzati Hassan, Mohd Sukor Su'ait, M. Angels Pelach Serrae Jos, Alberto Mendez Gonzalez, Luis Angel Granda and Khairiah Haji Badri. 2019. Photo-activated self-healing bio-based polyurethanes. *Industrial Crops and Products* 140: 111613.

Xie, Siyuan, Jianfeng Zhao, Bowu Zhang, Ziqiang Wang, Hongjuan Ma, Chuhong Yu, Ming Yu, Linfan Li and Jingye Li. 2015. Graphene oxide transparent hybrid film and its ultraviolet shielding property. *ACS Applied Materials & Interfaces* 7(32): 17558-17564.

Yang, Seong-Ryul, Oh-Jin Kwon, Dae-Hyun Kim and Jong-Shin Park. 2007. Characterization of the polyurethane foam using alginic acid as a polyol. *Fibers and Polymers* 8(3): 257-262.

Yang, X.F., C. Vang, D.E. Tallman, G.P. Bierwagen, S.G. Croll and S. Rohlik. 2001. Weathering degradation of a polyurethane coating. *Polymer Degradation and Stability* 74(2): 341-351.

Ye, Liyi, Jingmiao Zhang, Jie Zhao and Song Tu. 2014. Liquefaction of bamboo shoot shell for the production of polyols. *Bioresource Technology* 153: 147-153.

Yeganeh, Hamid, Pejman Hojati Talemi and Sadegh Jamshidi. 2007. Novel method for preparation of polyurethane elastomers with improved thermal stability and electrical insulating properties. *Journal of Applied Polymer Science* 103(3): 1776-1785.

Zain, Norazwani Muhammad, Eida Nadirah Roslin and Sahrim Ahmad. 2016. Preliminary study on bio-based polyurethane adhesive/aluminum laminated composites for automotive applications. *International Journal of Adhesion and Adhesives* 71: 1-9.

Zare, Yasser, Kyong Yop Rhee and David Hui. 2017. Influences of nanoparticles aggregation/agglomeration on the interfacial/interphase and tensile properties of nanocomposites. *Composites Part B: Engineering* 122: 41-46.

Zdrahala, Richard J. and Ivanka J. Zdrahala. 1999. Biomedical applications of polyurethanes: A review of past promises, present realities, and a vibrant future. *Journal of Biomaterials Applications* 14(1): 67-90.

Zhang, Chaoqun, Hongchao Wu and Michael R. Kessler. 2015. High bio-content polyurethane composites with urethane modified lignin as filler. *Polymer* 69: 52-57.

Zhang, Liqiang, Meng Zhang, Lihong Hu and Yonghong Zhou. 2014. Synthesis of rigid polyurethane foams with castor oil-based flame retardant polyols. *Industrial Crops and Products* 52: 380-388.

Zhao, Yusheng, Fu Zhong, Ali Tekeei and Galen J. Suppes. 2014. Modeling impact of catalyst loading on polyurethane foam polymerization. *Applied Catalysis A: General* 469: 229-238.

Zheng, Ning, Zizheng Fang, Weike Zou, Qian Zhao and Tao Xie. 2016. Thermoset shape-memory polyurethane with intrinsic plasticity enabled by transcarbamoylation. *Angewandte Chemie International Edition* 55(38): 11421-11425.

Zia, Khalid Mahmood, Mehdi Barikani, Mohammad Zuber, Ijaz Ahmad Bhatti and Munir Ahmad Sheikh. 2008. Molecular engineering of chitin based polyurethane elastomers. *Carbohydrate Polymers* 74(2): 149-158.

CHAPTER
12

Biopolymeric Sensors

Payam Zarrintaj[1*], Saeed Manouchehri[1], Mohammad Davachi[2], Mohsen Khodadadi Yazdi[3], Joshua D. Ramsey[1], Mohammad Reza Saeb[4] and Mohammad Reza Ganjali[3,5]

[1] School of Chemical Engineering, Oklahoma State University, Stillwater, OK, USA
[2] Department of Food Science, College of Agriculture and Life Sciences, Cornell University, Ithaca, NY, USA
[3] Center of Excellence in Electrochemistry, School of Chemistry, College of Science, University of Tehran, Tehran, Iran
[4] Departments of Resin and Additives, Institute for Color Science and Technology, P.O. Box 16765-654, Tehran, Iran
[5] Biosensor Research Center, Endocrinology and Metabolism Molecular-Cellular Sciences Institute, Tehran University of Medical Sciences, Tehran, Iran

1. Introduction

Sensors play an essential role in our daily life by alerting our bodies to toxicity and danger in different ways based on the mechanisms on which they are founded. Sensors make it possible to monitor various components or contaminants in the environment, food and biological fluid. Biosensors are a class of sensors obtained from bio-based materials with very low threat to the environment. The complex structure and high cost of biosensors are major challenges in using them as versatile materials. The design and development of inexpensive sensors has therefore attracted significant attention in the past three decades. Sensitivity, rapid detection and low cost are the main parameters considered in designing sensors for various uses such as medicine, food, environment and agriculture (Mehrotra and research 2016).

Polymers, because of their unique features, are obviously suitable as sensors. Ionic electro-active polymers such as polymer gels and biopolymer/conductive material composites as electro-responsive polymers exhibit electrical conductivity because of the migration of ions or molecules. For example, collagen and gelatin (denatured form of collagen) as an ampholyte exhibit conductivity around 10^{-5} S/m depending on water content, pH

*Corresponding author: payam.zarrintaj@gmail.com

and temperature. Water content affects the biopolymers' performance, but it can be regulated by crosslinking or additives to control their swelling/deswelling behavior. Such polymers have been used for fabrication of optical, electrochemical and other sensors. For example, Tiwari synthesized gum acacia-polyaniline-based water-soluble hybrid that exhibited pH-switching electrical conductivity. Increase in pH resulted in decrease in conductivity. The range of conductivity was reported to be about 10^{-5} S/cm to 10^{-2} S/cm with ohmic behavior with current flow. Such a hybrid can be used potentially in sensor applications (Krebsz et al. 2017, Tiwari 2008). Biopolymer-based sensors have been used for detection of various components. Table 1 shows a brief view of biopolymeric sensors and their uses.

Table 1. Biopolymer sensor for analyte detection

Component	*Biopolymer sensor*	*Transducer*	*Reference*
Ammonia	SWCNT–cellulose	Chemiresistor	Han et al. 2014
Water, methanol, ethanol, acetone, chloroform, THF, toluene, *n*-hexane	MWCNT–cellulose	Aerogel; Chemiresistor	Qi et al. 2015
Ammonia	MWCNT–polypeptide	Chemiresistor	L.-C. Wang et al. 2015
Chloroform, methanol, toluene, water	MWCNT–PLA	Chemiresistor	Kumar et al. 2012c
Ethyl acetate, carbon tetrachloride, methanol, ethanol, xylene, *n*-hexane	CB–poly(lactic acid)	Chemiresistor	K. Li et al. 2013
Water, methanol, toluene	CB–chitosan	Chemiresistor	Bouvree et al. 2009
Formaldehyde	Polyethyleneimine–chitosan	PZ(QCM)–gold	N. Wang et al. 2014
Methylamine, dimethylamine, ethanol	Polyaniline–chitosan	PZ(QCM)–gold	Ayad et al. 2014b
Hydrogen	Polyaniline–chitosan	Chemiresistor	W. Li et al. 2011
Water	SWCNT–carrageenan MWCNT–carrageenan	Chemiresistor	Aldalbahi et al. 2012
Water, methanol, ethanol, isopropanol, toluene, chloroform	MWCNT–chitosan	Chemiresistor	B. Kumar et al. 2012b, Kumar et al. 2010
Water, methanol, ethanol, isopropyl alcohol	MWCNT–chitosan–polyvinyl alcohol	Nanohybrid, chemiresistor	Molla-Abbasi and Ghaffarian 2014, Molla-Abbasi et al. 2016

Water, chloroform, ethanol, isopropyl alcohol	MWCNT–chitosan–polyethylene oxide	Nanohybrid, chemiresistor	Molla-Abbasi and Ghaffarian 2014
Methylamine, dimethylamine, diethylamine, methanol, ethanol, isopropyl alcohol	Chitosan	PZ(QCM)–gold	Ayad et al. 2014a
Chloroform Water	Chitosan	Chemiresistor	Chandrasakaran et al. 2014, Mironenko et al. 2014
Ammonia Ammonia, hydrochloric acid	Chitosan–carrageenan	Optical sensor	Mironenko et al. 2013
Water, methanol, toluene	Amylose Starch	Chemisensor	B. Kumar et al. 2012a
Methanol, propionic acid, trimethylamine, dinitrotoluene, dimethyl methylphosphonate	SWCNT–ss-DNA	FET	Khamis et al. 2010
Toluene, acetonitrile, *n*-hexane, *n*-heptane, chloroform, ethanol, methanol, cyclohexane, 2-propanol, dichloromethane, diethylether, tetrahydrofuran, cyclohexanone, triethylamine, anisole, 1-propanol, 1-butanol, 1-pentanol, butyraldehyde, styrene	DNA	PZ(QCM)–gold	Fu et al. 2015
Toluene, gasoline	Polyglycolic acid	PZ(silicon)	Hajjam and Pourkamali 2011
Acetic acid 1-Hexanol, 1-pentanol	Polypeptide SAM	PZ(QCM)–gold	Sankaran et al. 2011
Acetic acid, butyric acid, ammonia, dimethylamine, benzene, chlorobenzene, their mixtures	Oligopeptide SAM	PZ(QCM)–gold	H.-H. Lu et al. 2009
Trimethylamine, ammonia, acetone, methanol, ethyl acetate, *o*-xylene, acetic acid, *p*-xylene, toluene	Oligopeptide SAM	PZ(QCM)–gold	Wu et al. 2001

2. Optical sensors

Optical sensors are especially useful for easy detection, which is favored in chemical and biological applications. Detection of ammonia is crucial because of its wide use in industry and toxic effect on humans and animals. Hence, various techniques such as electrochemical, pH and optical sensors have been developed to monitor ammonia concentration with high and reversible sensitivity (Mader and Wolfbeis 2010, Onida et al. 2004). Most of these techniques are based on gas detection and are not useful for detecting ammonia concentration in water. A few investigations have been performed for ammonia solution detection. Optical sensors based on the surface plasmon resonance (SPR) feature of metal-based nanoparticles exhibit proper detection of liquid ammonia and have the advantages of low cost, simplicity and proper sensitivity (Abaker et al. 2011, Rahman et al. 2011).

Silver nanoparticles (Ag NP) have been widely used in sensors because of their remarkable chemical, optical and electronic properties. Ag NPs exhibit unique performance characteristics such as SPR and fluorescence. SPR is related to the coupled oscillation of free electrons on the conduction band with locally increased electromagnetic field that is sensitive to the conditions. Exposing the NPs to the electromagnetic field causes oscillation of the electrons at a frequency similar to that of the incident wave. Ag spectral feature depends on shape, size and inter-particle spacing. One attractive component for preparing a green nanoparticle is biopolymers. Raveendran et al. (2003) used β-d-glucose and starch as reducing and capping agent, respectively. Ag NPs were prepared using other biopolymers such as chitosan, heparin, acacia, and gum arabic (Kemp et al. 2009, Modrzejewska et al. 2010). Pandey et al. (2012) synthesized an optical sensor based on localized SPR using guar gum/Ag NP to detect ammonia. Guar gum was used as a capping and reducing agent to reduce Ag NPs instead of hydrazine and sodium borohydride. Such a method is considered green because biopolymers were used in the synthesis process instead of toxic materials. The importance of this method is that there is no need to maintain the system pH in the synthesis process. NP dispersion in biopolymer plays an important role in detection performance. The low limit detection was reported to be 1 ppm (Pandey et al. 2012).

Hyperbranch polymers because of the high density of the functional groups (Ghiyasi et al. 2018) have been used to synthesize various nanoparticles including Ag NPs, Au NPs, and ZnO NPs (Gao and Yan 2004), which resulted in narrow size distribution and proper stability. Amino-terminated hyperbranch polymers exhibited a proper ability to capture and reduce the metal ions to produce NPs (Zhang et al. 2009). Sherbiny et al. synthesized core–shell amino-terminated hyperbranched NPs based on Ag and chitosan for optical detection of ammonia. For this purpose, dendritic polyamidoamine (PAMAM) branches were synthesized (Tomalia et al. 1985) and coupled with chitosan nanoparticle. Such NPs were used as a template to fabricate the Ag-based NPs. As shown in Fig. 1, the Ag^{+} solution was

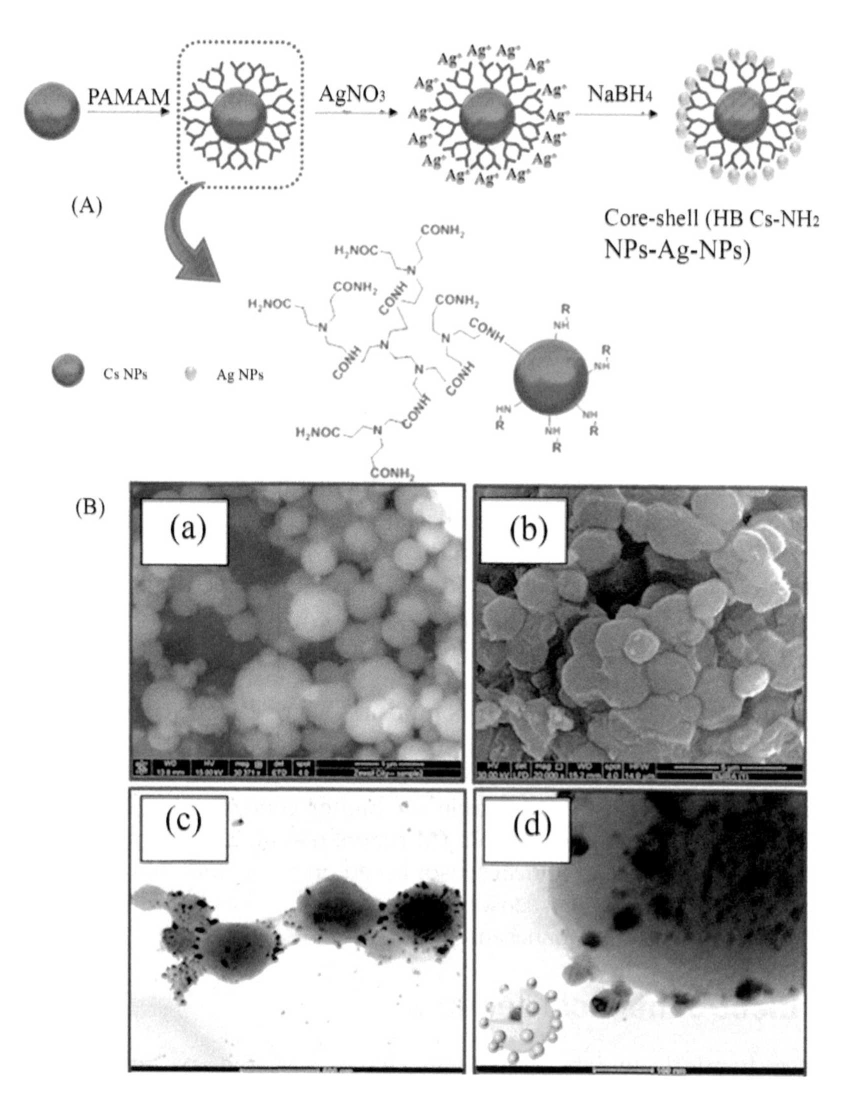

Fig. 1. (A) Schematic illustration for the formation of AgNPs onto the surface of the newly prepared core–shell (HBCs-NH_2) NPs. (B) (a, b) SEM images of Cs NPs and (HBCs-NH_2) NPs, (c, d) the typical HRTEM images of the developed (HBCs-NH_2) NPs-AgNPs. Reproduced with permission from El-Sherbiny et al. (2016).

exposed to synthetic NP and sonicated to achieve a stable solution. The addition of the reducing agent $NaBH_4$ reduced the ions to the zero valent Ag and resulted in the core-shell formation (El-Sherbiny et al. 2016).

For some time, the existence of lead in the ecosystem has been a huge worry due to its enhanced discharge and probable toxic impacts on humans (J. Liu et al. 2008). Hence, various methods have been proposed to monitor lead ions (Pb^+). Razali et al. used chitosan-coated optical fiber tip sensor to detect Pb^+. It was shown that the chitosan coating enhanced the sensitivity seven times.

It is proposed that such a sensor can be used potentially for environmental and wastewater monitoring (Razali et al. 2020). Fen at al. immobilized p-tert-butylcalix[4]arene-tetrakis (BCAT) in chitosan film, which detected Pb2+ by measuring the SPR signal. Pb^{+2} concentration was measured from 30 ppb to 5 ppm, which exhibited the linear relationship between the shift of SPR angle and Pb^{+2} concentration up to 1 ppm with 0.045 ppm^{-1} sensitivity. This sensor can serve as a selective detector in heavy metal ions mixture (Fen et al. 2012). Abdullah et al. synthesized an optical sensor based on chitosan in which the tyrosinase was immobilized. The sensor was used to detect phenol. Enzyme oxidized phenol and produced quinone, which reacted with 3-methyl-2-benzothiazolinone hydrazone to generate a maroon color adduct. The sensor exhibited optimum performance at 45°C and pH 6–7 with linear response between 2.5 and 70.0 µM for 4-chlorophenol with 0.9 µM LOD, 2.5 and 100.0 µM for m-cresol with 1.0 µM LOD, and 12.5 and 400.0 µM for p-cresol with 3.0 µM. Moreover, the sensor exhibited the proper stability for two months (Abdullah et al. 2006). Dubas et al. fabricated an optical sensor for alcohol detection based on chitosan. Layer-by-layer deposition was used to fabricate the multilayer polyelectrolyte sensor to detect ethanol in water. Chitosan immobilized anionic dye (Nylosan) as a polyelectrolyte that was sensitive to the 10-45% ethanol content in solution (Dubas et al. 2006). Mironenko et al. fabricated gas sensor for H_2S detection based on chitosan/gold and chitosan/silver. Such composites were fabricated using *in situ* reduction of metal precursors. Chitosan/Au composite is suitable for low concentration detection and detected ethanol in the range of 0.1-100 ppm. Chitosan/Ag composite is suitable for higher concentration and detected ethanol in the range of 5-300 ppm (Mironenko et al. 2016). Voznesenskiy et al. synthesized chitosan optical sensor based on waveguide films for relative humidity (Table 2). Such a sensor is low in cost and has high sensitivity and a short response time (Voznesenskiy et al. 2013).

3. Electrochemical sensors

Electrochemical analytical methods have garnered particular interest because they are highly sensitive, inexpensive, user-friendly, and simple, and they can be used to analyze the real sample. Electrochemical techniques can assess chemical reaction with goal elements by generating signals related to the concentration in environmental samples or biological fluids in which such technique can recognize markers produced during disease. For instance, urinalysis discloses different compounds at various concentrations reflecting the health and nutritional position. Any substantial variation of a compound from the usual level implies a metabolic syndrome. Hence, the use of electrochemical procedures for the evaluation of biological fluids composition is crucial for disease monitoring because early detection can guarantee effective therapy. Chitosan as a well-known biopolymer has been used in electrochemical techniques because of its biocompatibility, biodegradability, lack of toxicity and low cost. Chitosan composites such as metal composites exhibit chelating structure and enzyme characteristics that enhance the chemical reaction activity. This is attributable to the acidity of the primary amine pendant units in the chitosan (p*K*a 6.3), which can be categorized as an anionic chelating resin and has been used as ionophore in developing surface-modified sensors for the detection of various anions, such as bromide or an iron(III)-oxalate complex (G. Lu et al. 2001).

Diouf et al. fabricated an electrochemical sensor based on chitosan/Au NP for aspirin detection to control its dosage. The aspirin dosage was evaluated in urine, saliva and pharmaceutical tablet employing an electrochemical sensor. The sensor was manufactured by self-assembling chitosan capped with Au NPs with proper selectivity. It displayed a logarithmic-linear correlation between response and the aspirin concentration in the range 1 pg/mL to 1 µg/mL, with 0.03 pg/ml reported as a low detection limit (Diouf et al. 2020). Pauliukaite et al. fabricated an electrochemical sensor based on graphite-epoxy resin/chitosan. Various crosslinking agents (glutaraldehyde, glyoxal, epichlorohydrin and 1-ethyl-3-(3-dimethylaminopropyl) carbodiimide (EDC) together with N-hydroxysuccinimide (NHS)) were used to immobilize carbon nanotube (CNT). The electrode based on chitosan/CNT/EDC–NHS was sensitive to alterations in dipyrone and hydroquinone concentration. With an external glucose oxidase layer, electrochemical impedance spectroscopy confirms that the glucose biosensor features remain constant in the presence of the enzyme (Pauliukaite et al. 2010). Darder et al. fabricated an electrochemical sensor based on chitosan/nanoclay that can be used potentially for potentiometric evaluation of anionic species (Darder et al. 2005). Feng et al. fabricated an electrochemical sensor based on chitosan/Ag NP as a glucose sensor that exhibited broad linearity, straightforward operation and proper stability. It showed linearity range between 0.0004 and 0.01 $molL^{-1}$ with LOD about 0.0003 (Feng et al. 2009). B. Liu et al. (2012) synthesized molecularly imprinted polymer (MIP) electrochemical sensor based on graphene/chitosan to detect dopamine. MIPs are known as ideal components to fabricate sensors with particular binding to template molecule. Direct synthesis of the MIP on the electrode surface is better than conventional synthesis method because of thickness control, removal of toxic organic solvent, and use of aqueous systems with prolonged stability (Zangmeister et al. 2006). The sensor was used for detection of dopamine in wide ranges (10^{-9} to 10^{-8} M and 10^{-7} to 10^{-4} M) and exhibited complete recovery (B. Liu et al. 2012). Lian et al. synthesized MIP electrochemical sensor based on chitosan-platinum/graphene-gold for erythromycin detection. This sensor exhibited high selectivity, proper stability and appropriate reproducibility for detection of erythromycin, and it was effectively used to detect erythromycin in spiked samples (Lian et al. 2012). Huang et al. synthesized MIP electrochemical sensor based on chitosan/Au NP/CNT for tyramine detection. CNT-Au NPs were used for increment of sensitivity and electronic transmission. Chitosan served as a conduit for the imprinted layer and the NPs. Tyramine was used as the template molecule, and silicic acid tetraethyl ester and triethoxyphenylsilane were used as the functional monomers in MIP synthesis. MIP film showed outstanding selectivity for tyramine, demonstrating linear range between 10^{-7} and 1×10^{-5} mol/L, with LOD 5.7×10^{-8} mol/L (Huang et al. 2011).

4. Humidity sensors

Humidity sensors have been used in various applications as a component of home heating, air conditioning and ventilation systems in offices, cars, industries, museums and hospitals. Sick people are mostly affected by humidity; thus, monitoring the humidity is critical for them. Portable health monitoring devices are expected to be the future technology of personal health care. Respiration condition is a crucial indicator of sleep, rest, and exercise, and unusual conditions such as cardiovascular distress, flu, pneumonia, and asthma can be monitored using a wearable respiration device. Alterations in breathing rate can lead to local variations in humidity across the mouth and nose stimulated by exhaling and inhaling, so respiratory conditions can be precisely assessed by evaluating resistive or capacitive factors of humidity-sensitive substances (Nathan et al. 2012). Polymer nanocomposites using metal oxide are attracting increasing attention because of their use in electronics, memory, sensor and microwave devices. Inorganic NPs increase polymer properties such as physical and mechanical properties. Humidity sensor based on polymer metal oxide nanocomposites is usually fabricated by chemical processing techniques to achieve porous bodies. Such a surface provides better water permeability than a non-porous form; water vapor can simply pass through the pore openings and capillary condensation occurs in the capillary porous structures (Nagaraju et al. 2014).

Luo et al. fabricated a humidity sensor (captive type) based on silk fibroin (SF)/Ag as a wearable humidity sensor for respiration observation. Breathing rate and intensity were distinguished using the capacitance changes generated by vapor permeation. An inexpensive and easy patterning technique was used to fabricate the Ag nanowire interdigitated electrodes. Such sensors exhibited quick responses to circumstantial humidity changes and different breathing terms (breathing rate >4 Hz). They exhibited the proper mechanical and electrical stabilities (Fig. 2) (Luo et al. 2020).

Hashim et al. fabricated a humidity sensor based on polyvinyl alcohol/ carboxy methyl cellulose/ titanium carbide. The addition of titanium carbide to the biopolymer blend increased the conductivity, real/imaginary dielectric constants, absorption coefficient, refractive index, extinction coefficient, and optical conductivity. This nanocomposite exhibited high sensitivity to humidity (Hashim and Hadi 2018). Chen et al. synthesized chitosan fiber humidity sensor based on Fabry–Perot interferometry configuration. Sensing method was based on the swelling ratio of chitosan coating in the presence of humidity, which stimulated optical path modulation. The sensor exhibited 0.13 nm/%RH sensitivity with linear range from 20%RH to 95%RH and 380 ms response time (Chen et al. 2012). Mathew et al. fabricated fiber optic

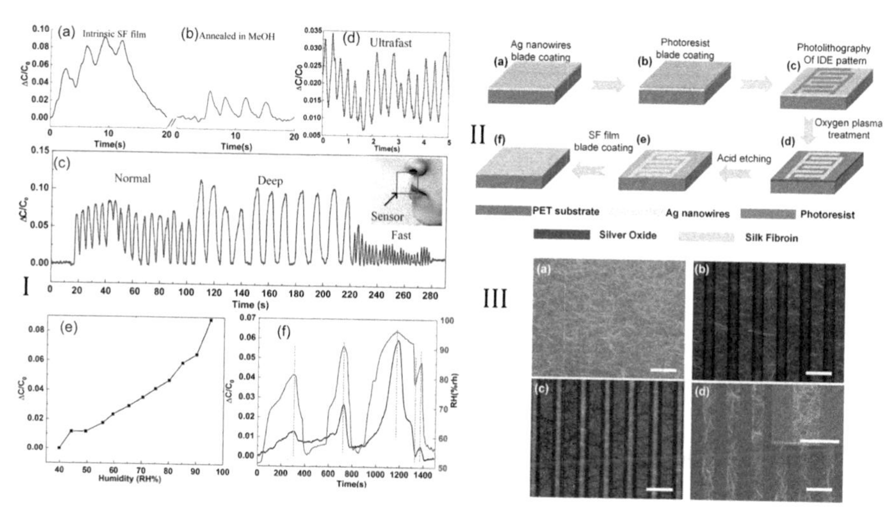

Fig. 2. (**I**) response curve of (a) silk fibroin film and (b) SF film annealed in methanol. (c) Capacitance response of the sensor for different breathing rates, and (d) ultra-fast breathing. (e) Capacitance alterations of the sensor under different RH levels. (f) Sensor responses to environmental humidity with cyclically varying RH. (**II**) The SF film-based humidity sensor preparation. (**III**) Optical microscopy images of (a) blade-coated Ag NWs. (b)-(d) Process of Ag NWs patterning: (b) photolithography process; (c) oxygen plasma treatment and (d) silver oxide and resin removal. Inset: SEM image of a patterned line. Reproduced with permission from Luo et al. (2020).

humidity sensors based on biopolymers. Chitosan and agarose were used as a biopolymer. Swelling behavior of such polymers was used as a humidity indicator. Biopolymers coated on an optical fiber swell by absorbing the water molecules and are confirmed to be useful as a reversible, stable and quickly responsive humidity sensor. Such sensors can be used for real-time humidity monitoring as a breathing monitor. Both exhibited 0.001 dB/RH sensitivity. Chitosan response time was lower than agarose response time with higher linear response (Mathew et al. 2007).

Table 2. Comparative characteristics of various types of RH sensors (Voznesenskiy et al. 2013)

Chitosan form	*RH range, %*	*Sensitivity, dB/%RH*	*Lag time, sec (adsorption)*	*Lag time, sec (desorption)*
Quartz fiber coated with chitosan acetate	17-95	0.001	2	2
Quartz fiber coated with agar	40-95	0.001	3	3
[Ru(phen)2(dppz)] 2 + -doped PTFE membrane	4-100	-	<80	<70

Table 2. (*Contd.*)

Chitosan form	*RH range, %*	*Sensitivity, dB/%RH*	*Lag time, sec (adsorption)*	*Lag time, sec (desorption)*
Phenol red doped PMMA film	20-80	-	5	<70
Agarose on tapered single-mode optical fiber	30-80	0.13	5	55
Ag-polyaniline nanocomposite clad planar optical waveguide	20-92	0.065	8	55
Nanoporous TiO_2	1-70	-	60	60
Agarose-gel with a periodic silicon-nitride film	20-80	-	8	55
Polyvinyl alcohol deposited on single-mode side-polished fiber	55-88 72-87	0.066 0.533	<60	<60
CH-Ac-N	20–45 45-90	0.0047 0.03	<1	1
CH-Cit-N	20-55 55-90	3.1 4.9	2	1
CH-Ac	15-50 50-90	0.018 0.57	<1	1
CH-Cit	50-70	0.2	1	200

Chitosan solution in acetic (CH-Ac) and citric (CH-Cit) acids, neutral form (CH-Ac-N) and (CH-Cit-N).

5. Conclusion

Because of their unique features, biopolymers have attracted significant attention in various applications, including sensors. Biopolymers along with metallic and ceramic nanoparticles exhibit electrical and mechanical properties suitable for this purpose. Moreover, their adsorption and swelling properties can be adjusted by addition of nanoparticles. Biopolymers are biocompatible and biodegradable, and they show good film-forming properties. They are plentiful in the environment and can be produced in large quantities from renewable and inexpensive sources. Various types of sensors such as optical and electrochemical can be fabricated to detect different types of analytes. In this chapter, the major types of sensors in demand are discussed to pave the way to fabricating a low-cost and highly sensitive sensor.

References

Abaker, M., Ahmad Umar, S. Baskoutas, G.N. Dar, S.A. Zaidi, S.A. Al-Sayari, A. Al-Hajry, S.H. Kim and S.W. Hwang. 2011. A highly sensitive ammonia chemical sensor based on α-Fe_2O_3 nanoellipsoids. *Journal of Physics D: Applied Physics* 44(42): 425401.

Abdullah, Jaafar, Musa Ahmad, Nadarajah Karuppiah, Lee Yook Heng and Hamidah Sidek. 2006. Immobilization of tyrosinase in chitosan film for an optical detection of phenol. *Sensors and Actuators B: Chemical* 114(2): 604-609.

Aldalbahi, Ali, Jin Chu and Peter Feng. 2012. Conducting composite materials from the biopolymer kappa-carrageenan and carbon nanotubes. *Beilstein Journal of Nanotechnology* 3(1): 415-427.

Ayad, Mohamad M., Nehal Salahuddin and Islam M. Minisy. 2014. Detection of some volatile organic compounds with chitosan-coated quartz crystal microbalance. *Designed Monomers and Polymers* 17(8): 795-802.

Ayad, Mohamad M., Nehal A. Salahuddin, Islam M. Minisy and Wael A. Amer. 2014. Chitosan/polyaniline nanofibers coating on the quartz crystal microbalance electrode for gas sensing. *Sensors and Actuators B: Chemical* 202: 144-153.

Bouvree, Audrey, Jean-François Feller, Mickaël Castro, Yves Grohens and Marguerite Rinaudo. 2009. Conductive polymer nano-biocomposites (CPC): Chitosan-carbon nanoparticle a good candidate to design polar vapour sensors. *Sensors and Actuators B: Chemical* 138(1): 138-147.

Chandrasakaran, Devi Shantini, Irwana Nainggolan, Nazree Derman and Tulus Ikhsan. 2014. Chloroform gas sensor based on chitosan biopolymer. pp. 45-49. *In*: Mohd Mustafa Al Bakri Abdullah, Liyana Jamaludin, Muhammad Faheem Mohd Tahir and Mohd Najmuddin Mohd Hassan (eds.). Applied Mechanics and Materials, vol. 679. Trans Tech Publications Ltd.

Chen, Li Han, Tao Li, Chi Chiu Chan, R. Menon, Papusamy Balamurali, Mutukumaraswamy Shaillender, Björn Neu, X.M. Ang, P. Zu, W.C. Wong and K.C. Leong. 2012. Chitosan based fiber-optic Fabry–Perot humidity sensor. *Sensors and Actuators B: Chemical* 169: 167-172.

Darder, Margarita, Montserrat Colilla and Eduardo Ruiz-Hitzky. 2005. Chitosan-clay nanocomposites: Application as electrochemical sensors. *Applied Clay Science* 28(1-4): 199-208.

Diouf, Alassane, Mohammed Moufid, Driss Bouyahya, Lars Österlund, Nezha El Bari and Benachir Bouchikhi. 2020. An electrochemical sensor based on chitosan capped with gold nanoparticles combined with a voltammetric electronic tongue for quantitative aspirin detection in human physiological fluids and tablets. *Materials Science and Engineering C* 110: 110665.

Dubas, Stephan T., Chularat Iamsamai and Pranut Potiyaraj. 2006. Optical alcohol sensor based on dye-chitosan polyelectrolyte multilayers. *Sensors and Actuators B: Chemical* 113(1): 370-375.

El-Sherbiny, Ibrahim M., Amr Hefnawy and Ehab Salih. 2016. New core-shell hyperbranched chitosan-based nanoparticles as optical sensor for ammonia detection. *International Journal of Biological Macromolecules* 86: 782-788.

Fen, Yap Wing, W. Mahmood Mat Yunus and Nor Azah Yusof. 2012. Surface plasmon resonance optical sensor for detection of Pb^{2+} based on immobilized p-tert-butylcalix [4] arene-tetrakis in chitosan thin film as an active layer. *Sensors and Actuators B: Chemical* 171: 287-293.

Feng, Duan, Fang Wang and Zilin Chen. 2009. Electrochemical glucose sensor based on one-step construction of gold nanoparticle-chitosan composite film. *Sensors and Actuators B: Chemical* 138(2): 539-544.

Fu, Kan and Brian G. Willis. 2015. Characterization of DNA as a solid-state sorptive vapor sensing material. *Sensors and Actuators B: Chemical* 220: 1023-1032.

Gao, Chao and Deyue Yan. 2004. Hyperbranched polymers: From synthesis to applications. *Progress in Polymer Science* 29(3): 183-275.

Ghiyasi, S., M.G. Sari, M. Shabanian, M. Hajibeygi, P. Zarrintaj, M. Rallini, L. Torre, D. Puglia, H. Vahabi, M. Jouyandeh and F. Laoutid. 2018. Hyperbranched poly(ethyleneimine) physically attached to silica nanoparticles to facilitate curing of epoxy nanocomposite coatings. *Progress in Organic Coatings* 120: 100-109.

Goyer, Robert A. and Thomas W. Clarkson. 1996. Toxic effects of metals. *Casarett and Doull's Toxicology: The Basic Science of Poisons* 5: 696-698.

Hajjam, Arash and Siavash Pourkamali. 2011. Fabrication and characterization of MEMS-based resonant organic gas sensors. *IEEE Sensors Journal* 12(6): 1958-1964.

Han, Jin-Woo, Beomseok Kim, Jing Li and M. Meyyappan. 2014. A carbon nanotube based ammonia sensor on cellulose paper. *RSC Advances* 4(2): 549-553.

Hashim, Ahmed and Qassim Hadi. 2018. Structural, electrical and optical properties of (biopolymer blend/titanium carbide) nanocomposites for low cost humidity sensors. *Journal of Materials Science: Materials in Electronics* 29(13): 11598-11604.

Huang, Jiadong, Xianrong Xing, Xiuming Zhang, Xiaorui He, Qing Lin, Wenjing Lian and Han Zhu. 2011. A molecularly imprinted electrochemical sensor based on multiwalled carbon nanotube-gold nanoparticle composites and chitosan for the detection of tyramine. *Food Research International* 44(1): 276-281.

Kemp, Melissa M., Ashavani Kumar, Shaymaa Mousa, Evgeny Dyskin, Murat Yalcin, Pulickel Ajayan, Robert J. Linhardt and Shaker A. Mousa. 2009. Gold and silver nanoparticles conjugated with heparin derivative possess anti-angiogenesis properties. *Nanotechnology* 20(45): 455104.

Khamis, Samuel M., Robert R. Johnson, Zhengtang Luo and A.T. Charlie Johnson. 2010. Homo-DNA functionalized carbon nanotube chemical sensors. *Journal of Physics and Chemistry of Solids* 71(4): 476-479.

Krebsz, M., T. Pasinszki, T.T. Tung and D. Losic. 2017. Development of vapor/gas sensors from biopolymer composites. pp. 385-403. *In*: K.K. Sadasivuni, D. Ponnamma, J. Kim, J.J. Cabibihanand M.A. AlMaadeed (eds.). Biopolymer Composites in Electronics. Elsevier.

Kumar, Bijandra, Mickaël Castro and Jean-François Feller. 2012. Tailoring the chemo-resistive response of self-assembled polysaccharide-CNT sensors by chain conformation at tunnel junctions. *Carbon* 50(10): 3627-3634.

Kumar, Bijandra, Mickaël Castro and Jean-François Feller. 2012. Controlled conductive junction gap for chitosan-carbon nanotube quantum resistive vapour sensors. *Journal of Materials Chemistry* 22(21): 10656-10664.

Kumar, Bijandra, Mickaël Castro and Jean-François Feller. 2012. Poly (lactic acid)–multi-wall carbon nanotube conductive biopolymer nanocomposite vapour sensors. *Sensors and Actuators B: Chemical* 161(1): 621-628.

Kumar, Bijandra, Jean-François Feller, Mickaël Castro and Jianbo Lu. 2010. Conductive bio-polymer nano-composites (CPC): Chitosan-carbon nanotube transducers assembled via spray layer-by-layer for volatile organic compound sensing. *Talanta* 81(3): 908-915.

Li, Ke, Kun Dai, Xiangbin Xu, Guoqiang Zheng, Chuntai Liu, Jingbo Chen and Changyu Shen. 2013. Organic vapor sensing behaviors of carbon black/poly (lactic acid) conductive biopolymer composite. *Colloid and Polymer Science* 291(12): 2871-2878.

Li, Wei, Dong Mi Jang, Sea Yong An, Dojin Kim, Soon-Ku Hong and Hyojin Kim. 2011. Polyaniline–chitosan nanocomposite: High performance hydrogen sensor from new principle. *Sensors and Actuators B: Chemical* 160(1): 1020-1025.

Lian, Wenjing, Su Liu, Jinghua Yu, Xianrong Xing, Jie Li, Min Cui and Jiadong Huang. 2012. Electrochemical sensor based on gold nanoparticles fabricated molecularly imprinted polymer film at chitosan–platinum nanoparticles/graphene–gold nanoparticles double nanocomposites modified electrode for detection of erythromycin. *Biosensors and Bioelectronics* 38(1): 163-169.

Liu, Bin, Hui Ting Lian, Jing Fen Yin and Xiang Ying Sun. 2012. Dopamine molecularly imprinted electrochemical sensor based on graphene–chitosan composite. *Electrochimica Acta* 75: 108-114.

Lu, Guanghan, Xin Yao, Xiaogang Wu and Tong Zhan. 2001. Determination of the total iron by chitosan-modified glassy carbon electrode. *Microchemical Journal* 69(1): 81-87.

Lu, Hsin-Hsien, Yerra Koteswara Rao, Tzong-Zeng Wu and Yew-Min Tzeng. 2009. Direct characterization and quantification of volatile organic compounds by piezoelectric module chips sensor. *Sensors and Actuators B: Chemical* 137(2): 741-746.

Luo, Yu, Yuechen Pei, Xueming Feng, Hao Zhang, Bingheng Lu and Li Wang. 2020. Silk fibroin based transparent and wearable humidity sensor for ultra-sensitive respiration monitoring. *Materials Letters* 260: 126945.

Mader, Heike S. and Otto S. Wolfbeis. 2010. Optical ammonia sensor based on upconverting luminescent nanoparticles. *Analytical Chemistry* 82(12): 5002-5004.

Mathew, Jinesh, K.J. Thomas, V.P.N. Nampoori and P. Radhakrishnan. 2007. A comparative study of fiber optic humidity sensors based on chitosan and agarose. *Sensors & Transducers Journal* 84(10): 1633-1640.

Mehrotra, Parikha. 2016. Biosensors and their applications – A review. *Journal of Oral Biology and Craniofacial Research* 6(2): 153-159.

Mironenko, A. Yu, A.A. Sergeev, A.E. Nazirov, E.B. Modin, S.S. Voznesenskiy and S. Yu Bratskaya. 2016. H2S optical waveguide gas sensors based on chitosan/Au and chitosan/Ag nanocomposites. *Sensors and Actuators B: Chemical* 225: 348-353.

Mironenko, A. Yu, A. A. Sergeev, S.S. Voznesenskiy, D.V. Marinin and S. Yu Bratskaya. 2013. pH-indicators doped polysaccharide LbL coatings for hazardous gases optical sensing. *Carbohydrate Polymers* 92(1): 769-774.

Mironenko, Alexander Y., Alexander A. Sergeev, Sergey Voznesensky and Svetlana Y. Bratskaya. 2014. Thin chitosan films for optical gas sensors. pp. 536-539. *In*: Evangelos Hristoforou and Dimitrios S. Vlachos (eds.). Key Engineering Materials, vol. 605. Trans Tech Publications Ltd.

Modrzejewska, Zofia, Roman Zarzycki and Jan Sielski. 2010. Synthesis of silver nanoparticles in a chitosan solution. *Progress on Chemistry and Application of Chitin and its Derivatives* 15: 63-72.

Molla-Abbasi, P. and S.R. Ghaffarian. 2014. Decoration of carbon nanotubes by chitosan in a nanohybrid conductive polymer composite for detection of polar vapours. *RSC Advances* 4(58): 30906-30913.

Molla-Abbasi, Payam, Seyed Reza Ghaffarian and Erfan Dashtimoghadam. 2016. Wrapping carbon nanotubes by biopolymer chains: Role of nanointerfaces in detection of vapors in conductive polymer composite transducers. *Polymer Composites* 37(9): 2803-2810.

Nagaraju, S.C., Aashis S. Roy, J.B. Kumar, Koppalkar R. Anilkumar and G. Ramagopal. 2014. Humidity sensing properties of surface modified polyaniline metal oxide composites. *Journal of Engineering* 2014.

Nathan, A., A. Ahnood, M.T. Cole, S. Lee, Y. Suzuki, P. Hiralal, F. Bonaccorso, T. Hasan, L. Garcia-Gancedo, A. Dyadyusha and S. Haque. 2012. Flexible electronics: The next ubiquitous platform. *Proceedings of the IEEE* 100(Special Centennial Issue): 1486-1517.

Onida, B., S. Fiorilli, L. Borello, Guido Viscardi, D. Macquarrie and E. Garrone. 2004. Mechanism of the optical response of mesoporous silica impregnated with Reichardt's dye to NH3 and other gases. *The Journal of Physical Chemistry B* 108(43): 16617-16620.

Pandey, Sadanand, Gopal K. Goswami and Karuna K. Nanda. 2012. Green synthesis of biopolymer-silver nanoparticle nanocomposite: An optical sensor for ammonia detection. *International Journal of Biological Macromolecules* 51(4): 583-589.

Pauliukaite, Rasa, Mariana E. Ghica, Orlando Fatibello-Filho and Christopher M.A. Brett. 2010. Electrochemical impedance studies of chitosan-modified electrodes for application in electrochemical sensors and biosensors. *Electrochimica Acta* 55(21): 6239-6247.

Qi, Haisong, Jianwen Liu, Jürgen Pionteck, Petra Pötschke and Edith Mäder. 2015. Carbon nanotube–cellulose composite aerogels for vapour sensing. *Sensors and Actuators B: Chemical* 213: 20-26.

Rahman, Mohammed M., Aslam Jamal, Sher Bahadar Khan and Mohd Faisal. 2011. Characterization and applications of as-grown β-Fe_2O_3 nanoparticles prepared by hydrothermal method. *Journal of Nanoparticle Research* 13(9): 3789-3799.

Raveendran, Poovathinthodiyil, Jie Fu and Scott L. Wallen. 2003. Completely "green" synthesis and stabilization of metal nanoparticles. *Journal of the American Chemical Society* 125(46): 13940-13941.

Razali, Nazirah Mohd, Nurul Farah Adilla Zaidi, Puteri Nadiah Syamimi Said Ja'afar, Azura Hamzah, Fauzan Ahmad and Sumiaty Ambran. 2020. Optical fibre tip sensor coated with chitosan for lead ion detection. *In*: AIP Conference Proceedings 2203(1): 020035. AIP Publishing LLC.

Sankaran, Sindhuja, Suranjan Panigrahi and Sanku Mallik. 2011. Olfactory receptor based piezoelectric biosensors for detection of alcohols related to food safety applications. *Sensors and Actuators B: Chemical* 155(1): 8-18.

Tiwari, Ashutosh. 2008. Synthesis and characterization of pH switching electrical conducting biopolymer hybrids for sensor applications. *Journal of Polymer Research* 15(4): 337-342.

Tomalia, Donald A., H. Baker, J. Dewald, M. Hall, G. Kallos, S. Martin, J. Roeck, J. Ryder and P. Smith. 1985. A new class of polymers: Starburst-dendritic macromolecules. *Polymer Journal* 17(1): 117-132.

Voznesenskiy, S.S., A.A. Sergeev, A. Yu Mironenko, S. Yu Bratskaya and Yu N. Kulchin. 2013. Integrated-optical sensors based on chitosan waveguide films for relative humidity measurements. *Sensors and Actuators B: Chemical* 188: 482-487.

Wang, Li-Chun, Tseng-Hsiung Su, Cheng-Long Ho, Shang-Ren Yang, Shih-Wen Chiu, Han-Wen Kuo and Kea-Tiong Tang. 2015. A bio-inspired two-layer sensing structure of polypeptide and multiple-walled carbon nanotube to sense small molecular gases. *Sensors* 15(3): 5390-5401.

Wang, Na, Xianfeng Wang, Yongtang Jia, Xiaoqi Li, Jianyong Yu and Bin Ding. 2014. Electrospun nanofibrous chitosan membranes modified with polyethyleneimine for formaldehyde detection. *Carbohydrate Polymers* 108: 192-199.

Wu, Tzong-Zeng, Yen-Ren Lo and Err-Cheng Chan. 2001. Exploring the recognized bio-mimicry materials for gas sensing. *Biosensors and Bioelectronics* 16(9-12): 945-953.

Zangmeister, Rebecca A., Jung J. Park, Gary W. Rubloff and Michael J. Tarlov. 2006. Electrochemical study of chitosan films deposited from solution at reducing potentials. *Electrochimica Acta* 51(25): 5324-5333.

Zhang, Feng, Xiaolan Wu, Yuyue Chen and Hong Lin. 2009. Application of silver nanoparticles to cotton fabric as an antibacterial textile finish. *Fibers and Polymers* 10(4): 496-501.

CHAPTER

13

Stimuli-responsive Polymers and Their Biomedical Applications

Dinesh K. Patel and Ki-Taek Lim*
Department of Biosystems Engineering, College of Agriculture and Life Science, The Institute of Forest Science, Kangwon National University, Chuncheon 24341, Republic of Korea

1. Introduction

On-demand release of loaded or active materials from the carrier has received a significant amount of interest in the field of medical science. For this, researchers are focusing on developing materials that can release the loaded materials in the targeted zone. The on-demand release system can overcome the limitations of traditional therapeutic systems and can enhance the lifespan and stability of the carrier in the targeted diseased organs or tissues to avoid the unnecessary accumulation of loaded materials in the vicinity of healthy tissues (Zhou et al. 2019a). Stimuli-responsive materials (SRMs), also known as smart materials, are commonly applied for this purpose. These materials have the potential to alter their structure in response to external or internal stimuli such as electric field (Xie et al. 2010), magnetic field (He et al. 2011), temperature (Zhang et al. 2011, Miaudet et al. 2007), pH (Harris et al. 2005), light (Yu and Ikeda 2011, Yu 2014), pressure (Ilievski et al. 2011), solvent (Lee et al. 2012), and moisture (Zakharchenko et al. 2010). These materials are actively used for the development of different electrochemical and biomimetic devices (Xie and Xiao 2008), medical instruments and auxiliaries (Lendlein and Langer 2002), sensors and actuators (Bar-Cohen and Zhang 2008), sound-absorbing materials, and other devices (Hu et al. 2012). Various polymers and their copolymers such as poly(*N*-isopropyl acrylamide), poly(*N N*″-diethyl acrylamide), poly(ethylene oxide), and poly(propylene oxide) are often applied for the fabrication of intelligent devices for different applications (Alarcon et al. 2005). Polymeric materials

*Corresponding author: ktlim@kangwon.ac.kr

are considered a suitable option for the development of stimuli-responsive systems because of the presence of different active groups that can be easily tailored. These materials have the ability to return their original structure when a counter-trigger is applied (Cabane et al. 2012). Biopolymers from living organic systems such as proteins and nucleic acids are SRMs and exhibit conformational changes at given critical points (Kumar et al. 2007). However, these materials remain stable over a wide range of stimuli. Natural SRMs inspire the development of synthetic polymers that can mimic their properties for broad applications.

The properties of SRMs can be easily modified by altering their chemical structures or combining suitable fillers in their matrices. Various kinds of fillers such as metal and their oxides, clay, nanocellulose, zeolites, fullerenes, carbon nanotubes, and graphite are commonly used to enhance the properties of pure polymers (Lee and Lin 2006, Wang et al. 2011, Ding et al. 2018). The loss of secondary forces, including hydrogen bonding, hydrophobic effects, electrostatic interactions, and progressive ionization in polymer units, are significant factors for the development of these properties in smart materials on application of external stimuli. The most widely studied and understood response for SRMs is to temperature. The lower critical solution temperature (LCST) is the lowest temperature of the polymers at which temperature-inspired phase change has occurred (Wei et al. 2017). Above the LCST, a transition from random coil to compact confirmation has occurred. Figure 1 represents the different types of stimuli-responsive polymers and changes in the confirmation above and below the LCST. For biomedical purposes, the material should not express any toxicity or adverse immune response in biological environments. The materials should be biodegradable and maintain the suitable mechanical potential to facilitate the cellular activity for biomedical applications (Oliveira et al. 2017). Biodegradable materials have additional advantages over non-degradable materials. They do not need to be removed or degraded from the applied conditions. In this chapter, we discuss stimuli-responsive systems with polymers as smart materials

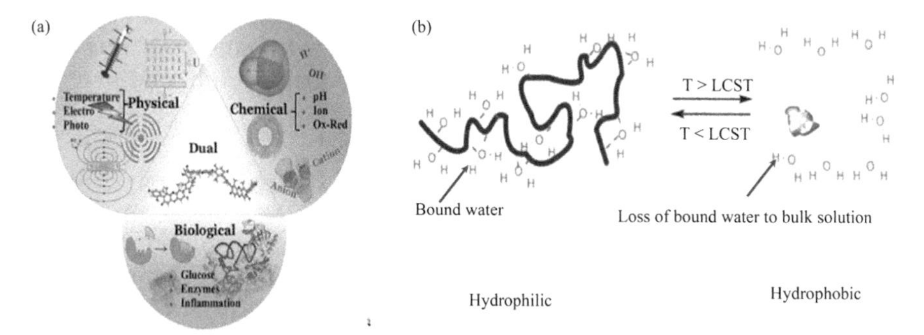

Fig. 1. (a) Classification of stimuli of stimuli-responsive polymers (Cabane et al. 2012). (b) Schematic of smart polymer response to temperature. Reprinted with permission from Alarcon et al. (2005).

by considering selected yet diverse recent research demonstrating their potential applications.

2. Types of stimuli

The properties of the polymers can be easily tuned by changing their wettability, solubility, and confirmation. External or internal factors play a significant role in dramatic physiochemical changes in the polymers. These factors are physical, chemical, and biological (Gil and Hudson 2004). Physical factors include temperature, light, electric and magnetic fields, mechanical forces, and ultrasound. It has been seen that physical factors alter the polymer chain dynamics. Temperature-responsive polymers show a transition above the LCST known as upper critical solution temperature (UCST). Conductive polymers such as polythiophene or sulfonated-polystyrene are more prone to electrical and electrochemical factors (Jones et al. 1998). Light- or photo-responsive polymers have a light-absorbing structure known as chromophores, which can absorb the light in a particular wavelength and consequently show the transition. The nature and strength of the solvent, pH, and electrochemical properties are classified as chemical factors. Chemical factors influence the interactions between the polymer chains or between polymer chains and the solvents (Liechty et al. 2010). pH is a critical stimulus factor and can be used for direct response at the cellular level. For this, the polymer should have weak acidic or basic groups that can easily be ionized by changing the conditions of the medium (Schmaljohann 2006). The ionic strength of the solution also influences the polymer solubility and fluorescence potential by altering the coulombic interactions between oppositely ionized species. Enzymes and receptors are considered biological factors (Delcea et al. 2011). pH-sensitive polymers exhibit conformational changes and, consequently, enzyme activity. Enzyme-responsive polymers have additional advantages over others in terms of their high selectivity and do not need external factors for stimulation. However, polymers responsive to more than one factor are also noticed. Multi-responsive polymers have received much attention in biomedical applications because of their multiple response activity under different conditions. Some commonly used polymers for stimuli-responsive applications and their stimuli conditions are summarized in Table 1.

3. Applications of stimuli-responsive polymers

The development of accurate and non-invasive devices is important for the early detection of disease. Polymers can exist in bulk or supramolecular assemblies, which provide various opportunities to modify their structure and form for smart applications. Smart materials are widely applied in controlled drug delivery, imaging, sensing, regenerative medicine, transport, and microfluidics (Stuart et al. 2010, Mano 2008, Christian et al. 2009, Sun and Qing 2011). Here we highlight stimuli-responsive polymeric systems in

Table 1. Some stimuli-responsive polymers and their stimulus conditions

Stimulus condition	*Material*	*Reference*
Temperature	N-(2-hydroxypropyl)methacrylamide) (HPMA), Poly(N-isopropylacrylamide) (PNIPAAM), Poly(2-isopropyl-2-oxazoline) (PiPOx)	Kost and Langer, 2012, Alfurhood et al. 2016, Skvarla et al. 2017
pH	Poly(acrylic acid), poly(methacrylic acid) (PMAA), poly(ethylene imine), poly(L-lysine), and poly(N,N-dimethyl aminoethyl methacrylamide)	Swift et al. 2016, Bal et al. 2016
Electric field	Poly(vinyl alcohol) and poly(acrylic acid-co-2-acrylamido-2-methyl propyl sulfonic acid)	Lin et al. 2008
Ultrasound	Poly(lactic acid-co-glycolic acid)	Zhang et al. 2014
Temperature and electric field	Poly(pyrrole)	Ge et al. 2012

the field of drug delivery, tissue engineering, and antibacterial applications by considering some selected, significant studies.

3.1. Intelligent drug delivery

Intelligent drug delivery systems have benefits over burst release where controlled delivery of the loaded materials is required. Most low-molecular-weight drugs are hydrophobic, and their solubility in the polar solvents is the key concern for optimum therapeutic applications. Stimuli-responsive polymers are applied in various forms, including polymeric micelles, polymersomes, or dendrimers. These structures have advantages for the delivery of hydrophobic drugs. The hydrophobic drug can easily be trapped in their structure and be stimulated by several factors for the desired release. Doxorubicin (DOX), paclitaxel, camptothecin (CPT), cisplatin, *N*-acetyl cysteine, indomethacin, and dexamethasone are the most intensively studied drugs in the delivery systems. It is well known that pH difference occurs between normal cells and tumor cells. This difference plays a significant role in the delivery of loaded antitumor drugs. The cleavage of pH-sensitive bonds occurs in stimuli-responsive polymer systems, and the release of loaded drugs is achieved (MacEwan et al. 2010, Bae et al. 2003, Alani et al. 2010). Surnar et al. have synthesized carboxylic functionalized poly(caprolactone) (PCL)/poly(ethylene glycol) block copolymer vesicles for the pH-responsive delivery of ibuprofen (IBU) and CPT drugs. These synthesized block copolymers form self-organized vesicles in water with a diameter of 100-250 nm. These pH-responsive vesicles were stable in strongly acidic media ($pH < 2.0$, stomach) and showed rapid release of the loaded drug under neutral or basic pH ($7.0 \leq pH$, small intestine) due to the rupture of the vesicle structure (Surnar and Jayakannan 2013). The cumulative release profiles of the IBU- and CPT-loaded drug from the synthesized block copolymers are shown in Fig. 2. Under the simulated gastrointestinal tract, the individual loaded drug vesicle showed a combination of diffusion and erosion release kinetics, whereas the dual-drug loaded block copolymer vesicle exhibited diffusion-controlled release kinetics. This custom-designed PCL-based block copolymer vesicle enables a new direction for the release of the loaded drug from the pH-responsive polymer for biomedical applications. Thermo-responsive polymers or copolymers are also explored for the release of the loaded drug. Quan and co-workers prepared thermo-responsive micelles of poly(*N*-acryloxysuccinimide)-*b*-poly(*N*-isopropylacrylamide)-*b*-poly(caprolactone) (PNAS-*b*-PNIPAAm-*b*-PCL) triblock copolymers for the delivery of anti-cancerous DOX for HeLa cell lines. It is well established that thermo-responsive polymers exhibit a transition above the LCST. It was interesting to see that 97% of the loaded drug was released above the LCST of PNIPAAm (Quan et al. 2009). Polymersomes have also been used as a carrier for low-molecular-weight loaded drugs. Aluri et al. developed the multi-responsive L-tyrosine-based amphiphilic poly(ester-urethane) carrier for the delivery of the anti-cancerous DOX and CPT drugs. The developed vehicle showed thermo-responsive behavior at 42-44°C, which is close to cancer tissue temperature. The prepared carrier was stable under physiological conditions (37°C, pH 7.4) and ruptured at cancer tissue temperature (42°C) with 90% release of the loaded drug, showing its thermo-responsive potential for

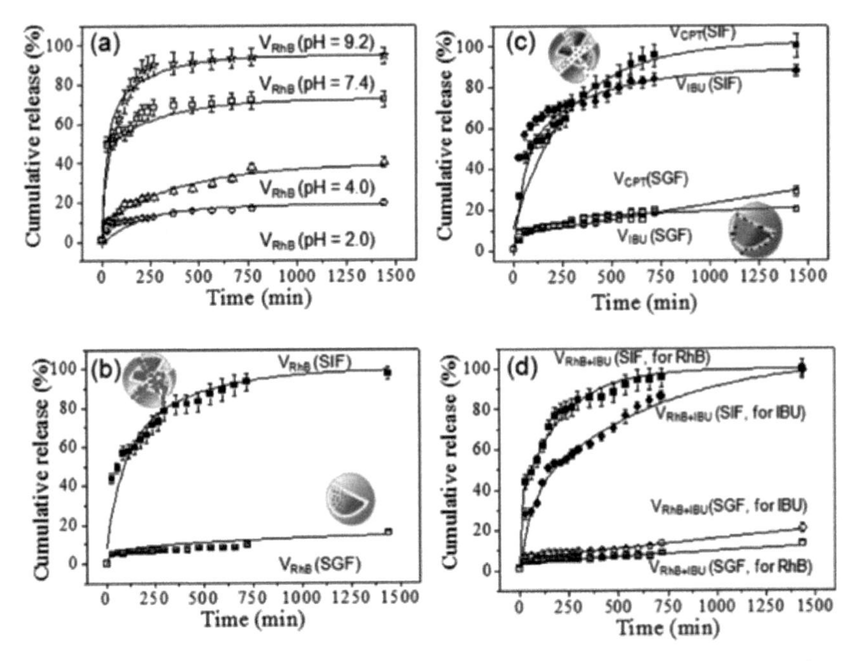

Fig. 2. (a) Cumulative release of V_{Rh-B} in various pH buffers at 37°C. (b) Cumulative release of V_{Rh-B} in SIF and SGF at 37°C. (c) Cumulative release of V_{IBU} and V_{CPT} (c) in SIF and SGF at 37°C. (d) Cumulative release of dual-loaded vesicle $V_{Rh-B+IBU}$ in SIF and SGF at 37°C. Reprinted with permission from Surnar and Jayakannan (2013).

intelligent drug delivery application (Aluri et al. 2018). Some other polymer-based stimuli-responsive drug delivery systems are summarized in Table 2.

Table 2. Stimuli-responsive polymers and their drug delivery applications

Stimulus condition	*Material*	*Application*	*Reference*
pH	Poly(ethylene oxide)-block-poly (ε-caprolactone) (PEO-b-PCL)	Targeted cellular and subcellular delivery of doxorubicin	Xiong et al. 2010
	Levulinic acid (LEV) or 4-acetyl benzoic acid (4AB) modified poly(ethylene glycol)-block-poly(aspartate-hydrazide) (PEG-p(Asp-Hyd)	Controlled release of anti-cancerous paclitaxel (PTX)	Alani et al. 2010
	Poly(lactic-co-glycolic acid) (PLGA)	Inflammation-induced drug release	Chung et al. 2014
Optical and ultrasound	Poly(lactic-co-glycolic acid) (PLGA)	Cancer targeting and imaging	Xu et al. 2010
Ultrasound	Chitosan	Targeted delivery of doxorubicin	Zhou et al. 2019b
Light	Spiropyran (SP) with polyglycerols (SP-hb-PG)	Smart drug delivery	Son et al. 2014
	Hydrogel of 2-hydroxyethyl methacrylate (HEMA) and ethylene glycol dimethacrylate (EGDMA)	On-demand transdermal drug delivery	Hardy et al. 2016
	Monomethoxy(polyethylene glycol)-poly(lactic-co-glycolic acid) (mPEG-PLGA)	Controlled release of doxorubicin	Fan et al. 2016
Near-infrared (NIR) light	Diselenide-crosslinked poly(methacrylic acid) (PMAA)-based nanogels	On-demand drug delivery	Tian et al. 2015
	Silica-coated lanthanum hexaboride (LaB_6@SiO_2) incorporated into poly(caprolactone) (PCL)		Chen et al. 2015
Temperature	Poly(N-isopropylacrylamide) crosslinked with di(ethylene glycol) (P-(NIPAAm-co-DEGDVE)	Controlled drug release	Werzer et al. 2019
	Poly(acrylamide-co-acrylic acid) hydrogel		(Dai et al. 2006)

3.2. Tissue engineering

The characteristics of fabricated scaffolds are the key concerns for tissue engineering applications. The scaffolds should be biocompatible and biodegradable. Scaffolds applied for tissue engineering should also have the adequate mechanical strength to support the developing tissue. Stimuli-responsive polymers are widely explored in the field of tissue engineering for the rapid development of new tissues or damage of cancerous cells. Wu et al. synthesized near-infrared (NIR) light-responsive polymer hydrogel for cancer treatment through controlled release. The hydrogels were synthesized using a semiconductor polymer poly(diketopyrrolopyrrole-*alt*-3,4-ethylenedioxythiophene) with the polymerization of *N*-isopropyl acrylamide. No significant cytotoxicity was expressed by the fabricated hydrogel, indicating their biocompatibility. The cell viability data for HeLa cultured on the fabricated hydrogels is given in Fig. 3a. However, a drastic decrease in cell viability was noted in DOX-encapsulated hydrogels under 0-2 W/cm^2 laser irradiation (808 nm). The hydrogel killed 30% of cancerous cells after 12 h of treatment. This efficiency was enhanced with time. Approximately 60% of HeLa cells were killed after 24 h treatment with laser irradiation. This is due to the enhanced release of the loaded anti-cancerous (DOX) drug by the shrinking of the hydrogel on laser treatment (Wu et al. 2017). The change in the cell viability of HeLa upon laser irradiation is given in Fig. 3b. Confocal fluorescence images of HeLa cells treated with the synthesized hydrogel upon irradiation of laser light at different time intervals are shown in Fig. 3c. The cells exhibited the normal morphology and low fluorescence of DOX without laser irradiation. However, strong fluorescence intensity with shrinkage morphology was noted under the NIR irradiation in the cytoplasm region of the HeLa cells. Roberts and co-workers prepared stimuli-responsive heparin-based hydrogel for cell encapsulation with poly(vinyl alcohol) (PVA). The developed PVA-heparin hydrogel was biocompatible and provided mild cell encapsulation (Roberts et al. 2015). Stimuli-responsive hydrogel microfibers and microtubes have gained much interest in biomedical applications owing to their similarity to the native extracellular matrix. Hydrogels are physically or chemically crosslinked polymer structures with a high water content that allows them to maintain their three-dimensional network structure. Kim et al. prepared the pH- and temperature-responsive hydrogel using a microfluidic device on alginate polymer through photopolymerization. For this, they irradiated the *N*-isopropylacrylamide (NIPAm) and sodium acrylate (SA) or allyl amine (AA) under UV light for *in situ* polymerization. Better cellular activities were noticed with positively charged microfibers, and these activities were further enhanced by supplying nutrient in the media, indicating their potential for tissue engineering (Kim et al. 2018).

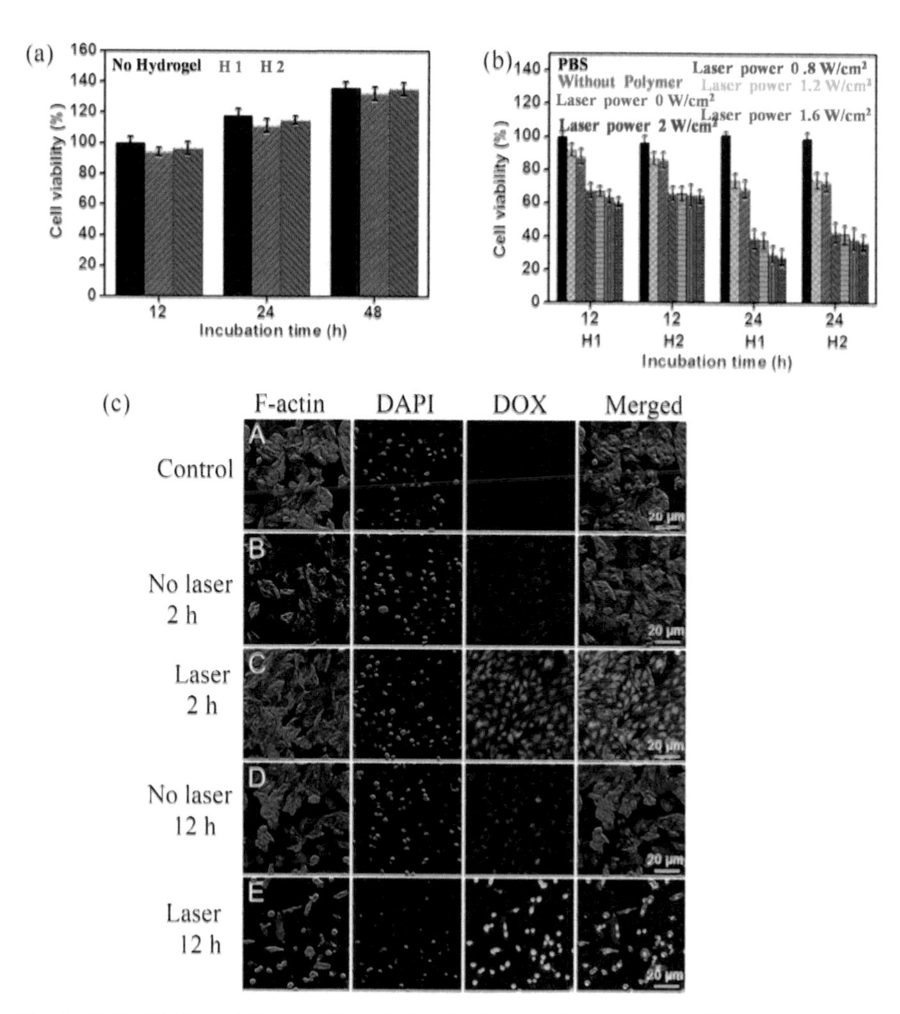

Fig. 3. Cell viability of HeLa cells tested using PrestoBlue assay. (a) Cytotoxicity assay after co-culture with H1 and H2 hydrogels over 12, 24, and 48 h. (b) Cell viability after treatment with DOX-encapsulated hydrogels under 0–2 W/cm2 laser irradiation (808 nm). (c) Confocal fluorescence images of HeLa cells incubated with the H1 hydrogel under different laser irradiation conditions. (A) Initial state of the cell. (B–E) HeLa cells treated with the H1 hydrogel after 20 min laser irradiation and incubation for 2 h and 12 h. Red, F-actin-stained cytomembrane; blue, DAPI-stained nucleus; green, pseudo label of DOX localization; and right, merged images. Reprinted with permission from Wu et al. (2017).

3.3. Sensors

Stimuli-responsive polymers are considered suitable for the fabrication of sensing and actuator devices owing to their ability to respond to stimuli such as temperature, pH, carbon dioxide, light, and electricity. These devices can be applied to detect the analytics, as well as monitor health (Hu et al. 2019).

The quantitative detection of toxic materials is important for both biological and environmental systems. Among different stimuli-responsive materials, thermo-responsive polymers are most studied because of their large-scale change in solubility, volume, and confirmation in response to temperature (Roy et al. 2013). Chen and co-workers developed temperature-responsive sensing film for the detection of the animal growth promoter ractopamine (RAC). The temperature-responsive sensing film was developed on the surface of glassy carbon electrode using reduced graphene oxide, fullerene (C_{60}), and the temperature-sensitive polymer poly(2-(2-methoxyethoxy) ethyl methacrylate) ($PMEO_2MA$). A broad oxidation peak was noted above the LCST of $PMEO_2MA$, and this peak disappeared below the LCST of $PMEO_2MA$. The developed sensor detected RAC concentrations from 0.1 to 3.1 µM, with an 82 nM detection limit under optimum conditions. The temperature-responsive polymer-modified glassy carbon electrode was successfully used to detect RAC in spiked pork samples (Chen et al. 2018). The schematic representation of the reversible temperature-controlled on/off electrochemical behavior of RAC at $PMEO_2MA/C_{60}$-rGO film and their electrochemical responses are shown in Fig. 4.

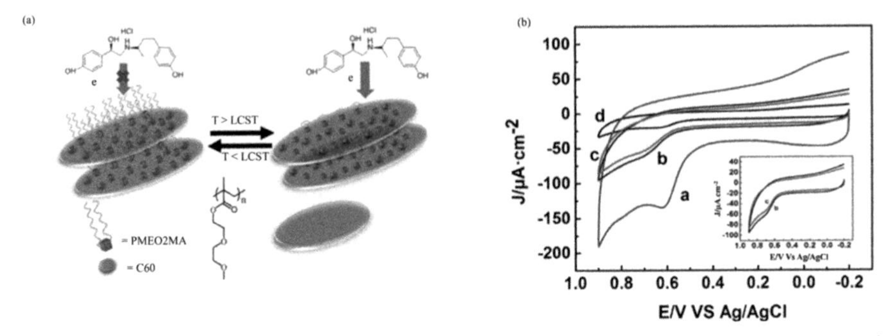

Fig. 4. (a) Schematic representation of reversible, temperature-controlled on/off electrochemical behavior of RAC at PMEO2MA/C60-rGO composited film. (b) Cyclic voltammograms of (a) PMEO 2MA/C60-rGO, (b) PMEO2MA/C60-GO, (c) PMEO 2MA/GO, (d) bare GCE with 10 µM RAC in 0.1 M PBS (pH = 7.0). Solution temperature: 30°C. Scan rate: 50 mV/s. Inset: enlarged view of (b) and (c). Reprinted with permission from Chen et al. (2018).

Wearable health-care devices provide very important information related to health care and store physical activity, body temperature, pH of sweat, and other data (Yamamoto et al. 2016) Among them, temperature is the most important sign, and it can provide critical information about human body conditions. Choe and co-workers developed temperature-responsive stretchable patches for smart wearable sensors and soft robotics applications. They prepared the patch by incorporating gold nanoparticles (AuNPs) on poly(*N*-isopropyl acrylamide) (PNIPAM), which show reversible and strain-sensitive color change (red ↔ grayish violet) in response to a temperature change. The transition temperature of the developed microgel can be easily altered by additives and comonomers (Choe et al. 2018). Wan et al. synthesized multi-responsive electron-deficient borinic acid polymers (PBAs) for the detection of alizarin red S (ARS), 8-hydroxyquinoline (HQ), and fluoride ions (Wan et al. 2017). It was noted that the thermo-responsive property and detection potential of PBAs is profoundly affected by the steric hindrance of the substituent. The thermo-responsive behavior of mesityl substituted PBA in DMSO, and the colorimetric observation of PBAs in the presence of different chemical moieties (ARS, HQ, and F^-) under irradiation with UV lamp at 365 nm are presented in Fig. 5. Gold nanoparticles are commonly applied for sensing applications because of their tunable optical properties (Saha et al. 2012). Stimuli-responsive polymers with NPs also have received widespread attention because NPs can easily be tuned by changing the conformation of a responsive polymer through external stimuli. Ma and co-workers prepared carbon dioxide (CO_2)-responsive polymer using poly(*N*-(3-amidino)-aniline) (PNAAN) with AuNPs by direct reduction of $HAuCl_4$ with *N*-(3-amidino)-aniline monomer. The protonation of PNAAN amidine group occurred in the hydrophilic amidinium group through dissolved CO_2 (dCO_2), causing the swelling of PNAAN and detachment of AuNPs. The detached AuNPs were aggregated, and change in the color was noted. The developed sensor exhibited a limit of detection (LOD) of 0.0024 hPa with a linear range of 0.0132 to 0.1584 hPa (Ma et al. 2016).

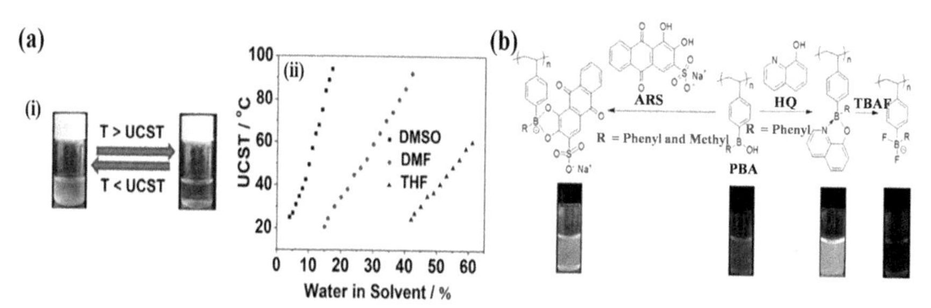

Fig. 5. (a) Digital photos of thermo-responsive behavior of mesityl PBA in (i) DMSO, (ii) plots illustrating mesityl PBA in DMSO, DMF, and THF ([PBA] = 1 mg/mL) in the presence of different amounts of water (v/v). (b) Proposed mechanism of structural response of substituted PBAs to diol compound, ARS and electron-rich compounds, HQ and fluoride ion, and the corresponding digital illustration under irradiation with UV lamp at 365 nm. Reprinted with permission from Wan et al. (2017).

4. Conclusions

In this chapter, we have discussed only a small part of the ongoing research into stimuli-responsive polymers. Stimuli-responsive polymers can undergo reversible changes in confirmation, volume, and solubility in response to different factors. They have demonstrated intelligent release of loaded drug at the desired site under the response of different stimuli. Intelligent drug release has its advantages over regular release. Stimuli-responsive polymers exhibit great potential in tissue engineering, especially in cancer cells. Better therapeutic activity was noticed with these materials. Polymeric materials with suitable design can detect biologically and environmentally toxic substances. Smartly designed materials can overcome the long response times for sensing and actuation applications where fast response is needed. Stimuli-responsive polymers with nanoparticles showed vast potential for sensing applications. Future efforts are needed towards the development of highly NIR-sensitive stimuli-responsive materials for targeted drug delivery through blood circulation and the living tissue where NIR is less absorbed. The development of low-cost stimuli-responsive polymers is an important task for sensing applications. The cyclability of these materials is also a critical issue for specific groups such as the azo group.

Acknowledg ements

The authors express their sincere gratitude to the Ministry of Education (No. 2018R1A6A1A03025582 & 2019R1D1A3A03103828), Republic of Korea, for providing financial support in the Basic Science Research Program through the National Research Foundation of Korea.

References

Alani, A.W., Y. Bae, D.A. Rao and G.S. Kwon. 2010. Polymeric micelles for the pH-dependent controlled, continuous low dose release of paclitaxel. *Biomaterials* 31: 1765-1772.

Alarcon, C.D.H., S. Pennadam and C. Alexander. 2005. Stimuli responsive polymers for biomedical applications. *Chemical Society Reviews* 34: 276-285.

AlfurhoodL, J.A., P.R. Bachler and B.S. Sumerlin. 2016. Hyperbranched polymers via RAFT self-condensing vinyl polymerization. *Polymer Chemistry* 7: 3361-3369.

Aluri, R., S. Saxena, D.C. Joshi and M. Jayakannan. 2018. Multistimuli-responsive amphiphilic poly (ester-urethane) nanoassemblies based on L-tyrosine for intracellular drug delivery to cancer cells. *Biomacromolecules* 19: 2166-2181.

Bae, Y., S. Fukushima, A. Harada and K. Kataoka. 2003. Design of environment-sensitive supramolecular assemblies for intracellular drug delivery: Polymeric micelles that are responsive to intracellular pH change. *Angewandte Chemie International Edition* 42: 4640-4643.

Bal, A., B. Ozkahraman and Z. Ozbas. 2016. Preparation and characterization of pH responsive poly(methacrylic acid-acrylamide-N-hydroxyethyl acrylamide) hydrogels for drug delivery systems. *Journal of Applied Polymer Science* 133.

Bar-Cohen, Y. and Q.M. Zhang. 2008. Electroactive polymer actuators and sensors. *Mrs Bulletin* 33: 173-181.

Cabane, E., X.Y. Zhang, K. Langowska, C.G. Palivan and W. Meier. 2012. Stimuli-responsive polymers and their applications in nanomedicine. *Biointerphases* 7.

Chen, C., M. Zhang, C. Li, Y. Xie and J. Fei. 2018. Switched voltammetric determination of ractopamine by using a temperature-responsive sensing film. *Microchimica Acta* 185: 155.

Chen, M.-C., M.-H. Ling, K.-W. Wang, Z.-W. Lin, B.-H. Lai and D.-H. Chen. 2015. Near-infrared light-responsive composite microneedles for on-demand transdermal drug delivery. *Biomacromolecules* 16: 1598-1607.

Choe, A., J. Yeom, R. Shanker, M.P. Kim, S. Kang and H. Ko. 2018. Stretchable and wearable colorimetric patches based on thermoresponsive plasmonic microgels embedded in a hydrogel film. *NPG Asia Materials* 10: 912-922.

Christian, D.A., S. Cai, D.M. Bowen, Y. Kim, J.D. Pajerowski and D.E. Discher. 2009. Polymersome carriers: From self-assembly to siRNA and protein therapeutics. *European Journal of Pharmaceutics and Biopharmaceutics* 71: 463-474.

Chung, M.F., W.T. Chia, H.Y. Liu, C.W. Hsiao, H.C. Hsiao, C.M. Yang and H.W. Sung. 2014. Inflammation-induced drug release by using a pH-responsive gas-generating hollow-microsphere system for the treatment of osteomyelitis. *Advanced Healthcare Materials* 3: 1854-1861.

Dai, H., Q. Chen, H. Qin, Y. Guan, D. Shen, Y. Hua, Y. Tang and J. Xu. 2006. A temperature-responsive copolymer hydrogel in controlled drug delivery. *Macromolecules* 39: 6584-6589.

Delcea, M., H. Mohwald and A.G. Skirtach. 2011. Stimuli-responsive LbL capsules and nanoshells for drug delivery. *Advanced Drug Delivery Reviews* 63: 730-747.

Ding, Q.Q., X.W. Xu, Y.Y. Yue, C.T. Mei, C.B. Huang, S.H. Jiang, Q.L. Wu and J.Q. Han. 2018. Nanocellulose-mediated electroconductive self-healing hydrogels with high strength, plasticity, viscoelasticity, stretchability, and biocompatibility toward multifunctional applications. *ACS Applied Materials & Interfaces* 10: 27987-28002.

Fan, J., Q. He, Y. Liu, F. Zhang, X. Yang, Z. Wang, N. Lu, W. Fan, L. Lin and G. Niu. 2016. Light-responsive biodegradable nanomedicine overcomes multidrug resistance via NO-enhanced chemosensitization. *ACS Applied Materials & Interfaces* 8: 13804-13811.

Ge, J., E. Neofytou, T.J. Cahill, 3RD, R.E. Beygui and R.N. Zare. 2012. Drug release from electric-field-responsive nanoparticles. *ACS Nano* 6: 227-233.

Gil, E.S. and S.M. Hudson. 2004. Stimuli-reponsive polymers and their bioconjugates. *Progress in Polymer Science* 29: 1173-1222.

Hardy, J.G., E. Larraneta, R.F. Donnelly, N. Mcgoldrick, K. Migalska, M.T. Mccrudden, N.J. Irwin, L. Donnelly and C.P. Mccoy. 2016. Hydrogel-forming microneedle arrays made from light-responsive materials for on-demand transdermal drug delivery. *Molecular Pharmaceutics* 13: 907-914.

Harris, K.D., C.W.M. Bastiaansen, J. Lub and D.J. Broer. 2005. Self-assembled polymer films for controlled agent-driven motion. *Nano Letters* 5: 1857-1860.

He, Z.W., N. Satarkar, T. Xie, Y.T. Cheng and J.Z. Hilt. 2011. Remote controlled multishape polymer nanocomposites with selective radiofrequency actuations. *Advanced Materials* 23: 3192.

Hu, J.L., H.P. Meng, G.Q. Li and S.I. Ibekwe. 2012. A review of stimuli-responsive polymers for smart textile applications. *Smart Materials and Structures* 21.

Hu, L., Q. Zhang, X. Li and M.J. Serpe. 2019. Stimuli-responsive polymers for sensing and actuation. *Materials Horizons* 6: 1774-1793.

Ilievski, F., A.D. Mazzeo, R.E. Shepherd, X. Chen and G.M. Whitesides. 2011. Soft robotics for chemists. *Angewandte Chemie-International Edition* 50: 1890-1895.

Jones, D.P., J.L. Carlson, P.S. Samiec, P. Sternberg, Jr., V.C. Mody, Jr., R.L. Reed and L.A. Brown. 1998. Glutathione measurement in human plasma. Evaluation of sample collection, storage and derivatization conditions for analysis of dansyl derivatives by HPLC. *Clin Chim Acta* 275: 175-184.

Kim, D., A. Jo, K.B.C. Imani, D. Kim, J.-W. Chung and J. Yoon. 2018. Microfluidic fabrication of multistimuli-responsive tubular hydrogels for cellular scaffolds. *Langmuir* 34: 4351-4359.

Kost, J. and R. Langer. 2012. Responsive polymeric delivery systems. *Advanced Drug Delivery Reviews* 64: 327-341.

Kumar, A., A. Srivastava, I.Y. Galaev and B. Mattiasson. 2007. Smart polymers: Physical forms and bioengineering applications. *Progress in Polymer Science* 32: 1205-1237.

Lee, H.T. and L.H. Lin. 2006. Waterborne polyurethane/clay nanocomposites: Novel effects of the clay and its interlayer ions on the morphology and physical and electrical properties. *Macromolecules* 39: 6133-6141.

Lee, W.E., Y.J. Jin, L.S. Park and G. Kwak. 2012. Fluorescent actuator based on microporous conjugated polymer with intramolecular stack structure. *Advanced Materials* 24: 5604-5609.

Lendlein, A. and R. Langer. 2002. Biodegradable, elastic shape-memory polymers for potential biomedical applications. *Science* 296: 1673-1676.

Liechty, W.B., D.R. Kryscio, V. Slaughterb and N.A. Peppas. 2010. Polymers for drug delivery systems. *Annual Review of Chemical and Biomolecular Engineering* 1(1): 149-173.

Lin, S.-B., C.-H. Yuan, A.-R. Ke and Z.-L. Quan. 2008. Electrical response characterization of PVA–P (AA/AMPS) IPN hydrogels in aqueous Na_2SO_4 solution. *Sensors and Actuators B: Chemical* 134: 281-286.

Ma, Y., K. Promthaveepong and N. Li. 2016. CO_2-responsive polymer-functionalized Au nanoparticles for CO_2 sensor. *Analytical Chemistry* 88: 8289-8293.

Macewan, S.R., D.J. Callahan and A. Chilkoti. 2010. Stimulus-responsive macromolecules and nanoparticles for cancer drug delivery. *Nanomedicine* 5: 793-806.

Mano, J.F. 2008. Stimuli-responsive polymeric systems for biomedical applications. *Advanced Engineering Materials* 10: 515-527.

Miaudet, P., A. Derre, M. Maugey, C. Zakri, P.M. Piccione, Inoublir and P. Poulin. 2007. Shape and temperature memory of nanocomposites with broadened glass transition. *Science* 318: 1294-1296.

Oliveira, E., R.C. Assuncao-Silva, O. Ziv-Polat, E.D. Gomes, F.G. Teixeira, N.A. Silva, A. Shahar and A.J. Salgado. 2017. Influence of different ECM-like hydrogels on neurite outgrowth induced by adipose tissue-derived stem cells. *Stem Cells International*. 2017: 1-10.

Quan, C.-Y., D.-Q. Wu, C. Chang, G.-B. Zhang, S.-X. Cheng, X.-Z. Zhang and R.-X. Zhuo. 2009. Synthesis of thermo-sensitive micellar aggregates self-assembled from biotinylated PNAS-b-PNIPAAm-b-PCL triblock copolymers for tumor targeting. *The Journal of Physical Chemistry C* 113: 11262-11267.

Roberts, J.J., P. Naudiyal, L. Juge, L.E. Bilston, A.M. Granville and P.J. Martens. 2015. Tailoring stimuli responsiveness using dynamic covalent cross-linking of poly(vinyl alcohol)-heparin hydrogels for controlled cell and growth factor delivery. *ACS Biomaterials Science & Engineering* 1: 1267-1277.

Roy, D., W.L. Brooks and B.S. Sumerlin. 2013. New directions in thermoresponsive polymers. *Chemical Society Reviews* 42: 7214-7243.

Saha, K., S.S. Agasti, C. Kim, X. Li and V.M. Rotello. 2012. Gold nanoparticles in chemical and biological sensing. *Chemical Reviews* 112: 2739-2779.

Schmaljohann, D. 2006. Thermo- and pH-responsive polymers in drug delivery. *Advanced Drug Delivery Reviews* 58: 1655-1670.

Skvarla, J., R.K. Raya, M. Uchman, J. Zednik, K. Prochazka, V.M. Garamus, A. Meristoudi, S. Pispas and M. Stepanek. 2017. Thermoresponsive behavior of poly(N-isopropylacrylamide)s with dodecyl and carboxyl terminal groups in aqueous solution: pH-dependent cloud point temperature. *Colloid and Polymer Science* 295: 1343-1349.

Son, S., E. Shin and B.-S. Kim. 2014. Light-responsive micelles of spiropyran initiated hyperbranched polyglycerol for smart drug delivery. *Biomacromolecules* 15: 628-634.

Stuart, M.A.C., W.T. Huck, J. Genzer, M. Muller, C. Ober, M. Stamm, G.B. Sukhorukov, I. Szleifer, V.V. Tsukruk and M. Urban. 2010. Emerging applications of stimuli-responsive polymer materials. *Nature Materials* 9: 101-113.

Sun, T. and G. Qing. 2011. Biomimetic smart interface materials for biological applications. *Advanced Materials* 23: H57-H77.

Surnar, B. and M. Jayakannan. 2013. Stimuli-responsive poly(caprolactone) vesicles for dual drug delivery under the gastrointestinal tract. *Biomacromolecules* 14: 4377-4387.

Swift, T., L. Swanson, M. Geoghegan and S. Rimmer. 2016. The pH-responsive behaviour of poly(acrylic acid) in aqueous solution is dependent on molar mass. *Soft Matter* 12: 2542-2549.

Tian, Y., J. Zheng, X. Tang, Q. Ren, Y. Wang and W. Yang. 2015. Near-infrared light-responsive nanogels with diselenide-cross-linkers for on-demand degradation and triggered drug release. *Particle & Particle Systems Characterization* 32: 547-551.

Wan, W.-M., S.-S. Li, D.-M. Liu, X.-H. Lv and X.-L. Sun. 2017. Synthesis of electron-deficient borinic acid polymers with multiresponsive properties and their application in the fluorescence detection of alizarin red s and electron-rich 8-hydroxyquinoline and fluoride ion: Substituent effects. *Macromolecules* 50: 6872-6879.

Wang, X., Y.A. Hu, L. Song, H.Y. Yang, W.Y. Xing and H.D. Lu. 2011. In situ polymerization of graphene nanosheets and polyurethane with enhanced mechanical and thermal properties. *Journal of Materials Chemistry* 21: 4222-4227.

Wei, M.L., Y.F. Gao, X. Li and M.J. Serpe. 2017. Stimuli-responsive polymers and their applications. *Polymer Chemistry* 8: 127-143.

Werzer, O., S. Tumphart, R. Keimel, P. Christian and A.M. Coclite. 2019. Drug release from thin films encapsulated by a temperature-responsive hydrogel. *Soft Matter* 15: 1853-1859.

Wu, Y., K. Wang, S. Huang, C. Yang and M. Wang. 2017. Near-infrared light-responsive semiconductor polymer composite hydrogels: Spatial/temporal-controlled release via a photothermal "sponge" effect. *ACS Applied Materials & Interfaces* 9: 13602-13610.

Xie, T. and X.C. Xiao. 2008. Self-peeling reversible dry adhesive system. *Chemistry of Materials* 20: 2866-2868.

Xie, X.J., L.T. Qu, C. Zhou, Y. Li, J. Zhu, H. Bai, G.Q. Shi and L.M. Dai. 2010. An asymmetrically surface-modified graphene film electrochemical actuator. *Acs Nano* 4: 6050-6054.

Xiong, X.-B., Z. Ma, R. Lai and A. Lavasanifar. 2010. The therapeutic response to multifunctional polymeric nano-conjugates in the targeted cellular and subcellular delivery of doxorubicin. *Biomaterials* 31: 757-768.

Xu, J.S., J. Huang, R. Qin, G.H. Hinkle, S.P. Povoski, E.W. Martin and R.X. Xu. 2010. Synthesizing and binding dual-mode poly(lactic-co-glycolic acid) (PLGA) nanobubbles for cancer targeting and imaging. *Biomaterials* 31: 1716-1722.

Yamamoto, Y., S. Harada, D. Yamamoto, W. Honda, T. Arie, S. Akita and K. Takei. 2016. Printed multifunctional flexible device with an integrated motion sensor for health care monitoring. *Science Advances* 2: e1601473.

Yu, H.F. 2014. Recent advances in photoresponsive liquid-crystalline polymers containing azobenzene chromophores. *Journal of Materials Chemistry C* 2: 3047-3054.

Yu, H.F. and T. Ikeda. 2011. Photocontrollable liquid-crystalline actuators. *Advanced Materials* 23: 2149-2180.

Zakharchenko, S., N. Puretskiy, G. Stoychev, M. Stamm and L. Ionov. 2010. Temperature controlled encapsulation and release using partially biodegradable thermo-magneto-sensitive self-rolling tubes. *Soft Matter* 6: 2633-2636.

Zhang, X., Y. Zheng, Z. Wang, S. Huang, Y. Chen, W. Jiang, H. Zhang, M. Ding, Q. Li and X. Xiao. 2014. Methotrexate-loaded PLGA nanobubbles for ultrasound imaging and synergistic targeted therapy of residual tumor during HIFU ablation. *Biomaterials* 35: 5148-5161.

Zhang, X.B., C.L. Pint, M.H. Lee, B.E. Schubert, A. Jamshidi, K. Takei, H. Ko, A. Gillies, R. Bardhan, J.J. Urban, M. Wu, R. Fearing and A. Javey. 2011. Optically- and thermally-responsive programmable materials based on carbon nanotube-hydrogel polymer composites. *Nano Letters* 11: 3239-3244.

Zhou, R.X., S. Zhu, L.J. Gong, Y.Y. Fu, Z.J. Gu and Y.L. Zhao. 2019a. Recent advances of stimuli-responsive systems based on transition metal dichalcogenides for smart cancer therapy. *Journal of Materials Chemistry B* 7: 2588-2607.

Zhou, X., L. Guo, D. Shi, S. Duan and J. Li. 2019b. Biocompatible chitosan nanobubbles for ultrasound-mediated targeted delivery of doxorubicin. *Nanoscale Research Letters* 14: 24.

CHAPTER

14

Poly(siloxane)s, Poly(silazane)s and Poly(carbosiloxane)s

Claire E. Martin, Giovanni Fardella, Ricardo Perez and Joseph W. Krumpfer*
Department of Chemistry and Physical Sciences, Pace University,
861 Bedford Road, Pleasantville, NY – 10570

1. Introduction

Perhaps one of the most ubiquitous and universal polymers, polysiloxanes (Rochow 1946, Post 1949, Brook 2000, Clarson and Semlyen 1993) have found a role in a wide range of applications from the mundane, such as shampoos and cosmetics (Horii and Kannan 2008), to the outlandish, including astronaut boots (Warrick 1990) and water-repellent agents (Krumpfer and McCarthy 2010, Krumpfer and McCarthy 2011). Siloxane polymers, commonly referred to as silicones, are capable of such a wide variety of functions because of their unique chemical structure, which consists of a polymer backbone containing alternating silicon and oxygen atoms. This clearly differentiates them from traditional polyolefin or carbon-based polymers and gives rise to a number of interesting considerations. In addition to the above applications, these polymers are featured prominently in lubricants (Aziz et al. 2019), coatings (Zahid et al. 2019) and medical implants (Brook 2012). Furthermore, siloxane copolymers are important materials in foaming and anti-foaming agents (Baferani et al. 2018), pharmaceuticals and surfactants.

Likewise, derivatives of polysiloxanes, namely carbosiloxanes, silphenylenes and silazanes, possess a number of unique properties and applications, including higher reactivities, greater thermal stabilities, flame resistance and the ability to be converted to ceramic materials, without sacrificing many of the superior properties of siloxanes. The subtle differences in the structures of these polymers allow them to be useful in rather extreme conditions, such as rubbers for flame resistance (Li et al. 2019).

In this chapter, we examine the structures and properties of these materials to better understand the underlying chemical reasons for their

*Corresponding author: jwkrumpfer@gmail.com

universality and wide range of applications. Where appropriate, the historical significance of these materials is highlighted, as silicones and their derivatives have had a fascinating and divisive past. Additionally, traditional methods for the preparation of these materials are reviewed, and some of the more recent advances in their preparations are highlighted. Particularly, an emphasis is placed on industrially relevant processes, and important applications of these materials, such as surfactants, coating materials and preceramic polymers, are examined.

2. Structure and properties

In order to have a better grasp of the unique properties of these materials, it is important to examine their chemical structures: the fundamental aspect of these materials that clearly differentiates them from other commercial polymers. It should be noted that silicon-containing compounds have a fairly unique nomenclature, which has been previously summarized (Longenberger et al. 2017) and may prove beneficial for the reader, but is certainly not necessary to understand the chemistry presented here. Figure 1 provides the basic structures of several silicon-containing polymers discussed in this chapter. The most basic of these structures is the polysiloxane (Fig. 1A), which contains a polymer backbone comprising alternating silicon and oxygen atoms. The nature of this particular polymer backbone is important, as it gives rise to the intriguing properties of these polymers, which are discussed later in this chapter.

We can imagine that all other polymers in Fig. 1 are simply derivatives of polysiloxanes. Polysilazanes (Fig. 1B) arise when the oxygen atom in a polysiloxane is replaced with a nitrogen atom, and the specific structure shown in Fig. 1B is called poly(dimethylsilazane). Likewise, if the oxygen atom is replaced by some carbon-based group (R = alkyl, aryl), then polycarbosilanes are produced. For example, if the R group is simply a methylene group (-CH_2-), then the polymer shown in Fig. 1C would be called poly(trimethylsilane). The replacement of the oxygen atom leads to drastic changes in the physical properties of these polymers. For instance, polysilazanes show far greater reactivities towards water than polysiloxanes, and polycarbosilanes are fairly rigid solid materials.

A wider variety of polymers can be produced by replacing only every second oxygen atom of a polysiloxane. By doing so, we can incorporate new functionalities while retaining many of the desirable properties of siloxanes. In particular, if the R group is an alkyl or aryl group, the class of polymers known as polycarbosiloxanes is derived (Fig. 1D). It should be noted that polysiloxysilazanes, the structure in which nitrogen is incorporated, is not of practical use owing to its difficulty in preparation and its high reactivity/instability. Conversely, if the replacing group is an aromatic aryl group (-C_6H_4-), the class of silphenylenes is derived, which has improved chemical and thermal stability over traditional polysiloxanes.

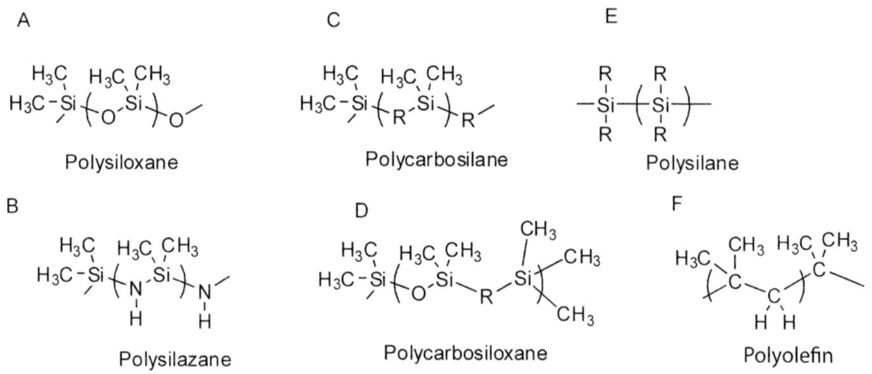

A Polysiloxane

C Polycarbosilane

E Polysilane

B Polysilazane

D Polycarbosiloxane

F Polyolefin

Fig. 1. Structures of silicon-containing polymers (A-E) and polyolefins (F).

While polysilanes are not discussed in this chapter, for completeness' sake, their structure is also provided (Fig. 1E). These polymers do not contain any heteroatoms between silicon atoms and are important in the preparation of silicon semiconductors (Kumar and Leitao 2020). This special property arises from the delocalized sigma electrons across Si-Si bonds (Orti et al. 1993, West et al. 1981), and there is still great interest in further exploring these materials. Additionally, there has been recent interest in other silicon-based polymeric structures worth noting, such as silicene, sp^2-hybridized silicon sheets, which are the silicon analogs to graphenes, and silicanes, sp^3-hybridized silicon sheets, which are analogs to 2D diamonds. Both of these structures have unique optical and electrical properties that have yet to be fully explored. However, the major limitation of these materials and polysilanes is their high reactivity with oxygen, requiring them to be handled within inert atmospheres.

Figure 1 also includes a traditional class of carbon-based polymers, polyolefins. Polyolefins are typical polymers used in conventional materials, such as plastic bags and styrofoam. The particular structure in Fig. 1F is poly(isobutylene) and the closest carbon-analog to poly(dimethylsiloxane) (PDMS) (Fig. 1A). Now, given that carbon (C) and silicon (Si) are both group 4 elements on the periodic table, it is commonly believed that they behave similarly. While that is true in some respects, the truth is that silicon is far different from carbon, which gives rise to many of the special properties found in silicon-based polymers.

To better understand how the physical properties manifest in silicon-containing polymers, we will start at the atomic level. Table 1 provides a few fundamental atomic properties relevant to these polymers. From a simple glance, stark differences between silicon and carbon can be noted. First, silicon (Si) is significantly larger than carbon (C), both in covalent radius and atomic mass. This larger radius results in longer bonds with other atoms, making silicon-based polymers less "compact" than carbon-based polymers. However, upon looking at the electronegativities, it is seen that silicon is far more electropositive than carbon (Rahm et al. 2019). In fact, silicon is even more electropositive than hydrogen (H). This has direct implications on the polarity of the bonds that form with silicon. For example, whereas the dipole moment in the hydrocarbon (C-H) bond is pointing towards the carbon atom, within the hydridosilane (Si-H) bond the dipole moment is pointing towards the hydrogen atom. This results in a much higher reactivity of the hydridosilane group than the hydrocarbon group, which is observed in a number of reactions, such as hydrosilylation (Stein et al. 1999) and reaction with inorganic surfaces (Fadeev and McCarthy 1999).

Table 1. Atomic properties of the elements of silicon-containing polymers (Chang and Overby 2019, Shriver et al. 2014)

		Atomic properties	
	Covalent radii (pm)	*Pauling's electronegativities* (χ)	*Atomic mass (amu)*
Si	118	1.8	28.09
C	77	2.5	12.01
O	66	3.5	16.00
N	71	3.0	14.01
H	37	2.2	1.008
B	84	2.0	10.81

Further, based on the electronegativities of these elements, silicon actually has more in common with boron (B) than any other element (Pauling 1960). This is taught to chemists as the "diagonal relationship" and arises from the similar electron densities, or the number of electrons per size of the atom, of these two elements. For this reason, we observe that silicon and boron behave similarly in a number of reactions (Thomas 2006) and form nearly covalent bonds. On the other hand, the electronegativity difference between silicon and oxygen is drastic. In fact, while the siloxane (Si-O) bond is often considered covalent, it actually has a stronger ionic nature (Weinhold and West 2011), having been calculated to be 51% ionic (Pauling 1980). Likewise, even the silicon carbide (Si-C) bond has a slightly ionic nature, being roughly 12% ionic.

As we can see, the atomic nature of these elements has a direct impact on the nature of the bonds they form. Table 2 shows the properties of several pertinent bonds. As expected, silicon-containing bonds are longer than carbon-containing bonds (Colvin 1981). Likewise, given the electropositivity of silicon, and thus the increased ionic nature of its bonds, the bond dissociation energy (BDE) of these bonds is higher than those with carbon (Beccera and Walsh 1998). This results in their higher thermodynamic stability. In fact, the Si-O bond has an extremely high BDE, which explains the excellent thermal properties of siloxane polymers. This BDE also explains why the vast majority of silicon found on the planet is directly tied to oxygen atoms in the form of silica (SiO_2) or silicates (SiO_4). Or, in more generic terms, silicon is largely found in dirt and sand and makes up approximately 25% of the earth's crust, while oxygen represents 50% (Rudnick and Fountain 1995).

Table 2. Properties of silicon- and carbon-containing bonds

	Bond length (pm)	*Bond dissociation energy (kJ/mol)*	*Bond angles*
Si-O	163	443	143 (Si-O-Si) 112 (O-Si-O)
Si-C	187	369	~ 120 (Si-C-Si) ~ 120 (C-Si-C)
Si-N	174	434	140 (Si-N-Si) 85.5 (N-Si-N)
Si-H	148	378	n/a
Si-Si	234	332	109.5 (Si-Si-Si)
C-C	153	346	109.5 (C-C-C)
C-H	107	420	n/a

Finally, we note that the bond angles containing silicon are rather large in all respects. This can be explained by the repulsive forces between the large silicon atoms, which widen the bond angles in silicon-containing compounds. The bond angles for the siloxane (Si-O) bond are unique in that they do not accurately describe the chemical structure of these materials. While the bond angles presented here are the traditionally accepted values, the truth is that the siloxane bond is easily able to swing through 180° with little barrier to the freedom of movement given its ionic nature, much like a ball-in-socket joint. This means that siloxanes are extremely flexible molecules. However, as can be seen in the silicon-silicon (Si-Si) bonds, without the presence of heteroatoms, silicon will conform similarly to carbon. These large bond angles have a direct impact on the physical properties of these polymers. Combined with the larger bond lengths, the increased bond angles suggest a greater extent of "open space", or more accurately free volume, between constituents, which will be an important consideration later.

Before discussing the physical properties of the polymer materials, it is worth considering one additional aspect related to bonds containing silicon atoms. While much of our previous explanation of these chemical bonds has centered around electronegativity and size differences, the location of silicon on the periodic table should also be mentioned. As silicon is a period 3 element (whereas carbon is a period 2 element), it has easier access to higher atomic orbitals for bonding, namely *d*-orbitals, while carbon does not. This further explains some of the bonding properties of this element and suggests that bonding with silicon atoms is very different from bonding with carbon atoms (Zeigler and Fearon 1989). For example, whereas carbon can form up to tetravalent (four bonds) compounds, silicon can form as high as hexavalent (six bonds) compounds (Holloczki and Nyulaszi 2009). While this property falls outside the scope of this chapter, it is worth noting.

Table 3 provides physical properties for several polymers. We note that the properties of these materials are extremely different and, in fact, a direct comparison between them is not entirely valid. In fact, many of the values in this table have been extrapolated from the nature of the materials, and others are approximations. For example, the degradation temperature (T_d) of polysilazanes are not true degradation temperatures, as they are known to form ceramic materials around this temperature. Rather, temperature range reported is when this ceramitization process typically begins. Regardless, the difference in properties showcased in Table 3 illustrates just how unique the class of polymers encompassing poly(siloxane)s, poly(silazane)s and poly(carbosiloxane)s is, and a few general observations can be derived from the previous chemical comparisons.

Looking at PDMS, we can instantly observe that it has a remarkably low glass transition temperature (T_g), in comparison to the carbon-based polymers, poly(isobutylene), poly(ethylene), and poly(styrene). This is due to several aspects. First, the large bond angles and lengths lead to a high degree of free volume in these materials (Hurd 1946). Second, the weak or nearly repulsive interactions between the methyl (-CH_3) groups and the siloxane backbone with itself make it unfavorable for the siloxane chains to pack easily into crystalline forms (Mark 2004). This also directly affects the low melting temperature of these materials. As an interesting aside, PDMS does not freeze upon cooling to its glass transition temperature but rather undergoes a "cold crystallization" at -80°C upon heating from the T_g, followed by a melting transition at -40°C. Finally, the high degree of vibrational and rotational freedom in these polymers, whereby siloxanes can "slide by" each other very easily, makes kinetic trapping of these polymers difficult. All of these factors result in an extremely low T_g value for nearly all siloxane polymers (Owen 1981).

Another direct result of the bonding nature in siloxanes is a relatively high thermal degradation temperature (T_d), which is attributed to the ionic nature of these bonds and the high BDE. On the other hand, polysilanes and polycarbosilanes, which are more covalent in nature, have much lower thermal stabilities due to the decreased bond stability (BDE). The major mechanism for degradation of siloxane polymers is not precisely direct bond-cleavage but the formation of cyclic monomers through a back-biting mechanism (Fig. 2) (Camino et al. 2002). A common strategy to increase thermal stability is to prevent reactive chain ends through the use of end-capping monomers, which is discussed later in this text. However, while Fig. 2 shows this as a chain end reacting with the polymer backbone, it is also possible for interchain scission to occur in a similar manner to form cyclic products.

Table 3. Physical properties of silicon-containing polymers and traditional polyolefins (Mark 1999, Bicerano 2009)

Polymer	*Acronym*	*Repeat unit*	T_g (°C)	T_m (°C)	T_d (N_2, °C)	ρ (g/cm^3)
Poly(dimethylsiloxane)	PDMS	$-(Me_2SiO)-$	-120	-40	~ 350	0.970
Poly(methylphenylsiloxane)	PMPS	-(MePhSiO)-	-86	35	371	1.11
Poly(hydridomethylsiloxane)	PHMS	-(MeHSiO)-	-119	~ -40	n/a	0.99
Poly(1,1-dimethylsilazane)	PDMZ	$-(Me_2SiNH)-$	-82	96.9	300-500	1.04
Poly(N-methyl-dimethylsilazane)	PMDMZ	-(MeHSiNMe)-	-38	227	300-500	1.0
Poly(silphenylene-siloxane)	n/a	$-(Me_2Si-Ph-Me_2Si-O)-$	-61	148	400	1.10
Poly(dimethylsilylene)	PDMSi	$-(Me_2Si)-$	60	n/a	n/a	0.971
Poly(silylenemethylene)	PSM	$-(H_2SiCH_2)-$	-100	-22	~ 200	~ 1.0
Poly(isobutylene)	PIB	$-[(CH_3)_2CHCH_2]-$	-71	300	300	0.917
Poly(ethylene), high density	HDPE	$-(CH_2CH_2)-$	150	~ 400	370-450	0.996
Poly(styrene)	PS	$-(CH_2CH_2Ph)-$	100	240	300	1.1

*Me = CH_3, Ph = C_6H_6

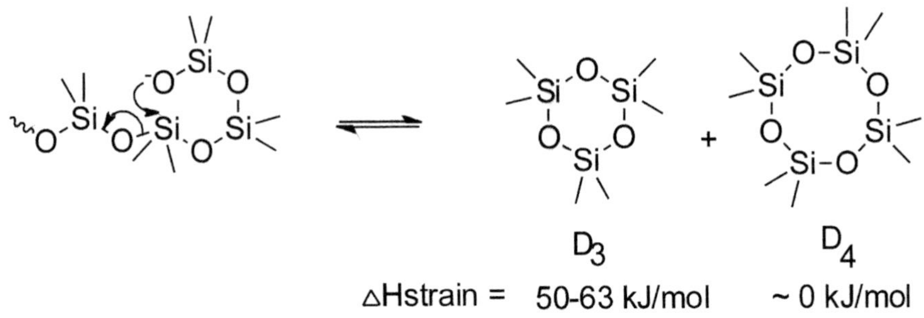

D_3 D_4

ΔH_{strain} = 50-63 kJ/mol ~ 0 kJ/mol

Fig. 2. Back-biting degradation mechanism of polysiloxanes and the formation of cyclic monomer species.

The two most prominent thermal degradation products are hexamethylcyclotrisiloxane (D_3) and octamethylcyclotetrasiloxane (D_4), with higher cyclic species also possible. An interesting aspect of the cyclic rings formed are the apparent ring strains for each of these compounds, particularly the absence of ring strain in the D_4 structure. While one is tempted to assume D_4 would be the most abundant product given that it is thermodynamically favored, it is rather the kinetic product, D_3, that forms in greater abundance (Camino et al. 2002).

Given that this is the major mechanism for thermal degradation, it is unsurprising that most silicon-based polymers have rather high thermal degradation temperatures. However, it is interesting to make two observations from the given data, which concern poly(methylphenylsiloxane) (PMPS) and poly(silphenylene)s. We note that these two polymers have higher degradation temperatures than PDMS, which is due to a disruption in the degradation process, albeit in different ways. For PMPS, the addition of the large phenyl side group provides steric interference during the back-biting cyclization process. We could imagine that poly(diphenylsiloxane) (not shown) would have an even higher degradation temperature, although this homopolymer is impractical to polymerize, owing to the bulkiness of the monomers. However, this brings up a popular strategy for increasing and controlling the thermal stability of silicone polymers, namely the preparations of copolymers containing dimethylsiloxane, phenylmethylsiloxanes, and/or diphenylsiloxanes through sterically hindering the degradation mechanism.

Silphenylenes, and carbosiloxanes in general, disrupt the degradation process in a different manner (Yang et al. 2011). Rather than directly interfere with the back-biting reaction, they limit the flexibility of the polymer backbone to a certain degree. In this way, the polymers are not able to bend sufficiently to form cyclic products. Rather, the degradation mechanism occurs primarily through interchain scission at elevated temperatures. As we can see in these two examples, the incorporation of phenyl moieties into either the side group or the polymer backbone has the effect of increasing thermal stability. However, it also increases the oxidative stability of these polymers, which is discussed below.

It is important to consider all of the discussed thermal properties in terms of thermal use ranges. In other words, these polymers are well known to be stable over an incredibly wide temperature range of approximately 400°C. Additionally, the extremely low glass transitions mean that these polymers are typically viscous liquids across these temperatures. And, in truth, these polymers are oils whose viscosity can range from "water-like" to "thick honey" consistencies depending on their molecular weights. However, when many discuss these polymers they are actually referring to crosslinked materials and not the simple homopolymers. Typically, these polymers require additional crosslinking and/or toughening agents in order to become viable mechanical materials.

As purely synthetic polymers with no naturally occurring analogs, siloxanes are surprisingly environmentally benign. In terms of environmental impact, it is important to discuss the oxidative degradation of these polymers, in comparison to carbon-based polymers. It is well established that the oxidative products of hydrocarbons are simply water and carbon dioxide, which has questionable effects for the future of this planet. On the other hand, the major oxidation products of siloxanes and carbosiloxanes in the presence of oxygen are silica (SiO_2) with some CO_2 and H_2O, depending on the side chains. Silazanes also produce ammonia or other amines. The production of a solid waste material has diminished impact on the atmosphere, and the nature of this product, sand, has no impact on the environment. Within soil, intermediate silanediols are also formed, which inevitably form silica as well (Graiver et al. 2003, Lin et al. 2013). However, as previously stated, the introduction of phenyl groups also increases the oxidative stability (Walsh 1981). This is due to the increased strength of the silicon-phenyl bond, which is more resistant to reaction with oxygen and thereby increases their environmental duration. For this reason, the desired material and properties must be carefully considered in conjunction with their eventual waste materials.

Another property of these materials that is worth mentioning is their low surface tension (γ_{LV}) and overall interaction with water. Poly(dimethylsiloxane)s exhibit surface tensions of approximately 20 dyn/cm, significantly lower than most hydrocarbons, making them excellent candidates for hydrophobizing agents and surfactants. This low surface tension is due to the low intermolecular attraction found in these materials. This important property has led to their wide use as water-repellent materials, such as nautical sealants or windshield coating. However, although they are largely water repellent, silicones and silphenylenes show a high degree of water vapor permeability. In this way, they are also "breathable" materials, which allow gases to easily flow through them. It is important to note that silazanes are highly reactive with water through hydrolysis of the Si-N bond and will ultimately convert to siloxanes if reacted with water. The use of these polymers as coating materials is further discussed later in this chapter.

Finally, it is important to mention the inherent biocompatibility and non-toxicity of siloxane polymers, allowing them to be used as implants and major components in over the counter drugs. In general, for good biocompatibility, a material must be resistant to chemical reactions, produce no harmful byproducts in the case of possible reaction, produce no immunoresponse, cause no allergic or inflammatory responses, and make no alterations to the biological environment (e.g., denaturing of proteins/enzymes, deterioration of surrounding tissue). With such a long list of demands for any biomaterial, it is a wonder that any material could satisfy all of these criteria.

While there are no overwhelmingly definitive arguments for why silicones, a material with no natural analogs, are biocompatible other than that they simply are, one of the more convincing arguments for this property within the human body centers on evolution. As a species, we (like nearly all species on this planet) have evolved in an environment primarily consisting of these two elements and have learned to coexist well. For example, swallowing sand, while not recommended, is not particularly harmful to humans and the sand will ultimately pass benignly through the digestive tract. One could imagine that the human body recognizes silicones, being primarily composed of silicon and oxygen, in much the same way: simply, it is an environmental component that neither helps nor harms. This principle can be easily found in the application of silicones as "zero-calorie oil substitutes" in low-fat food alternatives, such as potato chips. In this example, silicones are used as oils for the preparation of potato chips. Since the human body does not digest silicones, it offers no caloric contribution to the foods themselves. This argument for the biocompatibility of silicones is supported by other species, such as diatoms, sponges, and mollusks, which incorporate siliceous materials into their organisms (Voronkov et al. 1977, Simon and Volcani 1981).

Clearly, siloxane polymers and their derivatives are host to many, many remarkable properties, allowing them to be used in a wide number of applications. While the presented properties are by no means exhaustive, they provide a good overview of many of the important considerations of silicones when tailoring them to specific applications and properties. In the next section, we highlight the major synthetic processes for preparing these polymers.

3. Syntheses and preparations

3.1. Monomer preparation

The preparation of siloxane polymers consists of well-established, commercial processes that have been in use for almost a century with very little alteration. The breakthrough that allowed siloxanes to become commercially relevant materials arose in 1940 through the work of Eugene Rochow at General Electric (Rochow 1941) and eventually became known as the Direct Process (Rochow 1945). As an aside, German chemist Richard Müller developed a nearly identical process independently only a few weeks later (Müller 1942). For this reason, the Direct Process is also referred to as the Rochow-Müller

Process. Regardless of its name, this process was pivotally important since it provided a simple and efficient method for the preparation of functional silanes, the precursors for all silicon-derived polymers, and is an excellent place to start.

The purification and isolation of metallurgical silicon has been well known for over a century, by reacting silica (sand, quartz) with carbon (coal, charcoal) at 1500-2000°C. In such a way, we can imagine that silicones and their derivatives are "renewable" materials, since they are derived from silica and ultimately degrade back to silica, although this might be an oversimplification. With the recent advent of the semiconductor industry, additional methods of producing silicon of "semiconductor grade" or higher purity have been developed (Delannoy 2012, Khalifia et al. 2012). Approximately 6 million tons of silicon are produced each year, with China by far the world's leading producer of this material.

This silicon is then typically thermally reacted with any number of reagents, such as chloromethane, to produce reactive chlorosilanes (Fig. 3). In many cases, a catalytic amount of a heterometal, such as copper (Cu), is added to increase the efficiency of this reaction. However, it is not entirely clear what role this catalyst plays in this reaction (Brookes et al. 2001). Here, we see a number of different silanes, which all serve different purposes (Plueddemann 1982). It must be noted that these reaction schemes are simplified and, in reality, a mixture of mono-, di-, tri- and tetrafunctional materials are made, which are separated via subsequent distillation. In addition, there exists some disilane residue after reaction, which has recently been suggested as a starting material for additional silanes (Santowski 2019). The monomer, dichlorodimethylsilane (Fig. 3A), is by far the most industrially important of these silanes, and it is the common starting material by which all the polymers discussed in this chapter are derived.

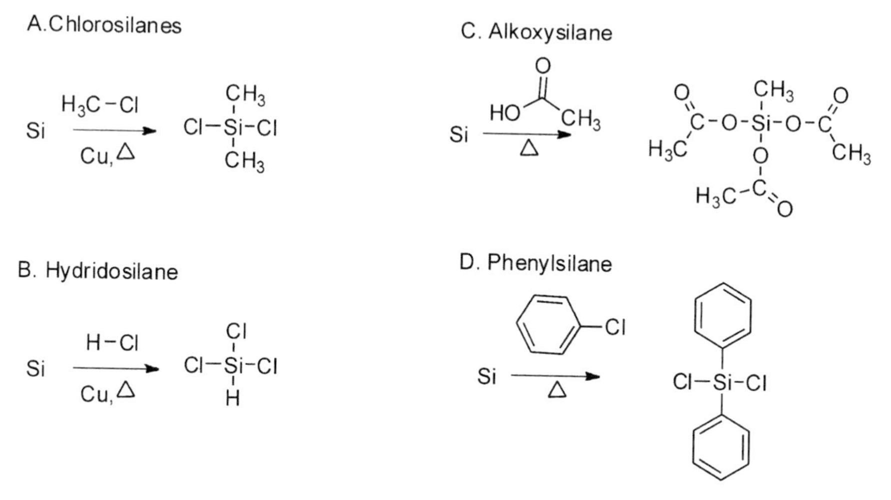

Fig. 3. The Direct (Rochow-Müller) Process for the preparation of functional silanes.

As we see, there are a number of additional monomers that can be derived from the Direct Process, and one can imagine a near endless possibility for controlling functional groups and reactivity. While we primarily focus on dimethyldichlorosilane throughout the chapter, it is also appropriate to discuss a few additional monomers. Figure 3B shows the preparation of hydridosilanes, which are important for the chemical vapor deposition of silicon semiconductors. Hydridosilanes also provide a starting point for more advanced monomers, via hydrosilylation (Fig. 4). This reaction is typically catalyzed using a platinum catalyst, such as Speier's Catalyst (H_2PtCl_6) or the more recent Karstedt's catalyst (divinylhexamethyldisiloxane platinum (0) complex) (Nakajima and Shimada 2015). The aminopropyl-, tridecafluorooctyl, and glycidoxysilanes are important functional groups for providing hydrophilicity, hydrophobicity and binding agents for adhesives, respectively, as just a few examples.

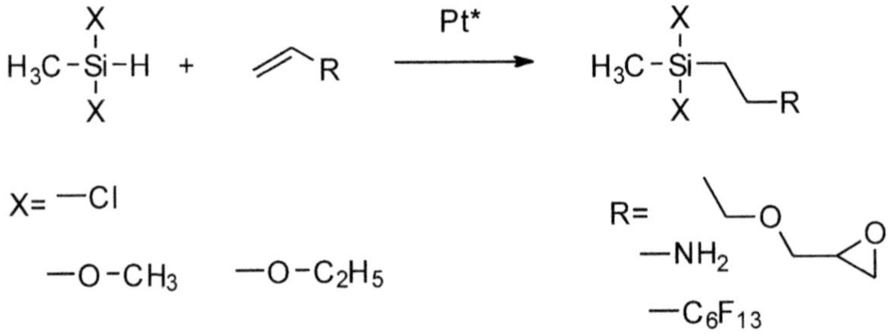

$$H_3C\text{-}SiX_2\text{-}H + CH_2{=}CH\text{-}R \xrightarrow{Pt^*} H_3C\text{-}SiX_2\text{-}CH_2CH_2\text{-}R$$

X = —Cl, —O-CH_3, —O-C_2H_5

R = —NH_2, —C_6F_{13}, —CH_2-O-CH_2-(epoxide)

Fig. 4. Hydrosilylation of hydridosilanes for the preparation of functional silanes.

While hydrosilylation is perhaps the most prominent method for preparing functional silanes, Grignard reagents are also a viable method. The preparation of silane monomers via Grignard reactions predates the Direct Process, and much of the early work on siloxane chemistry used this precursor method (Kipping and Lloyd 1901). Figure 5 shows the preparation of silanes via Grignard synthesis. While this technique is not used industrially, it is important for the preparation of unique monomers, particularly for polycarbosilanes and carbosiloxanes (Otomo et al. 2005).

Returning to Fig. 3, we note the final two structures have industrial importance. Figure 3C shows the preparation of an alkoxysilane, specifically methyltriacetylsilane. This particular silane is commonly used as a curing

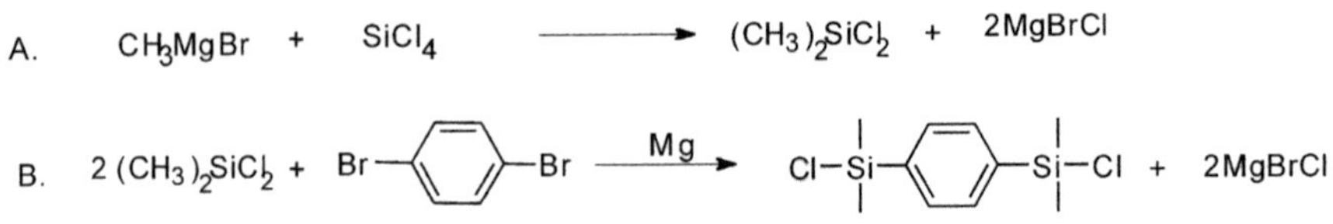

A. $$CH_3MgBr + SiCl_4 \longrightarrow (CH_3)_2SiCl_2 + 2MgBrCl$$

B. $$2\,(CH_3)_2SiCl_2 + Br\text{-}C_6H_4\text{-}Br \xrightarrow{Mg} Cl\text{-}Si(CH_3)_2\text{-}C_6H_4\text{-}Si(CH_3)_2\text{-}Cl + 2MgBrCl$$

Fig. 5. Preparation of (A) dichlorodimethylsilane and (B) 1,4-bis(chlorodimethylsilyl) benzene via Grignard synthesis.

agent in silicon sealants. The byproduct of this reaction, acetic acid, gives rise to the strong vinegar scent of these materials as they are setting. Finally, Fig. 3D showcases the preparation of phenylsilanes, which have previously been discussed as important comonomers for thermally stable silicone oils. Phenylsilanes are also often prepared through Grignard synthesis, as seen in Fig. 3B, which shows an important monomer for siphenylene-based polymers. While we have not discussed precursors for polysilazanes, they are easily derived from the aminolysis of chlorosilanes to directly produce polymeric materials, which are discussed below. In all the above cases for monomer preparations, anhydrous conditions are very important, as chlorosilanes are highly reactive with water.

3.2. Polymerization techniques

Just as the monomers for polysiloxanes serve as the precursors for other silicon-containing polymers, so do the polymerization mechanisms for polysiloxanes serve as the techniques for preparing the other classes of polymers. Mechanisms for the polymerization of functional silanes have been well studied since the early 1900s, and many of the techniques used to produce polysiloxanes have remained largely unchanged since then. These techniques were originally derived for PDMS but can be applied to other polymers as well. There are two primary mechanisms for preparing polysiloxanes: polycondensation of difunctional silanes and ring-opening polymerization (ROP) of cyclic monomers.

Figure 6 shows the polycondensation of dimethyldichlorosilane to produce PDMS. In the first step, hydrolysis of the chlorosilanes occurs to produce the silanediol intermediate. When using chlorosilanes, hydrochloric acid is formed as a byproduct. However, one can imagine that careful selection of the monomer, such as a diethoxysilane, can result in more benign byproducts, such as ethanol, while sacrificing monomer reactivity. Following this hydrolysis, the silanediol can then polymerize via a traditional polycondensation reaction (Fig. 6B). This also results in the reformation of water. We should note that the polymer produced is silanol terminated, which can very easily equilibrate through back-biting mechanism to form the cyclic monomers previously shown in Fig. 2. In fact, in order to achieve high molecular weight, an end-capper must be added to the equilibration to terminate the polymerization (Fig. 6C). This can either be a disiloxane or a monofunctional silane.

End-capping is an important aspect of introducing functionality to the polymer chain ends. For example, the use of chlorodimethylsilane, $(CH_3)_2SiHCl$, terminates the polymer chain with a hydridosilane, a functional group important in hydrosilylation reactions. Indeed, we will see this is one of the primary methods for preparing surfactants. Additionally, other monomers used to modify both chain ends, such as 3-aminopropyldimethylmethoxysilane, $(NH_2CH_2CH_2)(CH_3)_2Si(OCH_3)$, dimethylvinylchlorosilane, $(CH_3)_2CH_2CHSiCl$, or glycidoxysilanes, are

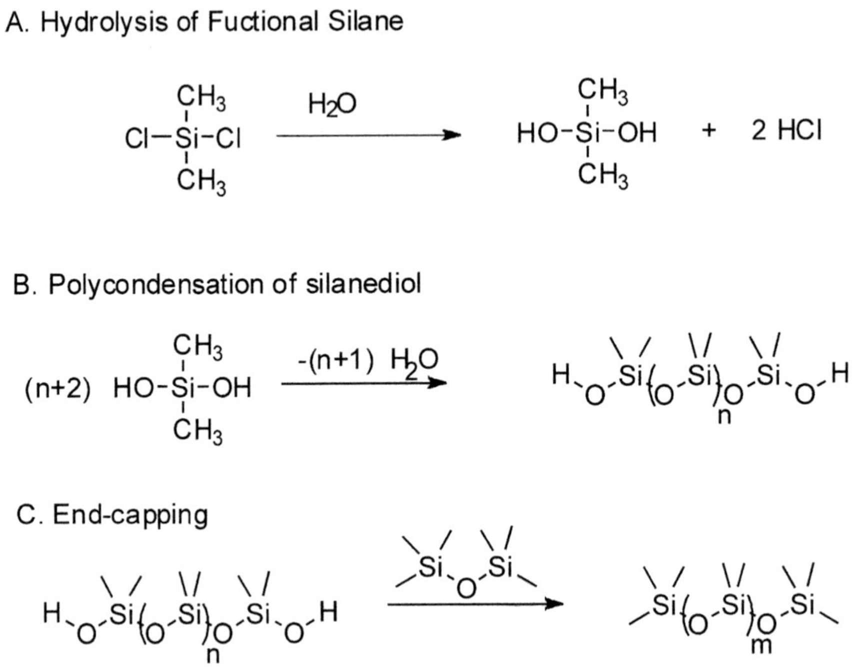

A. Hydrolysis of Fuctional Silane

$$Cl{-}Si(CH_3)_2{-}Cl \xrightarrow{H_2O} HO{-}Si(CH_3)_2{-}OH + 2\ HCl$$

B. Polycondensation of silanediol

$$(n+2)\ HO{-}Si(CH_3)_2{-}OH \xrightarrow{-(n+1)\ H_2O} H{-}O{-}Si(CH_3)_2{-}(O{-}Si(CH_3)_2)_n{-}O{-}Si(CH_3)_2{-}O{-}H$$

C. End-capping

$$H{-}O{-}Si(CH_3)_2{-}(O{-}Si(CH_3)_2)_n{-}O{-}Si(CH_3)_2{-}O{-}H \xrightarrow{(CH_3)_3Si{-}O{-}Si(CH_3)_3} (CH_3)_3Si{-}(O{-}Si(CH_3)_2)_m{-}O{-}Si(CH_3)_3$$

Fig. 6. Polycondensation of dichlorodimethylsiloxane for the synthesis of poly(dimethylsiloxane)s.

important crosslinking agents for polyurethanes, hydrosilylation networks, and epoxies, respectively. We note that this is a thermodynamic control over the molecular weight through affecting the equilibration process, which occurs late in the polymerization.

While the step-growth polymerization of polysiloxanes has widespread use, the equilibration of the polymer with its cyclic monomers ultimately limits the molecular weight that can be achieved through this method. In order to achieve higher molecular weight, the ROP of the cyclic monomers is used (Lee et al. 1969). Particularly, the ROP of the D_3 monomer is used, owing to its increased ring strain, although the polymerization is interesting in that it is one of the few entropically driven polymerizations known. This is due to the greater flexibility of the linear siloxane chain, which provides more states of freedom and motion, thereby increasing the entropy of the system over the cyclic monomers. Figure 7 shows a typical anionic polymerization of cyclic siloxane monomers. Typically, a Lewis base is used to catalyze this reaction, such as potassium hydroxide, potassium trimethylsilanolate, or n-butyl lithium (Suzuki 1989). This is also possible using a Lewis acid through a cationic polymerization.

In the first step of the ROP of D_3, the anionic initiator attacks a silicon atom, which results in the cleavage of a siloxane bond. This leads to a reactive silanolate chain end. For ROP, the choice of the initiator is extremely important in imparting functionality. In Fig. 7A, an initiator that results in

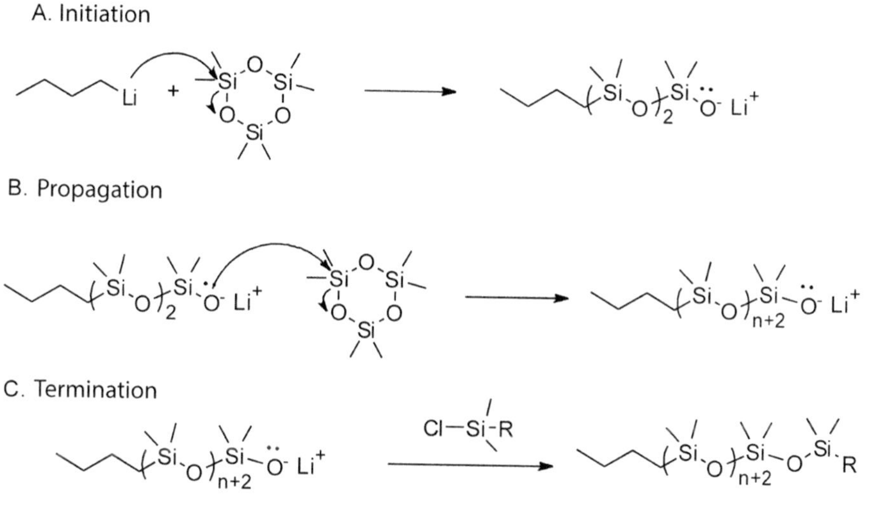

A. Initiation

B. Propagation

C. Termination

Cl—Si—R

Fig. 7. Ring-opening polymerization of poly(dimethylsiloxane).

an inert butyl chain end was chosen. However, if potassium hydroxide or a reactive silanolate, such as potassium 3-thiopropyldimethylsilanolate, is used as an end-capping agent, then a silanol or thiol functional end group, respectively, can be incorporated for additional reactions. Following initiation, propagation occurs whereby additional monomer units are added. If left unchecked, the polymer chain will also begin to equilibrate similarly to the polycondensation. However, molecular weight can be kinetically controlled by the addition of a terminating end-capping reagent. In general, ROP of D_3 can result in polymers of much higher molecular weight and lower polydispersity indices (Hölle and Lehnen 1975). Additionally, asymmetric chain ends are possible by varying the initiator groups and the terminating groups. This leads to a host of different potential architectures for future reactivity.

We can also imagine that the majority of polycarbosiloxanes are produced via polycondensation in a similar fashion to that of polysiloxanes with difunctional carbosiloxanes as the primary monomer. Figure 8A shows the preparation of a silphenylene-derived polymer through the same polycondensation process as PDMS (Dvornic et al. 1989). This process is in no way remarkable in comparison, but of course the polymerization kinetics are slightly affected by the decreased backbone flexibility. However, this ultimately does not have much of a practical effect on the polymerization itself. Figure 8B shows the same polymer prepared through Grignard synthesis. We can imagine these two as simply the same steps in different order. However, the efficiency of Grignard-based polymerization is far lower than that of traditional polycondensation. Nevertheless, the reactions presented in Fig. 8B have been used to synthesize not only polycarbosiloxanes, but also polycarbosilanes when a functional silane is used (Boury et al. 1989).

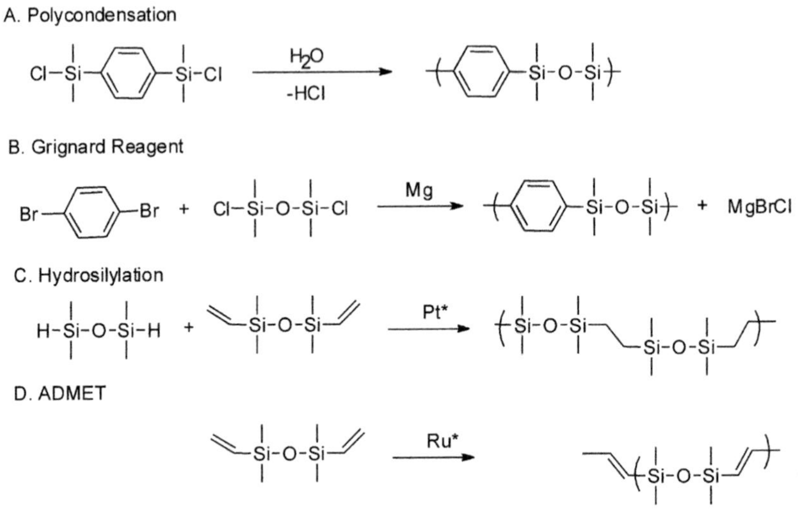

A. Polycondensation

H_2O / -HCl

B. Grignard Reagent

Br + Cl–Si–O–Si–Cl → Mg → + MgBrCl

C. Hydrosilylation

H–Si–O–Si–H + → Pt*

D. ADMET

Ru*

Fig. 8. Polymerizations of poly(carbosiloxane)s via (A) polycondensation, (B) Grignard reagents, (C) hydrosilylation, and (D) acyclic diene methathesis (ADMET).

The final two polymerization techniques require a transition metal catalyst. Figure 8C shows the polymerization of 1,1,3,3-tetramethyldisiloxane and 1,1,3,3-dimethyl-1,3-divinyldisiloxane catalyzed by a platinum catalyst. Given the high efficiency and exothermic nature of hydrosilylation reactions, such preparations must have very careful temperature control in order to prevent explosions. Acyclic diene metathesis (ADMET) is another route towards creating polycarbosiloxanes, especially those containing unsaturated bonds in the backbone. ADMET requires the use of common ruthenium catalysts, such as Grubb's II, to prepare polymer (Smith and Wagener 1993). Both hydrosilylation and ADMET can also be used in the preparation of polycarbosilanes through the use of difunctional silanes or monomers that do not contain siloxane bonds. Furthermore, the number of different addition techniques shown here opens up many avenues to orthogonal syntheses, particularly for dendrimer preparations, which are briefly discussed later in this chapter.

Just as many of the polymerization techniques of carbosiloxanes mimic those of polysiloxanes, so do the processes by which polysilazanes are prepared. Figure 9 shows both the polycondensation of dimethyldichlorosilane and the ROP of hexamethylcyclotrisilazane (Z_3). The polycondensation reaction is nearly identical; however, functional silazanes are prepared through the aminolysis of the chlorosilanes, whereby diaminosilanes are formed. These can further undergo reaction with additional chlorosilanes to form linear polymer. However, just like polysiloxanes, polysilazanes can undergo back-biting cyclization reactions and form an equilibrium with cyclic silazanes. For this reason, end-capping is again essential to achieve species of relatively high molecular weight. However, the resulting cyclic silazanes can

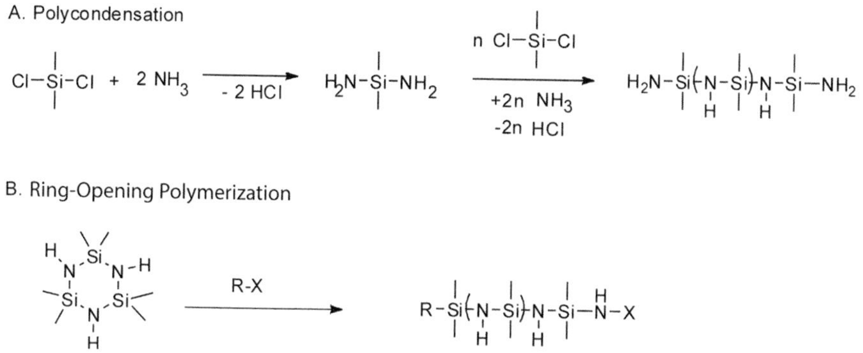

Fig. 9. Formation of polysilazanes via (A) aminolysis and polycondensation of dichlorodimethylsilane and (B) ring-opening polymerization of hexamethylcyclotrisilazanes.

be isolated for the ROP reactions to form higher-molecular-weight polymers with lower polydispersity index values (Bouquey et al. 1996).

While much of the chemistry shown here has been portrayed for the synthesis of monomers and polymers, many of these reactions are also important for the preparation of polymeric materials, as we will see in the following sections. Therefore, these reactions are traditional tenets of silicon polymers that have found universal use. Despite their simplicity, the number of and differences in these reactions can lead to a wide variety of structures and materials. However, although these are the primary reactions widely used through silicon chemistry, there have been a few recent advances worth noting.

Perhaps the most important recent reaction introduced to silicon chemistry is the Piers-Rubinsztajn reaction (Brook 2018). Although this reaction was first noted by Piers in 1996 (Parks and Piers 1996), it would not be until 2004 that Rubinsztajn and Cella recognized its importance in polymer chemistry (Rubinsztajn and Cella 2005). The reaction involves the direct formation of a siloxane bond from a hydridosilane and either a silanol or an alkoxysilane (Fig. 10). This reaction is catalyzed by tris(pentafluorophenyl) borane. This new cross-coupling reaction has seen increased use over the past several years, and additional examples of this chemistry are discussed later in this chapter.

Beyond this, some modern organic chemistry reactions have been applied to silicon polymers. Brook et al. have recently introduced the azide-alkyne "click" reaction to siloxane polymers (Gonzaga et al. 2009). This work

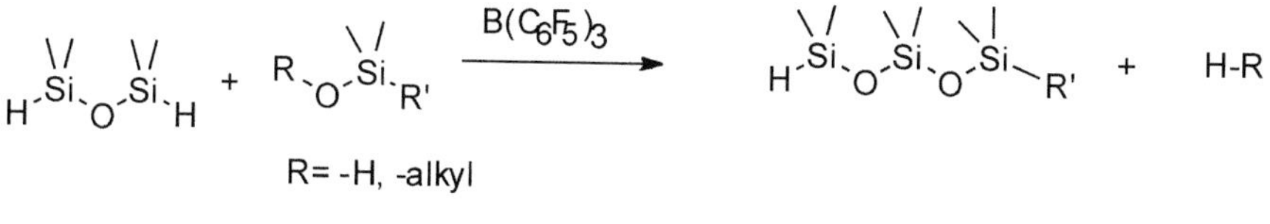

Fig. 10. The Piers-Rubinsztajn reaction.

has been extended to "metal-free" click reactions for the preparation of amphiphiles (Rambarran et al. 2013, Rambarran et al. 2015). Likewise, thiol-ene click reactions have been employed (Zhang et al. 2013). The use of these click reactions requires the incorporation of a functional organic side chain. Typically, this is done by preparing special monomers and copolymerizing them with dimethyldichlorosilane. In general, we can imagine this as typical organic chemistry, rather than silicon chemistry.

4. Applications of polysiloxanes, polysilazanes, and polycarbosiloxanes

Given the many desirable qualities of polysiloxanes, polysilazanes, and polycarbosiloxanes, it is no surprise to find them in a wide range of applications. It is difficult to cover in depth all of these applications, but in this section we explore a select handful of them and how their material properties and structures directly impact their use. First, we discuss their widespread use as surfactants and the new emerging interest in their dendrimer structures. Afterwards, their uses as coatings, particularly for hydrophobic and protective applications, are reviewed. Next, crosslinked bulk materials extend from the discussion of coating materials. Finally, we look at the use of these polymers to make ceramic materials.

4.1. Silicon polymer surfactants and dendrimers

Perhaps the most commonplace application of silicon polymers comes from their use as surfactants. As previously mentioned, silicones have remarkably low surface tensions, making them ideally suited for this application as surface modifiers and anti-foaming agents. Additionally, their biocompatibility/non-toxicity allows them to find further uses in cosmetics and pharmaceuticals. By far the most important surfactants are the class of poly(dimethylsiloxane)-poly(ethylene glycol) (PDMS-PEG) copolymers, owing to the biocompatibility of both materials, as well as the drastic differences in hydrophilicity between these two blocks (Yilgör and Yilgör 2014).

Figure 11 shows structures of several common silicone surfactants. In each of these cases, the polymerization of PDMS was end-capped with dimethylchlorosilane in order to achieve a hydridosilane chain end. This in turn was reacted with a terminal vinyl group via hydrosilylation (Kanner et al. 1967). Each of these different structures is largely used for different applications. For instance, the diblock copolymers can be found as emulsifying agents in pharmaceuticals and cosmetics. On the other hand, graph copolymers are largely used as foaming agents in the production of polyurethane foams (Zhang et al. 1999). The trisiloxane structure is unique in that it allows for rapid spreading across surfaces. Trisiloxanes are used widely in the application of aqueous pesticides on agricultural crops. In this way, the trisiloxane surfactant-coated droplets do not "bounce" off leaves, but quickly coat them to deliver the pesticide.

In each of these surfactants, the size of the respective blocks has significant impact on performance, unsurprisingly. For example, decreasing the size of the polysiloxane block, as seen in the superspreaders, increases the rate of their kinetic spreading (Sankaran et al. 2019), but also increases their irritation to human skin. On the other hand, increasing the size of polyethylene block increases the critical micelle concentration (CMC), or the concentration at which these surfactants operate as emulsifiers, requiring more material for surfactant applications. Furthermore, increasing the size of both blocks slows the kinetics of these surfactants (Gentle and Snow 1995). For this reason, there has been recent work on investigating additional structures to better control these properties.

The use of gemini surfactants has recently seen renewed interest. Lin et al. (2017) prepared gemini trisiloxane surfactants that have decreased irritation, lower CMC, and relatively fast dynamic properties. Likewise, Tong et al. (Chen et al. 2018) investigated branched, cationic gemini surfactants with different rigid spacer units. This study reported extremely low surface tensions and low CMC values, while also able to create a variety of self-assembled structures, from micelles and vesicles to worm-like tube structures.

Other approaches to better control surfactant performance have led to the use of additional polymer blocks. One potential new class of surfactant consists of PDMS-polyols. A typical example of this type of polymer is PDMS-poly(gylcerol) (Fig. 11D), which has exhibited surface tensions as low as 21 dynes/cm, but with significantly lower CMC values (Wang et

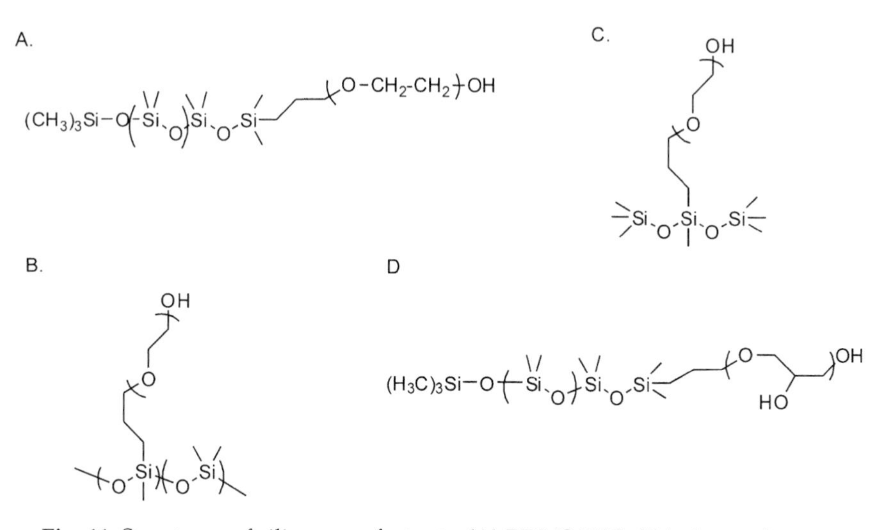

Fig. 11. Structures of silicone surfactants: (A) PDMS-PEG diblock copolymers, (B) PDMS-PEG graft copolymers, (C) trisiloxane superspreaders and (D) PDMS-poly(glycerol).

al. 2017). Other polyols, such as polysaccharides and "poly-oils", have also been explored (Furtwengler and Avérous 2018). One such surfactant derived from hyaluronan has seen application as a contact lens wetting agent owing to the biocompatibility of these materials (Paterson et al. 2014). Surfactants using palm oils have also been investigated in polyurethane foams (Nasir et al. 2016). Beyond copolymer surfactants, small molecule surfactants have been prepared by varying the architecture and hydrophilic chain end. Figure 12 shows a recent report by Huang et al. (2019). Clearly, the structure has a direct impact on the mechanisms of spreading for these molecules. Additionally, sulfonates are of particular interest owing to their remarkably high hydrophilicity.

Control of polymer architecture is also important for the field of dendrimers. Silicon-derived dendrimers are particularly interesting, owing to their biocompatibility and ability to be ceramitized for applications in drug delivery (Jimenez et al. 2019) and particle synthesis, respectively. Although there have been attempts prior to the past decade (Uchida et al. 1990, Lang and Luhmann 2001), recent advances in silicon chemistry have opened many additional routes to these structures and therefore renewed interest in them. Using the Piers-Rubinstajzn reaction, Brook et al. (Grande et al. 2014) were able to prepare up to a generation 3 (G3) polysiloxane dendrimer. In truth,

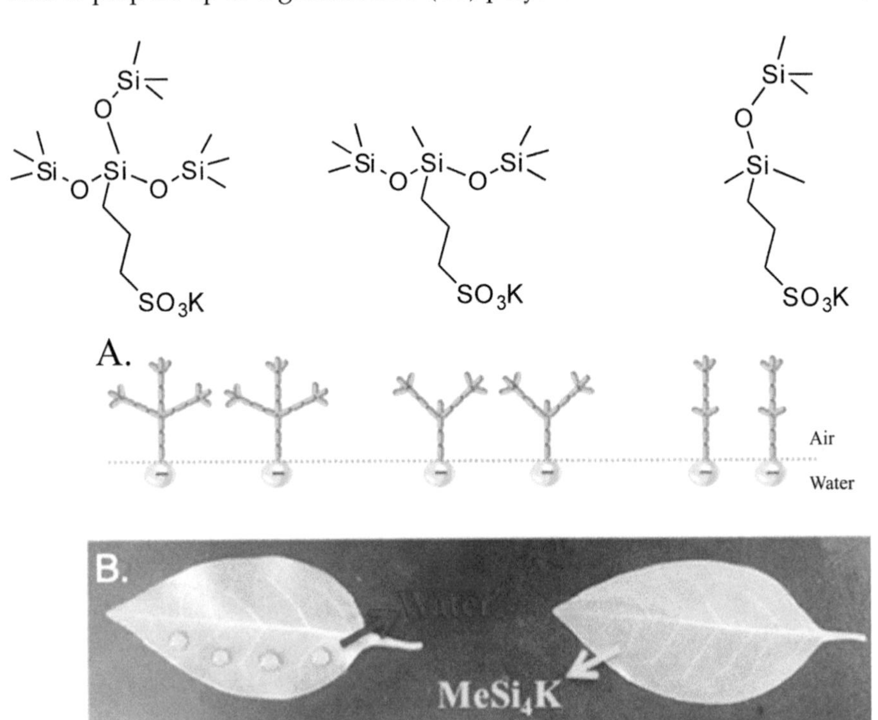

Fig. 12. (A) Structures of trisiloxane-sulfonate surfactants and (B) effect of surfactant coating on a paraffin leaf. Reprinted with permission from Jimenez et al. (2019), © 2019 the American Chemical Society.

these dendrimers are polycarbosiloxanes, owing to the alkyl spacers within them. The same strategy was followed in the preparation of hyperbranched silicones as novel surfactants (Morgan et al. 2017). Grignard syntheses and hydrosilylation have also found roles in carbosiloxane formation. Saxena et al. (2010) developed an epoxy-terminated dendrimer through hydrosilylation from a tetrakis(dimethylsiloxy)silane core. Likewise, thiol-ene click chemistry was used to prepare a G5 carbosiloxane dendrimer (Zhang et al. 2016).

In each of these dendrimer syntheses, a combination of reactions must be carefully selected (orthogonal synthesis). Typically, this requires the use of traditional condensation chemistry, along with an additional cross-coupling reaction. Furthermore, while there has been increased interest in the field of siloxane dendrimers, the potential and application of these materials has been largely untapped. With the recent development of new silicon chemistries, it is expected that these materials will continue to grow in importance with increased visibility in the literature.

4.2. Coating applications

While surfactants are considered coating materials, in this section we explore silicon-polymers as chemical modifiers for inorganic substrates, as well as traditional coatings of composites. We also explore some of the crosslinking techniques commonly found in siloxane materials, which can be applied to elastomers and bulk materials.

First, we begin with siloxanes and their reactions with inorganic oxide substrates. While silanes have long been used as surface-modifying agents in covalently attached monolayers and self-assembled monolayers (Krumpfer and Fadeev 2006), it was not until the last decade that the siloxane bond itself was considered a reactive functional group on its own (Krumpfer and McCarthy 2010, Krumpfer and McCarthy 2011). In retrospect, this is not altogether surprising, since it is well known that the siloxane bond is reactive with Lewis acids and bases, as seen in the ROP reactions. In addition to this, the fact that metal oxides have acid/base properties is well established in the fields of chromatography.

Figure 13 shows the reaction of PDMS with an inorganic oxide substrate, such as silica (Krumpfer and McCarthy 2011). While the exact reaction mechanism is not clear, it is likely a mixture of the mechanisms provided here. Given the acidic nature of many inorganic oxides, it is probable that the terminal hydroxyl groups can directly react with the siloxane bond, ultimately forming covalent attachment to the surface. On the other hand, inorganic surfaces are well known to be hydrated from atmospheric moisture. This surface-bound water can serve as a catalyst to cleave the siloxane bond, followed by traditional condensation with the surface. In this respect, we can see that the siloxane bond is actually an inherent functional group when used properly.

While this chemistry is relatively unsurprising, its implications and the resulting properties are not. Given that all inorganic oxides have some acid/

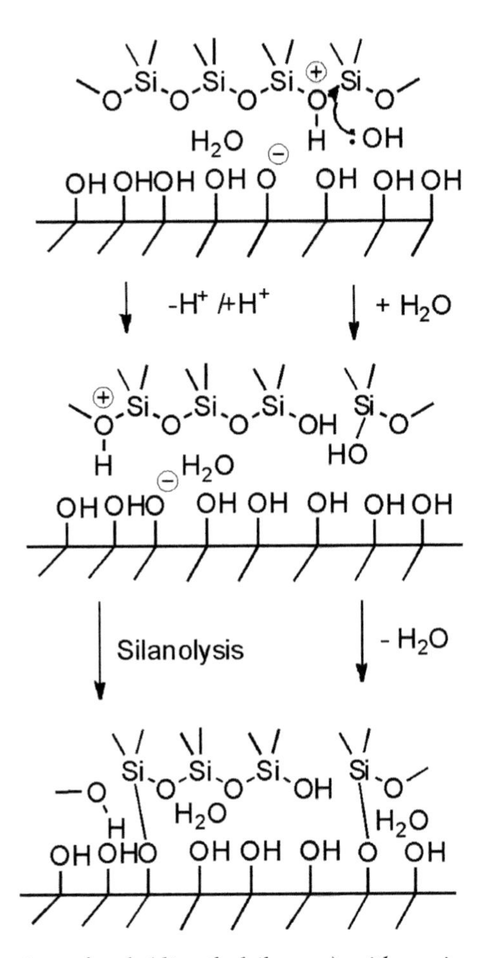

Fig. 13. Reaction of poly(dimethylsiloxane) with an inorganic oxide surface via direct silanolysis (left) and/or hydrolysis and condensation (right). Reprinted with permission from Krumpfer and McCarthy (2011), © 2011 the American Chemical Society.

base nature, this insinuates that the siloxane bond is actually a universal modifying agent regardless of the substrate. Indeed, Table 4 provides water contact angle data for a variety of substrates from acidic (SiO_2, TiO_2) to amphoteric (Al_2O_3) to basic (NiO). In each of these cases, we can see that hydrophobization occurs, although conditions for each substrate are not optimized. Another additional consideration for siloxanes as modifiers is that no caustic byproducts are produced. This is especially important as many byproducts can degrade inorganic surfaces, and furthermore, it makes this reaction a "green" reaction. The last and perhaps most interesting aspect of this reaction is the production of surfaces that exhibit negligible contact angle hysteresis. Physically, this means that droplets of various liquids will slide with almost no barrier to movement upon the slightest tilting. This implies that although these polymers covalently react with solid substrates, the interfaces themselves behave as liquids with a high dynamic motion, thereby minimizing contact line pinning (Cheng et al. 2013).

Table 4. Water contact angle data for surfaces reacted with PDMS (MW~2,000 g/mol) on various inorganic oxide surfaces (top) and contact angle data for various probe fluids of PDMS (MW~2,000 g/mol)-modified silica substrates (Reprinted with permission from Krumpfer and McCarthy (2011), © 2011 American Chemical Society)

	Contact angles (θ_A/θ_R), *deg*		
Temp, °C	*Tio*$_2$	Al_2O_3	*NiO*
25	77/34	75/35	84/37
60	89/54	87/44	90/57
100	97/67	92/51	95/74
150	101/84	106/100	103/85

		Contact angles (θ_A/θ_R), *deg*			
Time, h	*Temp, °C*	H_2O	CH_2I_2	$C_{16}H_{34}$	*Thickness, nm*
24	25	94/80	71/61	37/30	0.67
24	60	102/93	72/65	37/34	0.72
24	100	104/102	76/74	36/35	1.15
24	150	105/102	75/72	37/33	3.1

Hozumi et al. (2011) demonstrated a similar reaction using hydridomethylsiloxane polymers. However, with the incorporation of the hydridosilane group, covalent attachment is also possible through hydrolysis of the hydridosilane bond, followed by subsequent condensation. These polymers exhibited nearly identical contact angle behavior to that of pure PDMS, likewise suggesting that they produce liquid-like interfaces. In another report, cyclic siloxanes were used for the modification of magnesium powders to incorporate in solid-state explosives (Hastings et al. 2019).

While siloxanes are relatively new to the surface modification scene, silazanes have long been known to be excellent hydrophobizing agents. In fact, hexamethyldisilazane is one of the most popular hydrophobizing agents for titania and silica powders. It is particularly useful in modifying titania, since its byproduct is ammonia gas, which does not degrade the material. However, this concerns small molecule surface modifiers. Poly(silazane) s as surface-modifying agents are known to lack hydrolytic stability in comparison to poly(siloxane)s (Gengenbach and Griesser 1999).

Despite relatively new interest concerning the siloxane bond, silicones have long been used as coating materials, which most showers in the world will attest. They make excellent coating materials because of their low surface tension, giving them the ability to easily and conformally spread across nearly all substrates, while at the same time being resistant to adhesion to their surfaces. The most common form of water-repellent coating uses triacetitoxymethylsilane (Fig. 3C) and silanol-terminated PDMS (Fig. 6B) (Yoshimura et al. 1999, Kumar and Lee 2017). Acetitoxysilanes are particularly useful, since they show a slow reaction with moisture allowing enough time to apply and set, while also only producing a benign byproduct, acetic acid, for domestic commercial use. Since these formulations will gradually set by reacting with moisture in the air at room temperature, this class of coating is known as room-temperature vulcanization silicones.

On the other hand, most engineering siloxane materials are produced through hydrosilylation (Marciniec and Gulinski 1993). By mixing poly(hydridomethylsiloxane)s with vinyl-terminated PDMS in the appropriate mixtures along with a platinum catalyst, silicone materials can be developed with a wide range of mechanical properties, from elastomers to solid materials (Zheng and McCarthy 2010). Of particular interest is the use of these materials as composites. Such composites may find use in the preparation of silicone-based paints using classical pigments (Fig. 14) (Longenberger et al. 2017), or they can also impart more practical properties, such as electrical conductivity, in these coatings (Fig. 15) (Krumpfer 2012). Polycarbosiloxane and polysilazane coatings also follow the same principles, but they are largely used as preceramic materials, which are discussed in the following section.

4.3. Preceramic polymers

Finally, an important property of each of these polymers is the formation of solid residue upon ceramitization (Colombo et al. 2010, Birot et al. 1995).

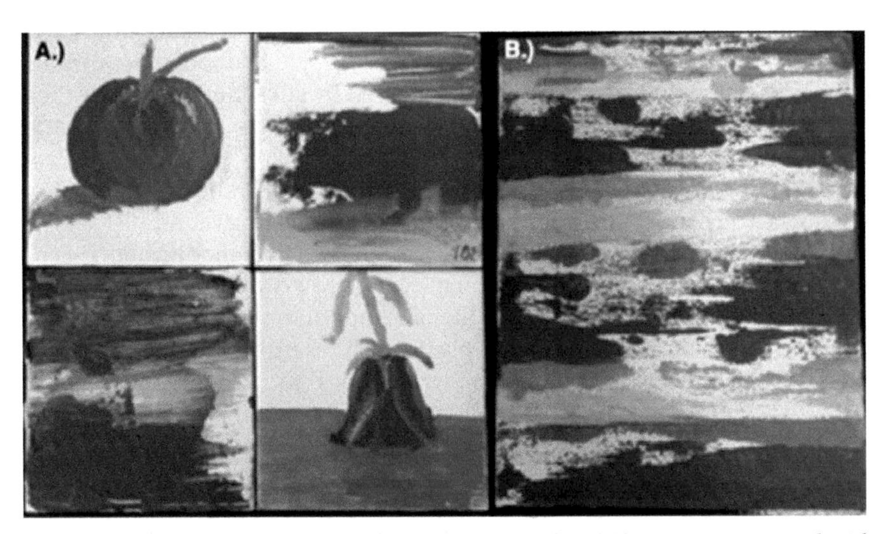

Fig. 14. Hydridomethylsiloxane and vinyl-terminated PDMS paints composited with Prussian blue, chrome yellow and iron oxide as artistic paint medium on ceramic tiles (A) and canvas (B). Reprinted from Longenberger et al. (2017) with permission, © 2017 the American Chemical Society.

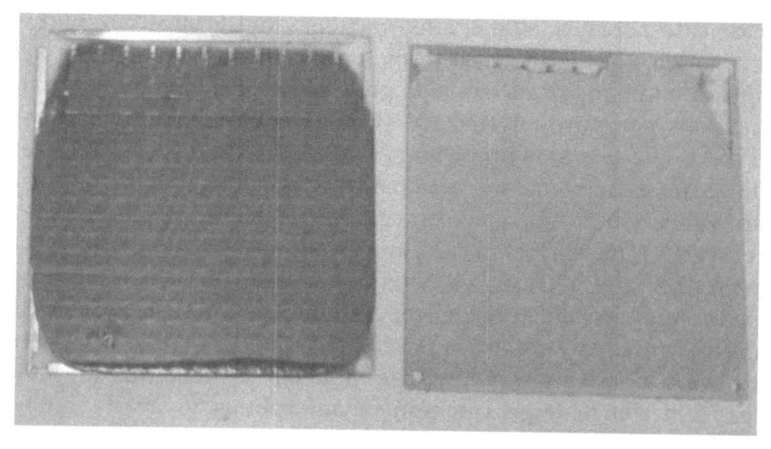

Fig. 15. Silicone composite film with nickel particles to impart electrical conductivity. Reprinted from Krumpfer (2012) with permission, © 2012 Joseph W Krumpfer.

Ceramics are particularly important materials owing to their lightweight, thermally stable and high-strength properties. The predisposition of silicon-polymers for this application is entirely due to the close proximity they have to traditional ceramics in atomic ratios. For example, PDMS has a Si:O ratio of 1:1, which is similar to silica, 1:2. Likewise, polysilazanes have a ratio of Si:N of 1:1, which is even closer to that of silicon nitride, Si_3N_4. By incorporating carbon into the polymer backbone, as is the case for carbosiloxanes, ternary ceramic materials with $Si_xC_yO_z$ can be prepared. With respect to this application, these polymers have largely cornered the market.

Because of their close atomic ratio to a well-known ceramic, polysilazanes were among the first preceramic polymers investigated (Lücke et al. 1997). Polymer precursors are highly favored over pure materials, since they significantly cut down on manufacturing costs. The process by which silazanes are converted to ceramic materials has been well understood for quite some time. Generally speaking, silazanes begin ceramitization at approximately 300-500°C and remain a metastable structure, $Si_{3+x}C_{x+y}N_4$, to temperatures up to 1400°C. However, above this temperature, the structure degrades into silicon nitride, Si_3N_4, silicon carbide, SiC, and carbon, C. Yet, one of the potential side reactions in this process is the evolution of nitrogen gas, which is thermodynamically favored. In order to limit this evolution, and therefore increase the stability of the resulting ceramic, this process must be performed under inert atmospheres, such as argon, or in the presence of ammonia, which can reintroduce nitrogen into the structure (Riedel et al. 1992). These resulting materials are highly resistant to oxidation below temperatures of 1600°C and show very little change in their physical properties below this temperature.

The use of polymers also opens up more avenues towards processing and can be deposited on surfaces via chemical vapor deposition, spun into fibers or crosslinked into powders for compression molding. In many of these cases, specific side chains, such as vinyl groups, are necessary to insure stability of the material prior to ceramitization (Riedel et al. 1995). To illustrate the importance of side groups, we can compare the pyrolyzed products of perhydropolysilazane (PHPZ) and polymethylhydrosilazane (PMHZ). PHPZ contains no carbon atoms and would therefore be an ideal candidate for directly making silicon nitride, whereas PMHZ has one methyl group, which would make silicon carbonitride ceramic. While PHPZ does make nearly pure silicon nitride, the resulting ceramic easily crystallizes between 1200 and 1300°C, which is lower than the necessary processing temperature. Crystallization of the ceramic material negatively impacts the desired physical properties (Vaahs et al. 1992). On the other hand, the introduction of a single methyl group helps prevent this crystallization process and the material can remain amorphous to nearly 1400-1500°C (An et al. 1998).

From these examples, we can see two important properties. It is important that neither the preceramic polymer, intermediates, nor ceramic readily crystallize in order to maintain the required properties. Additionally, the presence of a "heteroatom" such as carbon can help prevent crystallization. In a way, the carbon serves as a glue between the silicon and nitrogen atoms. Another heteroatom often used in the preparation of polymer-derived ceramics is boron (B). Borosilicates are well known to have enhanced properties over traditional silica glasses and for this reason are often added to prepare SiCNB ceramics. Boron compounds can also be used to crosslink these materials, aiding in increasing stability of the intermediate during the pyrolysis process.

Figure 16 shows the hydroboration reaction of vinylmethylsilazane and borane. In this reaction the borane may react with the vinyl group through the well-known hydroboration reaction. However, since it still contains B-H bonds, it may further react with nearby vinyl groups, thus crosslinking the intermediate material. SiCNB fibers prepared from this polymer followed by pyrolysis at 1400°C show remarkable properties, with tensile strength up to 1.3 Gpa and Young's modulus of 172 GPa. Furthermore, because of the heteroatom mixture in these materials, these fibers retain their amorphous character even up to temperatures as high as 1700°C (Bernard et al. 2005). Additionally, greater tailorability of ceramic properties can be ensured using other boranes, such as methylborane.

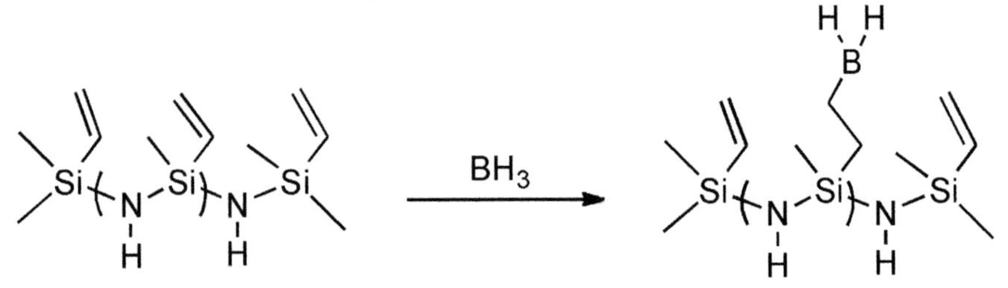

Fig. 16. Crosslinking of vinylsilazanes via hydroboration for the preparation of SiCNB ceramics.

5. Outlook and conclusion

Over the course of the last decade, there has been a "rediscovery" of siloxane polymers. Recent work has been performed to further study and understand the chemical potential of these polymers long relegated to engineering materials. The faulty belief that these polymers were chemically inert has largely faded from most chemists' minds, and many researchers have strived to enlarge the synthetic toolbox by imparting well-known organic moieties to engage them in a larger context. Given the wide range of excellent properties these polymers have, it is likely they will draw increased attention for the foreseeable future. Likewise, the potential of polycarbosiloxanes, polysilazanes and other derivatives is still only being explored; only the surface has been scratched. Particularly, better control of polymer architecture for the preparation of organic-inorganic polymers has exciting implications in any number of fields, including fire-resistant clothing, highly detailed ceramics, and pharmaceuticals.

Conflicts of interest

The authors declare no conflict of interest.

Acknowledgements

The authors thank the Pace University Office of the Provost for financial support.

References

An, L., R. Riedel, C. Konetschny, H. Kleebe and R. Raj. 1998. Newtonian viscosity of amorphous silicon carbonitride at high temperatur. *Journal of the American Ceramic Society* 81: 1349-1352.

Aziz, T., H. Fan, F.U. Khan, M. Haroon and L. Cheng. 2019 Modified silicone oil types, mechanical properties and applications. *Polymer Bulletin* 76: 2129-2145.

Baferani, A.H., R. Keshavarz, M. Asadi and A.R. Ohadi. 2018. Effects of silicone surfactant on the properties of open-cell flexible polyurethane foams. *Advanced Polymer Technology* 37: 71-83.

Beccera, R. and R. Walsh. 1998. The Chemistry of Organic Silicon Compounds, volume 2. Wiley: Chichester.

Bernard, S., M. Weinmann, P. Gerstel, P. Miele and F. Aldinger. 2005. Boron-modified polysilazane as a novel single-source precursor for SiBCN ceramic fibers: Synthesis, melt-spinning, curing and ceramic conversion. *Journal of Materials Chemistry* 15: 289-299.

Bicerano, J. 2009. Prediction of Polymer Properties. Marcel-Decker, Inc. New York.

Birot, M., J-P. Pillot and J. Bunogues. 1995. Comprehensive chemistry of polycarbosilanes, polysilazanes, and polycarbosilazanes as precursors of ceramics. *Chemical Reviews* 95: 1443-1477.

Bouquey, M., C. Brochon, S. Bruzaud, A.-F. Mingotaud, M. Schapper and A. Soum. 1996. Ring-opening polymerization of nitrogen containing cyclic organosilicon monomers. *Journal of Organometallic Chemistry* 521: 21-27.

Boury, B., R.J.P., Corriu and W.E. Douglas. 1991 Poly(carbosilane) precursors of silicon-carbide – The effect of crosslinking on ceramic residue. *Chemistry of Materials* 3: 487-489.

Brook, M.A. 2000. Silicon in Organic, Organometallic, and Polymer Chemistry. John Wiley & Sons: New York.

Brook, M.A. 2012. The chemistry and physical properties of biomedical silicones. pp. 52-67. *In*: W. Peters, H. Brandon, K.L. Jerina, C. Wolf and V.L. Young (eds.). Biomaterials in Plastic Surgery: Breast Implants. Woodward Publishing, Cambridge.

Brook, M.A. 2018. New control over silicone synthesis using SiH Chemistry: The Piers-Rubinsztajn reaction. *Chemistry – A European Journal* 24: 8458-8469.

Brookes, K.H., M.R.H. Siddiqui, H.M. Rong, R.W. Joyner and G.J. Hutchings. 2001. Effect of Al and Ca addition on the copper catalysed formation of silanes from Is and CH_3Cl. *Applied Catalysis A: General* 206: 257-265.

Camino, G., S.M. Lomakin and M. Lageard. 2002. Thermal polydimethylsiloxane degradation. Part 2: The degradation mechanisms. *Polymer* 43: 2011-2015.

Chang, R. and J. Overby. 2019. Chemistry, 13th edition. McCraw-Hill Education. New York.

Chen, C-P., G. Lu and Q-X. Tong. 2018. Three tetrasiloxane-tailed cationic gemini surfactants: The effect of different spacer rigidity on surface properties and aggregation behaviors. *Journal of Molecular Liquids* 266: 504-513.

Cheng, D.F., B. Masheder, C. Urata and A. Hozumi. 2013. Smooth perfluorinated surfaces with different chemical and physical natures: Their unusual dynamic wetting behavior toward polar and non polar liquids. *Langmuir* 29: 11322-11329.

Clarson, S.J. and J.A. Semlyen. 1993. Siloxane Polymers. Prentice Hill: Englewood Cliffs.

Colombo, P., G. Mera, R. Riedel and G.D. Sorarù. 2010. Polymer-derived ceramics: 40 years of research and innovation in advanced ceramics. *Journal of the American Ceramic Society* 93: 1805-1837.

Colvin, E.W. 1981. Silicon in Organic Synthesis. Buttersworths: London.

Delannoy, Y. 2012. Purification of silicon for photovoltaic applications. *Journal of Crystal Growth* 360: 61-67.

Dvornic, P.R., H.J. Perpall, P.C. Uden and R.W. Lenz. 1989. Exactly alternating silarylene siloxane polymers. 7. Thermal-stability and degradation behavior of para-silphenylene siloxane polymers with methyl, vinyl, hydride, and/or fluoroalkyl side groups. *Journal of Polymer Science A* 27: 3503-3514.

Fadeev, A.Y. and T.J. McCarthy. 1999. A new route to covalently attached monolayers: Reaction of hydridosilanes with titanium and other metal surfaces. *Journal of the American Chemical Society* 121: 12184-12185.

Furtwengler, P. and L. Avérous. 2018. Renewable polyols for advanced polyurethane foams from diverse biomass resources. *Polymer Chemistry* 9: 4258-4287.

Gengenbach, T.R. and H.J. Griesser. 1999. Post-deposition aging reactions differ markedly between plasmas polymers deposited from siloxane and silane monomers. *Polymer* 40: 5079-5094.

Gentle, T.E. and S.A. Snow. 1995. Absorption of small silicone polyether surfactants at the air/water surface. *Langmuir* 11: 2905-2910.

Gonzaga, F., G. Yu and M.A. Brook. 2009. Versatile, efficient derivatization of polysiloxanes via click technology. *Chemical Communications* 1730-1732.

Graiver, D., K.W. Farminer and R. Narayan. 2003. A review of the fate and effects of silicones in the environment. *Journal of Polymers and the Environment* 11: 129-136.

Grande, J.B., T. Urlich, T. Dickie and M.A. Brook. 2014. Silicone dendrons and dendrimers from orthogonal SiH coupling reactions. *Polymer Chemistry* 5: 6728-6739.

Hastings, D.L., M. Schoenitz, K.M. Ryan, E.L. Dreizin and J.W. Krumpfer. 2019. Stability and ignition of a siloxane-coated magnesium powder. *Propellents, Explosives, Pyrotechnics* 45(4): 621-627.

Hölle, H.J. and B.R. Lehnen. 1975. Preparation and characterization of polydimethylsiloxanes with narrow molecular weight distribution. *European Polymer Journal* 11: 663-667.

Holloczki, O. and L. Nyulaszi. 2009. Stability and structure of carbene-derived neutral penta- and hexacoordinate silicon complexes. *Organometallics* 28: 4159-4164.

Hozumi, A., D.F. Cheng and M. Yagihashi. 2011. Hydrophobic/superhydrophobic oxidized metal surfaces showing negligible contact angle hysteresis. *Journal of Colloid and Interface Science* 353: 582-587.

Huang, Y., M. Guo and S. Feng. 2019. Synthesis and solution behavior of sulfonate-based silicone surfactants with specific, atomically defined hydrophobic tails. *Langmuir* 35: 9785-9793.

Hurd, C.B. 1946. Studies on siloxanes. I: The specific volume and viscosity in relation to temperature and constitution. *Journal of the American Chemical Society* 68: 364-370.

Horii, Y. and K. Kannan. 2008. Survey of organosilicone compounds, including cyclic and linear siloxanes, in personal-care and household products. *Archives of Environmental Contamination and Toxicology* 55: 701-710.

Jimenez, J.L., R. Gomez, V. Briz, R. Madrid, M. Bryszewsk, F.J. de la Mata and M.A. Munoz-Fernandez. 2012. *Journal of Drug Delivery Science and Technology* 22: 75-82.

Kanner, B., W.G. Reid and I.H. Petersen. 1967. Synthesis and properties of siloxane-polyether copolymer surfactants. *Industrial & Engineering Chemistry Process Design and Development* 6: 88-92.

Khalifia, M., M. Hajji and H. Ezzaouia. 2012. Purification of silicon powder by the formation of thin porous layer followed by photothermal annealing. *Nanoscale Research Letters* 7: 444.

Kipping, F.S. and L.L. Lloyd. 1901. XLVII.—Organic derivatives of silicon. Triphenylsilicol and alkyloxysilicon chlorides. *Journal of the Chemistry Society* 79: 449-459.

Krumpfer, J.W. 2012. Chemistry of siloxane-inorganic oxide surfaces. Dissertation.

Krumpfer, J.W. and A.Y. Fadeev. 2006. Displacement reactions of covalently attached organosilicon monolayers on Si. *Langmuir* 22: 8271-8272.

Krumpfer, J.W. and T.J. McCarthy. 2010. Contact angle hysteresis explained. *Faraday Discussions* 146: 103-111.

Krumpfer, J.W. and T.J. McCarthy. 2011. Rediscovering silicones: "Unreactive" silicones react with inorganic surfaces. *Langmuir* 27: 11514-11519.

Kumar, V. and D.-J. Lee. 2017. Effects of thinner on RTV silicone rubber nanocomposites reinforced with GR and CNTs. *Polymers for Advanced Technologies* 28: 1842-1850.

Kumar, V.B. and E.M. Leitao. 2020. Properties and applications of polysilanes. *Applied Organometallic Chemistry* 34(3): e5402.

Lang, H. and B. Luhmann. 2001. Siloxane and carbosiloxane based dendrimers: Synthesis, reaction chemistry, and potential applications. *Advanced Materials* 13: 1523-1540.

Lee, C.L., C.L. Frye and O.K. Johannson. 1969. Selective polymerization of reactive cyclosiloxanes to give non-equilibrium molecular weight distributions, monodisperse siloxane polymers. *Polymer Preprints: Polymer Division of the American Chemical Society* 10: 1361.

Li, Z.X., W.J. Liang, Y.F. Shan, X.X. Wang, K. Yang and Y.Y. Cui. 2019. Study of flame-retarded silicone rubber with ceramifiable property. *Fire and Materials* 44(4): 487-496.

Lin, J., W. Wang, W. Bai, M. Zhu, C. Zheng, Z. Liu, X. Cai, D. Lu, Z. Qiao, F. Chen, and J. Chen. 2017. A gemini-type superspreader: Synthesis, spreading behavior and superspreading mechanism. *Chemical Engineering Journal* 315: 262-273.

Lin, Y., L. Wang, J.W. Krumpfer, J.J. Watkins and T.J. McCarthy. 2013. Hydrophobization of inorganic oxide surfaces using dimethylsilanediol. *Langmuir* 29: 1329-1332.

Longenberger, T.B., K.M. Ryan, W.Y. Bender, A-K. Krumpfer and J.W. Krumpfer. 2017. The art of silicones: Bringing siloxane chemistry to the undergraduate curriculum. *Journal of Chemical Education* 94: 1682-1690.

Lücke, J., J. Hacker, D. Suttor and G. Ziegler. 1997. Synthesis and characterization of silazane-based polymers as precursors for ceramic matrix composites. *Applied Organometallic Chemistry* 11: 181-194.

Marciniec, B. and J. Gulinski. 1993. Recent advances in catalytic hydrosilylation. *Journal of Organometallic Chemistry* 446: 15-23.

Mark, J.E., ed. 1999. Polymer Data Handbook. Oxford University Press: New York.

Mark, J.E. 2004. Some interesting things about polysiloxanes. *Accounts of Chemical Research* 37: 946-953.

Morgan, J., T. Chen, R. Hayes, T. Dickie, T. Urlich and M. Brook. 2017. Facile synthesis of dendron-branched silicone polymers. *Polymer Chemistry* 8: 2743-2746.

Müller, R. 1942. German Patent No. C57411.

Nakajima, Y. and S. Shimada. 2015. Hydrosilylation reaction of olefins: Recent advances and perspectives. *Catalysis Communications* 4: 637-639.

Nasir, I.A.A., A. Alis, Z. Mohamad, S.H. Che Man and R.A. Majid. 2016. Rigid polyurethane foam from palm oil polyol-polyethylene glycol blend. *Journal of Polymer Materials* 33: 629-637.

Orti, E., R. Crespo, M.C. Piqueras, F. Tomas and J.L. Bredas. 1993. Electronic structure of polysilanes: Influence of substitution and conformation. *Synthetic Metals* 57: 4419-4424.

Otomo, Y., Y. Nagase and N. Nemoto. 2005. Synthesis and properties of novel poly(tetramethylsilnaphthylenesiloxane) derivatives. *Polymer* 46: 9714-9724.

Owen, M.J. 1981. Why silicones behave funny. *CHEMTECH* 11: 288.

Parks, D.J. and W.E. Piers. 1996. Tris(pentafluorophenyl)boron-catalyzed hydrosilation of aromatic aldehydes, ketones, and esters. *Journal of the American Chemical Society* 118: 9440-9441.

Paterson, S.M., L. Liu, M.A. Brook and H. Sheardown. 2014. Poly(ethylene glycol)- or silicone-modified hyaluronan for contact lens wetting agent applications. *J. Biomedical Materials Research A* 103A: 2602-2610.

Pauling, L. 1960. The Nature of Chemical Bonding, 3rd Edition. Cornell University Press, Ithaca.

Pauling, L. 1980. The nature of silicon-oxygen bonds. *American Mineralogist* 65: 321.

Plueddemann, E.P. 1982. Silane Coupling Agents. Plenum Press: New York.

Post, H.W. 1949. Silicones and Other Organosilicon Compounds. Reinhold: New York.

Rahm, M., T. Zeng and R. Hoffmann. 2019. Electronegativity seen as the ground state average valence electron binding energy. *Journal of the American Chemical Society* 141: 342-351.

Rambarran, T., F. Gonzaga and M.A. Brook. 2013. Multifunctional amphiphilic siloxane architectures using sequential, metal-free click ligations. *Journal of Polymer Science A: Polymer Chemistry* 51: 855-864.

Rambarran, T., F. Gonzaga, M.A. Brook, F. Lasowski and H. Sheardown. 2015. Amphiphilic thermoset elastomers from metal-free, click crosslinking of PEG-grafted silicone surfactants. *Journal of Polymer Science A: Polymer Chemistry* 53: 1082-1093.

Riedel R., G. Passing, H. SchoÈnfelder and R.J. Brook. 1992. Synthesis of dense silicon-based ceramics at low temperatures. *Nature* 355: 714.

Riedel R., A. Kienzle and M. Friess. J.F. Harrod, R.M. Laine (eds). 1995. Application of Organometallic Chemistry in the Preparation and Processing of Advanced Materials. Kluwer Academic Publishers, The Netherlands.

Rochow, E.G. 1941. U.S. Patent No. 2,380,995.

Rochow, E.G. 1945. The direct synthesis of organosilicon compounds. *Journal of the American Chemical Society* 67: 963-965.

Rochow, E.G. 1946. Introduction to the Chemistry of the Silicones. J. Wiley & Sons, Chapman & Hall: New York, London.

Rubinsztajn, S. and J.A. Cella. 2005. A new polycondensation process for the preparation of polysiloxane copolymers. *Macromolecules* 38: 1061-1063.

Rudnick, R.L. and D.M. Fountain. 1995. Nature and composition of the continental crust – A lower crustal perspective. *Reviews of Geophysics* 33: 267-309.

Sankaran, A., S.I. Karakashev, S. Sett, N. Grozev and A.L. Yarin. 2019. On the nature of the superspreaders. *Advances in Colloid and Interface Science* 263: 1-18.

Santowski, T., A.G. Sturm, K.M. Lewis, T. Felder, M.C. Holthausen and N. Auner. 2019. Synthesis of functional monoplanes by disilane cleavage with phosphonium chlorides. *Chemistry – A European Journal* 25: 3809-3815.

Saxena, K., C.S. Bisaria and A.K. Saxena. 2010. Studies on the synthesis and thermal properties of alkoxysilane-terminated organosilicone dendrimers. *Applied Organometallic Chemistry* 24: 251-256.

Shriver, D., M. Weller, T. Overton, J. Rourke and F. Armstrong. 2014. Inorganic Chemistry, 6th ed. W.H. Freeman and Company: New York.

Simon, T.L. and B. Volcani. 1981. Silicon and Siliceous Structures in Biological Systems. Springer: New York, NY, USA.

Smith, D.W. and K.B. Wagener. 1993. Acyclic diene metathesis (ADMET) polymerization – Design and synthesis of unsaturated poly(carbosiloxane)s. *Macromolecules* 26: 1633-1642.

Stein, J., L.N. Lewis, Y. Gao and R.A. Scott. 1999. In situ determination of the active catalyst in hydrosilylation reactions using highly reactive Pt(0) catalyst precursors. *Journal of the American Chemical Society* 121: 3693-3703.

Suzuki, T. 1989. Preparation of poly(dimethylsiloxane) macro monomers by the 'initiator method': 2. Polymerization mechanism. *Polymer* 30: 333-337.

Thomas, S.E. 2006. Organic Synthesis: The Roles of Boron and Silicon. Oxford University Press. Oxford.

Uchida, H., Y. Kabe, K. Yoshino, A. Kawamata, T. Tsumuraya and S. Masamune. 1990. General strategy for the systematic synthesis of oligosiloxanes. Silicone dendrimers. *Journal of the American Chemical Society* 112: 7077-7079.

Vaahs, T., M. Brück and W.D.G. Böcker. 1992. Polymer-derived silicon nitride and silicon carbonitride. *Advanced Materials* 4: 224-226.

Voronkov, M.G., G.I. Zelchan and E.J. Lukevits. 1977. Silicon and Life, 2nd ed. Zinatne Publishing: Vilnius, Lithuania.

Walsh, R. 1981. Bond dissociation energy values in silicon-containing compounds and some of their implications. *Accounts of Chemical Research* 14: 246-252.

Wang, G., Y. Zhu, Y. Zhai, W. Wang, Z. Du and J. Qin. 2017. Polyglycerol modified polysiloxane surfactants: Their adsorption and aggregation behavior in aqueous solution. *Journal of Industrial and Engineering Chemistry* 47: 121-127.

Warrick, E.L. 1990. Forty Years of Firsts. McGraw-Hill: New York.

Weinhold, F. and R. West. 2011. The nature of the silicon-oxygen bond. *Organometallics*, 30(21): 5815-5824.

West, R., L.D. David, P.I. Djurovich, K.L. Stearley, K.S.V. Srinivasan and H. Yu. 1981. Phenylmethylpolysilanes: Formable silane copolymers with potential semiconducting properties. *Journal of the American Chemical Society* 103: 7352-7354.

Yang, Z., S. Han, R. Zhang, S.Y. Feng, C.Q. Zhang and S.Y. Zhang. 2011. Effects of silphenylene units on the thermal stability of silicone resins. *Polymer Degradation and Stability* 96: 2145-2151.

Yilgör, E. and I. Yilgör. 2014. Silicone containing copolymers: Synthesis, properties and applications. *Progress in Polymer Science* 39: 1165-1195.

Yoshimura, N., S. Kumagai and S. Nishimura. 1999. Electrical and environmental aging of silicone rubber used in outdoor insulation. *IEEE Transactions on Dielectrics and Electrical Insulation* 6: 632-650.

Zahid, M., G. Mazzon, A. Athanassiou and I.S. Bayer. 2019. Environmentally benign non-wettable textile treatments: A review of recent state-of-the-art. *Advances in Colloid and Interface Science* 270: 216-250.

Zeigler, J.M. and F.W.G. Fearon. 1989. Silicon-based Polymer Science. ACS Publishing, Washington D.C.

Zhang, J., Y. Chen and M.A. Brook. 2013. Facile functionalization of PDMS elastomer surfaces using thiol-ene click chemistry. *Langmuir* 29: 12432-12442.

Zhang, X.D., C.W. Macosko, H.T. Davis, A.D. Nikolov and D.T. Wasan. 1999. Role of silicone surfactant in flexible polyurethane foam. *Journal of Colloid and Interface Science* 215: 270-279.

Zhang, Z., S. Feng and J. Zhang. 2016. Facile and efficient synthesis of carbosiloxane dendrimers via orthogonal click chemistry between thiol and ene. *Macromolecular Rapid Communications* 37: 318-322.

Zheng, P. and T.J. McCarthy. 2010. Rediscovering silicones: Molecularly smooth, low surface energy, unfilled, UV/Vis-transparent, extremely cross-linked, thermally stable, hard, elastic PDMS. *Langmuir* 26: 18585-18590.

Index

9780367655617